FAULT-TOLERANCE THROUGH RECONFIGURATION OF VLSI AND WSI ARRAYS

MIT PRESS SERIES IN COMPUTER SYSTEMS
Herb Schwetman, editor

Metamodeling: A Study of Approximations in Queueing Models, Subhash Chandra Agrawal, 1985

Logic Testing and Design for Testability, Hideo Fujiwara, 1985

Performance and Evaluation of LISP Systems, Richard P. Gabriel, 1985

The LOCUS Distributed System Architecture, edited by Gerald Popek and Bruce J. Walker, 1985

Analysis of Polling Systems, Hideaki Takagi, 1986

Performance Analysis of Multiple Access Protocols, Shuji Tasaka, 1986

Performance Models of Multiprocessor Systems, M. Ajmone Marsan, G. Balbo, and G. Conte, 1986

Microprogrammable Parallel Computer: MUNAP and Its Applications, Takanobu Baba, 1987

Simulating Computer Systems: Techniques and Tools, M. H. MacDougall, 1987

A Commonsense Approach to Theory of Error-Correcting Codes, Benjamin Arazi, 1987

Research Directions in Object-Oriented Programming, edited by Bruce Shriver and Peter Wegner, 1987

Object-Oriented Concurrent Programming, edited by Akinori Yonezawa and Mario Tokoro, 1987

Networks and Distributed Computation: Concepts, Tools, and Algorithms, Michel Raynal, 1988

Fault Tolerance Through Reconfiguration of VLSI and WSI Arrays, R. Negrini, M. G. Sami, and R. Stefanelli, 1989

FAULT-TOLERANCE THROUGH RECONFIGURATION OF VLSI AND WSI ARRAYS

R. Negrini, M.G. Sami, R. Stefanelli

The MIT Press
Cambridge, Massachusetts
London, England

Printed and bound in the United States of America.

Library of Congress Cataloging-in-Publication Data

Negrini, R.
    Fault tolerance through reconfiguration in VLSI and WSI arrays / R. Negrini, M. Sami, and R. Stefanelli.
       p. cm. -- (MIT Press series in computer systems)

    Includes bibliographies and index.
    ISBN 0-262-14044-6
    1. Fault-tolerant computing. 2. Integrated circuits -- very large scale integration. 3. Integrated circuits -- Wafer-scale integration.
    I. Sami, Mariagiovanna. II. Stefanelli, Renato. III. Title.
    IV. Series.
QA 76.9.F38N44   1989
004.2--dc19

88-23225
CIP

CONTENTS

SERIES FOREWORD

This series is devoted to all aspects of computer systems. This means that subjects ranging from circuit components and microprocessors to architecture to supercomputers and systems programming will be appropriate. Analysis of systems will be important as well. System theories are developing, theories that permit deeper understanding of complex interrelationships and their effects on performance, reliability, and usefulness.

We expect to offer books that not only develop new materials but also describe projects and systems. In addition to understanding concepts, we need to benefit from the decision making that goes into actual development projects; selection from various alternatives can be crucial to success. We are soliciting contributions in which several aspects of systems are classified and compared. A better understanding of both the similarities and the differences found in systems is needed.

It is an exciting time in the area of computer systems. New technologies mean that architectures that were at one time interesting but not feasible are now feasible. Better software engineering means that we can consider several software alternatives, instead of "more of the same old thing," in terms of operating systems and system software. Faster and cheaper communications mean that intercomponent distances are less important. We hope that this series contributes to this excitement in the area of computer systems by chronicling past achievements and publicizing new concepts. The format allows publication of lengthy presentations that are of interest to select readership.

Herb Schwetman

PREFACE

A book should first of all justify its own right to exist — in other words, what was the rationale for its creation (apart from personal satisfaction of the authors)? Let us therefore check the motivations that led to writing a full book devoted to a fairly specialized subject: the reconfiguration of processing arrays aiming at fault-tolerance.

Processing arrays are defined simply as computing structures consisting of:

- a set of identical Processing Elements (PEs) or *cells*;

- an interconnection network linking together the PEs in a regular scheme, so that each PE will exchange information with a limited subset of PEs (*neighbors*). PEs along the *border* of the array are interconnected by the same regular structure with external, interacting devices, and this is the *only* I/O mechanism foreseen for the whole array. The interconnection network can, and usually will, also be comprised of control devices (e.g., switches and the related control logic) routing data along the interconnection paths.

It is clear that such a definition is quite general and that it can be applied to a large number of very well-known, conventional devices. In fact, this same definition applies to the large class of *iterative networks*; for example, any ripple-carry adder could be considered as a array in its own right. Still, architectures referred to as *processing arrays* are usually made up of PEs of much higher complexity than a single full adder: they may even be full-fledged microcomputers and are in any event processing units capable of several functions (arithmetical or others). In recent years, interest in such structures has been growing as a result of two concurrent factors: development of algorithms for high-performance applications that can be well mapped upon processing arrays, and development of VLSI and even Wafer-Scale Integration (WSI) technologies that have enabled efficient and cost-effective implementation of these same arrays.

A typical application area for such architectures is represented by signal processing, where high processing speed on large masses of data is a general specification, particularly if real-time features are required. Signal processing algorithms, in turn, have been devised that can be efficiently mapped onto processing arrays, thus providing the necessary speed for real-time operation through high parallelism and pipelining. A number of VLSI and experimental WSI devices for signal processing applications have been implemented. The regularity and locality of the architectures make them particularly suited to that type of implementation, even when very large chips are envisioned — chips whose dimensions would certainly deter the implementation of *random* or irregular structures.

With regard to the importance of fault-tolerant design for processing arrays, it thus becomes apparent that both application areas and technology require the introduction of a capacity to survive (multiple) faults. While signal processing applications are often mission-critical ones (e.g., in the case of avionics or of spaceborne applications) it is also true that very large chips are characterized by decreasing yields that would make them too expensive to produce. Thus, the need to meet mission critical standards leads to developing forms of fault-tolerance to increase system lifetime, while the cost factor makes fault-tolerance mandatory during production so as to circumvent production defects and increase yield.

The extreme regularity of array architectures allows fault-tolerance to be achieved through reconfiguration in a particularly cost-effective way. Redundancy, in fact, need not be introduced on a massive scale, on the "system level" (as in the case of conventional structures when techniques such as TMR and similar ones are used), but rather through adoption of a limited number of spare PEs and of a suitably augmented, but still regular, interconnection network. It should be noted that proper definition of the reconfiguration techniques will, with such limited redundancy, allow survival in spite of *multiple* faults, in contrast to the single-fault assumption underlying system-level TMR. This fact has fostered considerable research on the subject of reconfiguration of processing arrays, leading to attractive theoretical formulations and to interesting and useful methodologies. Although far from exhaustive, this book is dedicated to an analysis of a number of such results.

The first section provides an introduction to the framework of fundamental information required to discuss the problem of reconfiguration and to evaluate its solutions. Structures and characteristics of some main classes of processing arrays are reviewed, emphasis being placed upon their applications. This allows a better understanding of the restrictions to be adopted later with regard to interconnection structures. The problem of fault models will then be analyzed. In addition to consideration of physical causes of defects, the functional fault models adopted by various authors will be discussed in relation to:

- faults caused by production defects *vs.* faults caused by run-time failures;
- relative relevance of PEs *vs.* interconnection paths and related control logic;
- chip dimensions and regularity of design.

Having defined the characteristics of the fault environment, we will discuss problems that must be taken into account when attempting to achieve fault-tolerance through reconfiguration. The factors of merit usually adopted to carry out a comparative evaluation of different algorithms, or even of different alternatives for one single algorithm, will also be presented.

In chapter four, technologies and operations required to support and physically implement the reconfiguration methods will be reviewed. Suitable technologies are necessary when *designing* a fault-tolerant array to allow for subsequent reconfiguration. Again, technologies will vary depending upon whether reconfiguration is performed once, *statically* at end of production, to overcome production defects, or *dynamically*, at run time, to overcome failures during device operation. Any fault-tolerance policy, moreover, requires as its starting point a full test of the system. Since they are amply documented in the literature, details will not be given about testing techniques that might be considered *conventional*, that is, techniques that are not specific of array devices but rather useful for any

VLSI structure. The entire problem of production-time testing will be dealt with briefly, since general techniques such as E-beam testing appear to be satisfactory. Rather, some major aspects of run-time testing will be analyzed — concurrent or semiconcurrent, self-driven or host-driven — since they are strongly related to the architecture and are also instrumental for dynamic reconfiguration.

The central problem of the book — reconfiguration techniques — will then be discussed. Considering the vast literature on this subject that has been published in the past few years, only a relatively small part is analyzed in detail here. The proposals chosen exemplify well some corresponding line of approach. With regard to a given reconfiguration philosophy, in some instances, we preferred to examine not necessarily the most efficient algorithm but rather the one better suited to didactic purposes. Since fault-tolerance through reconfiguration is the theme of this book, there is very little discussion of two other closely related themes: *functional reconfiguration* (which deals with mapping different target architectures, and therefore different algorithms, onto a given physical architecture) and *fault-tolerance through algorithm reconfiguration* (by which the very organization of the algorithm and its mapping onto the physical system are modified to suit fault distributions).

Presentation of the various reconfiguration techniques is organized by groups of algorithms corresponding to different basic philosophies. A simple, first subdivision concerns topology of the array interconnection structure. A particular instance is represented by a criterion that superimposes fault-tolerance upon a general philosophy whose aim is functional reconfiguration of various target architectures onto linear physical structures. This is the well-known *Diogenes approach*. The general problem of reconfiguring target linear arrays will be examined next. This is a subject widely dealt with in the literature, because linear *processing* arrays have many important applications (e.g., FIR filters etc.) and because a much wider class of digital structures can be described as linear arrays (e.g., serial memories).

Most published reconfiguration techniques deal with *rectangular* arrays, i.e., with structures in which each PE is connected with its four *nearest neighbors* in a plane. In fact, the techniques devised for such arrays could well be extended without relevant modifications to other planar structures with different connectivity. Moreover, even when *three-dimensional* arrays (such as *pyramidal* ones for image processing) are discussed, their spatial characteristics are actually only logical ones. The array is obviously mapped onto a plane physical one, so as to be implementable with a VLSI chip.

Among techniques suited for reconfiguration of rectangular arrays, we chose to examine first a class that can be defined as corresponding to a *mathematical* approach to the problem. In this case, emphasis is not so much on definition of structures and circuits supporting reconfiguration, as on a rigorous theoretical analysis of bounds that can be predetermined for various figures of merit. The results of such analysis lend authority and significance to the methods discussed in the following chapters.

Among the implementation-oriented proposals, a few are based on the concept of *local* reconfiguration: i.e., the problem is solved inside small subarrays by means of predetermined, fairly simple rules, and the global solution derives from the composition of the local ones. In contrast, the majority of approaches prefer to look for *global* reconfiguration, dealing with the whole array and with the complete fault distribution. These approaches can be subdivided according to different philosophies. A variety of techniques are used to isolate faulty areas and choose a policy for introduction of spares, also depending upon the percentage of fault-free PEs that will be excluded from the final operating array. This factor will be balanced against bounds on delays and interconnection paths, on algorithm simplicity and so on. Thus, *Row and column elimination* techniques give a premium to guaranteeing bounds for delays (i.e., they try to grant a balanced distribution of lengths for interconnection links all over the array). These techniques usually require a large number of spares and are better adopted for *production-time restructuring* of wafer-scale circuits. On the other hand, *Index-mapping* techniques choose to map the array functions onto the available fault-free PEs, trying to grant strict bounds for the *individual* interconnection between operating PEs and at the same time to achieve a high utilization of the available spares. Thus, such techniques are particularly suited to instances allowing only a reduced number of spares.

All these techniques refer to fault models that locate randomly distributed faults in PEs only, and assume interconnections even between faulty PEs to be always fully operational. In contrast, a few algorithms consider the possibility of large clusters of faults; this model is tuned to wafer-scale structures where defects are not only of electrical origin but may also derive from factors such as scratches in the photoresist etc., affecting whole areas independently of the functions performed by the devices in that area. To avoid propagation of signals inside faulty areas, corresponding techniques are based upon simple message exchanges using specific protocols. This allows the creation of rows and columns of the target array almost independently, requiring no information transfer with devices inside the faulty area.

All previous algorithm classes are designed for general plane arrays with a nearest-neighbor interconnection structure. As detailed in the chapter on array types and applications, a large number of signal processing algorithms are actually mapped upon simplified regular arrays also called *multipipeline*. Their peculiar structure permits the adoption of reconfiguration techniques that are both efficient and require less costly interconnection networks.

These approaches require a redundancy expressed in terms of spare PEs and augmented interconnection networks, while they strive to keep processing speed nominal (or, at least, as near as possible to nominal) by introducing bounds on the communication delays. Given the stringent requirements usually made in array applications, specifically, with regard to processing speed, we have paid most attention to such approaches. A number of alternative proposals have also been presented in which performance degradation in terms of lower speed is accepted when faults are present, while structure redundancy is as limited as possible. Most such techniques are related to the specific algorithm implemented by the array, being based upon modified mappings of the algorithm itself onto the reduced array where faulty PEs are present. Time redundancy is thus achieved through a *graceful degradation* tailored upon a specific application. Since in this book we abstract from application-specific examples involving well defined application algorithms, we do not discuss any such approach. Instead, a few suggestions for time redundancy solutions are discussed that are closely related to the earlier, *structure-redundant* approaches and that may be seen as transferring global philosophies from the space domain to the time domain.

Barring a very few exceptions, evaluation of reliability achieved by the various solutions will be deduced from results of statistical simulation. Complexity of the architectures envisioned (and, often, of the reconfiguration policies adopted) justifies this choice in preference of mathematical deduction of reliability expressions. Anyway, theoretical bases for such mathematical deduction are briefly recalled in the appendix.

An absolute definition of *optimality* independent of any other consideration cannot be deduced from the set of approaches and their performances. The existence of multiple factors of merit, often contrasting with each other, makes each choice the result of a trade-off among a number of considerations. This study, therefore, does not offer one *solution*, but presents possible alternatives for a number of possible environments that will allow guidelines to be made available. Mathematical techniques suggest bounds for at least some of the various factors of merit. Specific technological solutions may in some instances lead to implementations that perform even *better* than such bounds, by sharing resources (e.g., interconnection links) that were considered separate in the theoretical analysis. Thus, technological knowledge can complete the information deduced from abstract analysis of some problems. Similarly, information on functions and structure of the individual PEs can offer further guidelines when the balance between complexity due to PEs and to the interconnection network must be established. While we do not deal with an actual example of complete design involving all these factors simultaneously, we hope that the reader will find this book a useful tool for achieving such a design.

1 TYPICAL PROCESSING ARRAYS

This chapter analyzes typical structures found in the best known processing arrays. It is a partial survey, not to be considered as an introduction to methods for array design, but rather as an attempt to highlight real structures found in many useful arrays. Particular attention will be given to aspects that have some relevance to the main subject of this book.

The literature concerning array architectures and related design problems is by now quite vast. In addition to the large number of papers published in various journals, especially the *IEEE Transactions on Computers*, and in proceedings of conferences on computer architecture and on digital signal/image processing, many references are to be found in books specifically dedicated to this subject or to closely related ones (see, e.g., [CAN86, CAP84, COS86, LEG86, MOO85, MOO86, POT85, REI86, SAU86, SWA86]).

At the end of this chapter, we will concentrate on the physical architectures depicted in figure 1.1, where boxes are identical Processing Elements (PEs, or cells) that execute the same processing steps upon data flowing through the local, uniform paths that connect these PEs.

This chapter will:

(1) identify the most important and common architectural features found in practical arrays, such as dimensions and typical functions performed by the individual PE, number and topology of interconnections, etc.;

(2) implicitly define a practical, valid frame for the choice of requirements that may appropriately be attributed to the reconfiguration algorithms in the various applications (for example, costs of introducing spare paths and spare processing elements, or foreseen production yields for PEs);

(3) verify the usefulness of arrays in some practical applications.

The usefulness of arrays directly derives in fact from their main performance: high throughput. By means of dedicated arrays it is possible to build systems that are orders of magnitude more powerful than programmed computers, even special computers, having digital signal processing oriented CPUs.

Arrays can be the only means of implementation for high performance applications. They can be exploited in two main ways:

(1) inside powerful embedded digital systems that implement more or less fixed functions: in this case, arrays work as stand-alone units;

(2) inside programmable computers, as attached processors for implementing special functions when high throughput is required (e.g., for high performance matrix arithmetics). Collections of different arrays can thus be attached to standard machines, taking advantage both of the programming environment of the computer, and of high performances for the algorithms that are hard-

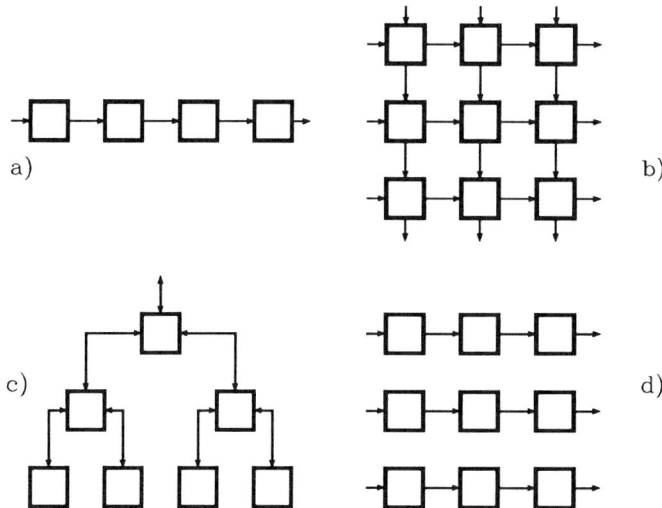

Figure 1.1

The four chosen array classes: a) linear, b) mesh, c) tree, d) multiple pipeline.

wired into the dedicated processing arrays.

In both cases the processing array often executes a fixed algorithm only or, at most, a very few similar algorithms. Processing Elements usually cannot be programmed. Rather, they can be personalized so as to be tailored according to dimensions of vectors and matrices under processing or to changes of parameters or coefficients in the given algorithm.

Even microprocessors, especially single chip microcomputers, can be usefully adopted as processing elements. This implies giving up orders of magnitude of throughput, exchanged against ease of implementation resulting from adoption of commercial, well-supported and inexpensive components. Tightly connected multimicroprocessors are thus obtained. Such structures will not be discussed here.

In this chapter, some well known as well as lesser known but typical examples of processing arrays will be described, that clearly show very different and significant structures. The arrays will be ranked by their most important application fields:

(1) digital signal processing (DSP)

(2) image processing (IP)

(3) non numeric processing

(4) cellular arithmetics

1.1. Digital Signal Processing (DSP) Arrays

Many typical dedicated arrays are found, because in these application areas classic Digital Signal Processing (DSP) algorithms show a very high degree of intrinsic parallelism, operate upon huge amounts of data, and require high performances; all of which are characteristics allowing and asking for dedicated arrays.

Furthermore, DSP structures often require very high reliability even in unmanned environments, reliability that can be achieved through reconfiguration for fault-tolerance. In practice, this is the application area that can gain the most from the application of reconfiguration algorithms discussed in this book.

Standard algorithms of this field are convolution and correlation, and various transforms (among which Fourier transforms are by far the most useful ones). Even signals occupying 40 Mhz bands can be treated by means of arrays of acceptable cost, implemented by integrated circuits of standard technology. By the sampling theorem, 40 Mhz signals can be sampled by Analog-to-Digital converters at a rate of two samples every 1/40 microseconds, i.e., 12.5 nanoseconds.

Classic DSP algorithms, moreover, require relatively simple PEs. Multipliers and adders for complex operands, or scalers and rotators, i.e., multipliers by a complex constant, are the usual necessary components, together with FIFO memories.

Interconnection structures are often very simple. It has been proven that two-dimensional arrays where each processing element is linked only to its four neighbors (often called *meshes*) satisfy the needs of many DSP algorithms applied to monodimensional signals.

Furthermore, algorithms for multiple-dimensional signals can often be implemented by cascading algorithms for monodimensional signals, a result of the fact that the algorithms are linear. A classic example is that of Fourier transform of two-dimensional signals (e.g., images) that can be obtained by means of two cascaded and independent monodimensional Fourier transforms, applied following two orthogonal directions in the image. This allows for adoption of simple cascaded arrays, with — interposed between them — a memory in which the two-dimensional matrix of signal samples can be written by rows and read by columns (a memory that is often dubbed *Corner Turning Memory*).

Dimensions of *signals* must not be confused with dimensions of *arrays*. This will become clear in the first example of array to be discussed: a two-dimensional mesh implementing the *Discrete Fourier Transform (DFT)* of a monodimensional signal.

Consider a time-varying signal that is sampled at N equally spaced times: the DFT of these samples gives N samples of the Fourier transform of the signal. These transform values are equally spaced in the frequency domain; the N time domain samples can be easily reconstructed starting from the N frequency domain values.

The underlying theory will not be necessary in the following (see, e.g., [OPP75] for a complete treatment); for our scope it will be sufficient to recall that the monodimensional N-point Discrete Fourier Transform of a vector of N complex data samples is given by:

$$X(k) = \sum_{n=0}^{N-1} x(n) \cdot W_N^{nk} \qquad [1.1]$$

where $X(k)$ are the N transform values, $x(n)$ are the N input data, and W_N^{nk} are constant weights, obtained by raising to the (nk)-th power the N-th radix of the unity $(W_N = e^{-j(2\pi/N)})$:

$$W_N^{nk} = e^{-j(2\pi/N)nk} \qquad [1.2]$$

As an example, when $N = 4$ equation [1.1] becomes:

$$\begin{bmatrix} X(0) \\ X(1) \\ X(2) \\ X(3) \end{bmatrix} = \begin{bmatrix} W_4^{00} & W_4^{10} & W_4^{20} & W_4^{30} \\ W_4^{01} & W_4^{11} & W_4^{21} & W_4^{31} \\ W_4^{02} & W_4^{12} & W_4^{22} & W_4^{32} \\ W_4^{03} & W_4^{13} & W_4^{23} & W_4^{33} \end{bmatrix} \cdot \begin{bmatrix} x(0) \\ x(1) \\ x(2) \\ x(3) \end{bmatrix}$$

By defining p as $p = nk$ modulo N (i.e., $nk = p + mN$, being m integer and $0 \le p < N$) then:

$$W_N^{nk} = e^{-j2\pi nk/N} = e^{-j2\pi p/N} \cdot e^{-j2\pi mN/N} = e^{-j2\pi p/N} = W_N^p$$

and the matrix of weights (for $N = 4$) can be written as:

$$\begin{bmatrix} W_4^0 & W_4^0 & W_4^0 & W_4^0 \\ W_4^0 & W_4^1 & W_4^2 & W_4^3 \\ W_4^0 & W_4^2 & W_4^0 & W_4^2 \\ W_4^0 & W_4^3 & W_4^2 & W_4^1 \end{bmatrix}$$

The flow graph corresponding to the above example is given in figure 1.2, where data paths and operators are complex.

If, inside equation [1.1], we substitute the matrix of weights with a matrix formed by N rows obtained by shifting N complex samples of a given reference function r (each row shifted one place with respect to the previous one), the same formula computes the convolution c between the reference function and the input samples x.

In case of $N = 4$, this corresponds to:

$$\begin{bmatrix} c(0) \\ c(1) \\ c(2) \\ c(3) \end{bmatrix} = \begin{bmatrix} r_0 & r_1 & r_2 & r_3 \\ r_1 & r_2 & r_3 & r_0 \\ r_2 & r_3 & r_0 & r_1 \\ r_3 & r_0 & r_1 & r_2 \end{bmatrix} \cdot \begin{bmatrix} x(0) \\ x(1) \\ x(2) \\ x(3) \end{bmatrix}$$

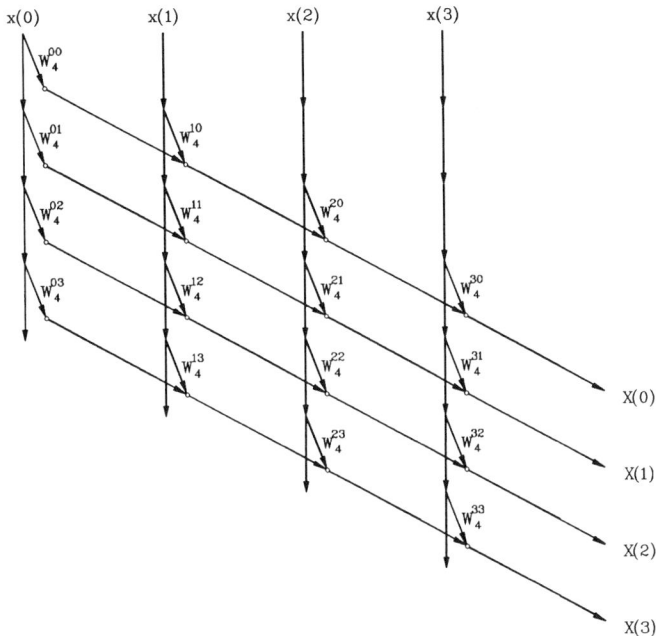

Figure 1.2
Flow graph of DFT.

1.1.1. The mono-dimensional array for DFT

The linear (i.e., monodimensional) array of figure 1.3.a executes equation [1.1], and its processing elements correspond one-to-one to the operators of a (sloping) row of the flow graph of figure 1.2: the k-th element receives at its external (upper) inputs the sequence W_N^{nk} of weights (i.e., a row of the weights matrix), and from its neighboring PE the sequence $x(n)$ of samples; these sequences of samples pass through the PE with no other modification than the single step delay.

Each sample $x(n)$ meets its weight W_N^{nk} at the k-th cell. There, the two terms are multiplied and the result is accumulated inside the cell, thus executing the summation of equation 1.1. When all the samples have passed through the cell, it contains $X(k)$.

Since the array is synchronous, the k-th PE receives $x(1)$ only k steps after this sample passed through the first cell; thus, the k-th row of the weight matrix must be fed to the k-th PE with a delay of k steps. Transform data $X(k)$ become available inside the k-th PE with the same relative delay.

An actual corresponding PE structure is given in figure 1.4: the summation is accumulated inside the latch that is fedback around the adder. A buffer, clocked by the propagated clock $CIN - COUT$, captures the final result $X(k)$, and presents it at the broadcast output line labelled $X(k)$ in figure. Note that the presence of

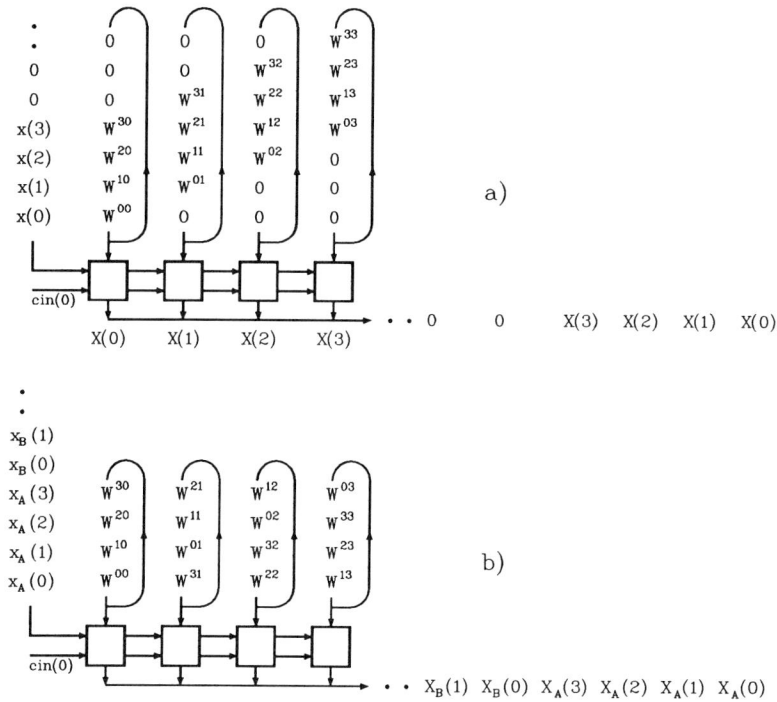

Figure 1.3
Pipeline for DFT implementation.

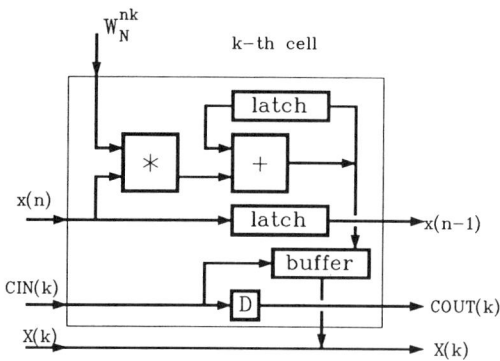

Figure 1.4
PE of the DFT pipeline.

one clock-step delay D guarantees that the buffers of the various PEs capture and output their result $X(k)$ at the right time step. If necessary, the broadcast output line can be substituted by a line that chains all the buffers of PEs, so that all the $X(k)$ flow through the chain of these buffers, instead of being broadcast.

During the sequential output phase of the results that have been stored in the buffers, the pipeline can begin operation on a new input sequence, so that no *idle* phases are introduced. This possibility is exploited by feeding the weights as shown in figure 1.3.b.

Many other monodimensional array structures could be adopted to compute equation [1.1]. In [ANT86] four typically different approaches are presented, but in any case we obtain a PE architecture that is not very different from that of figure 1.4.

1.1.2. Two-dimensional arrays for DFT

Another possibility might be to adopt a two-dimensional array of $N \times N$ cells: PE (n, k) contains entry W_N^{nk} of the weight matrix, receives the input sample $x(n)$ from its upper neighbor $(n, k - 1)$, and passes it to its lower neighbor $(n, k + 1)$. Inside the PEs of the n-th column, the n-th sample flows meeting all the W_N^{nk} weights.

All the PEs of the k-th row can thus compute the transform $X(k)$, simply by receiving from their left the partial sum, adding to it the local product $x(n) \cdot W_N^{nk}$, and passing the partial sum towards right. Again, sample $x(n)$ must be fed to the n-th column with a delay of n steps.

This array can support pipelined processing: continuously fed, one step after the other, by different rows of N input samples to be separately transformed, the two-dimensional array would give contemporarily N transformed values. For example, at a certain time step, the array gives:

- at the output of row N, the N-th transform value of the row of samples that have been input N times before;

- at the output of row $N - 1$, the $(N - 1)$-th transform value of the row of samples that have been input $N - 1$ times before;

- and so on, up to the upper row, where the first transform value of the lastly input row is output.

1.1.3. Systolic and wavefront arrays

Consider the class of processing arrays in which all cells (PEs):

(1) are exactly identical (position-independency is achieved);

(2) are independent from the number N of samples (size-independency);

(3) process all the data flowing through them always in the same way, i.e., they always repeat exactly the same computation beat at constant frequency (time-independency); this means that all PEs are synchronized by one common clock and that the two phases computation/data transfer alternate in a regular way;

(4) are connected only to neighboring cells, and always in the same way. All data flow through short links between neighbors (locality of interconnections, no broadcasting).

These properties are extremely useful, because they lead to arrays that are completely modular and expandable with no internal modifications, that are iterative, and characterized by extremely simple control networks inside cells.

In fact, the two arrays seen in §1.1.1 and §1.1.2 do not fully satisfy conditions (3) and (4). In particular, the array of figure 1.3.a requires extraction of results from the PEs at the end of the whole computation, while the array in 1.1.2 requires latching of the fixed terms $W_N^n k$ inside the PEs before starting computations.

Actually, as previously shown in the case of the monodimensional array for DFT (i.e., the chaining of the buffers of figure 1.4 by means of an output line), slight modifications of arrays would also eliminate the necessity of setting up coefficients or extracting results, thus adding a fifth property that enhances position independency, namely:

(5) the PEs do not contain private data.

This last point can be obtained by letting *all* data, coefficients and partial results flow inside the array. The DFT monodimensional array, for example, can be fed by input samples flowing from right to left, whereas partial results flow from left to right, and the order of input weights sequences is thus rearranged. A continuous operation can consequently be obtained without complex control networks.

Very high throughputs can thus be obtained as well because no idle phases for setting up internal data or extracting results are necessary. A further advantage regards I/O management at the array borders since, in the typical instance of DSP applications, data can be directly fed by analog-to-digital converters without need of rearrangement.

Arrays (and algorithms) having all the properties listed above are called *systolic*, and have received wide attention in recent years (see, e.g., [KUN80]). Methods have also been developed that allow transformation of many non systolic flow graphs into completely systolic flow graphs (see, e.g., [ULL84]).

From a hardware perspective, the global synchronization involved in systolic arrays incurs problems of *clock skew*; moreover, a further fault-tolerance problem is introduced by the clock itself, and it is much harder to overcome then the ones concerning PEs and interconnection networks. Finally, global sinchronization creates peak power requests in correspondence of clock edges. As an answer to all the previous questions, *wavefront arrays* have been proposed [KUN88]. They are characterized by self-timed, data driven computation, i.e., there is no global timing references in wavefront arrays: a PE works only when new data are available at its inputs and old results have yet been passed to the neighbors at its outputs.

Data sequence   Transform sequence

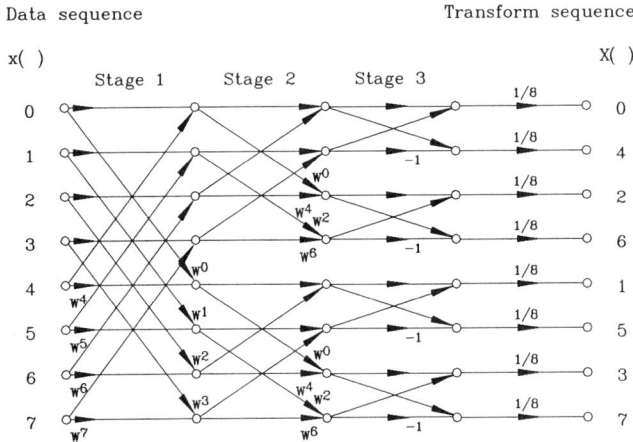

Figure 1.5
Flow graph of FFT.

In the wavefront architecture, the information transfer is by mutual convenience between a PE and its immediate neighbors: this scheme can be implemented by means of simple handshaking protocols [KUN82].

1.1.4. Commutators

From the above discussion, it might seem that only truly systolic algorithms, corresponding to systolic signal flow graphs, can be easily hardwired into arrays, i.e., mapped onto arrays where only neighboring processing elements are linked together. This is false. Fortunately, non local (and not truly systolic) algorithms can also correspond to simple arrays, characterized by local connections only. A clear example of this is given by the Fast Fourier Transform (FFT) that will now be examined in the case of monodimensional signals.

This algorithm corresponds to signal flow graphs like that of figure 1.5 where data produced by distant processing elements (i.e., PEs operating the multiplications and additions of the flow graph) must meet at some fixed point. For example, sample pairs 0-4, 1-5, 2-6, and 3-7 must be coupled and fed in this order to stage 1 in figure 1.5, whereas usually they arrive in the order 0,1,2,...,7 from an Analog-to-Digital converter.

This data switching can be achieved by means of special cells whose operation consists of changing the temporal ordering of data flowing through them. These cells, called *commutators*, can be built in various ways. An example of structure is shown in figure 1.6.a, where data coming along two lines are output in a different order by means of four switches and a FIFO memory that are controlled by a simple circuit (not shown in detail in figure 1.6.a).

In figure 1.6.b, time flows from left to right; the first two rows correspond to the two inputs: at the upper one, the input samples appear in their natural order 0,1,2,...,7. At the same time, the lower input could well receive other samples. They are marked as undefined in the figure ("-") because they are of no interest for the present analysis.

The third row of figure 1.6.b represents the *states* of the commutator at each time step:

- if it is "=" the commutator simply transfers the upper input value at the upper output, and the lower input value at the lower output (two switches directly connect inputs to outputs);

- if it is "X" the commutator exchanges the two input values at the two outputs (the upper input is connected to the lower output, and the lower input is connected to the upper output).

The fourth and the fifth rows show the two output sequences at points c and d; it is now sufficient to insert at the upper output of the commutator a delay of four time steps (i.e., a four-cell FIFO) to guarantee that the upper output sequence 0, 1, 2, 3 waits for the lower output sequence 4, 5, 6, 7 so that data pairs 0-4, 1-5, 2-6, 3-7 are fed in a timely way to the following stage of the array (see figure 1.7).

Obviously, cells such as the commutator above considered have some drawbacks:

- large quantities of FIFO memory are required;

- a latency is introduced, i.e., output from a commutator is delayed till the latest data of the group that must be processed together arrive at the input;

- length of FIFO memory and structure of circuits that control interconnection phases, depend upon specific parameters of the algorithm (e.g., number of samples to be treated, matrix or vector dimensions, etc.) and can be different from commutator to commutator, as we will see in few moments for the array of figure 1.7. Thus, cells *are not completely homogeneous*. This is a drawback that cannot be eliminated from algorithms that are not truly systolic.

It is now possible to pinpoint another general concept. Dimensions of a physical array implementing a flow graph can be smaller than the dimensions of the flow graph itself. For example, a two-dimensional graph can be implemented by a monodimensional linear array, as also occurred in the case of the DFT linear array discussed previously. A graph such as that of figure 1.5 can be implemented, other than by an array in which processing elements correspond one-to-one to operators in the graph by:

(1) a *column* of processing elements. The column starts executing, in the first time step, the first column of operations in the graph. Then, in the second step it executes the second column of operations, then the third and so on. At each step, intermediate results are stored in a memory, and then read — changing the structure of interconnection paths between processing elements and memory cells — as inputs for the following operational step.

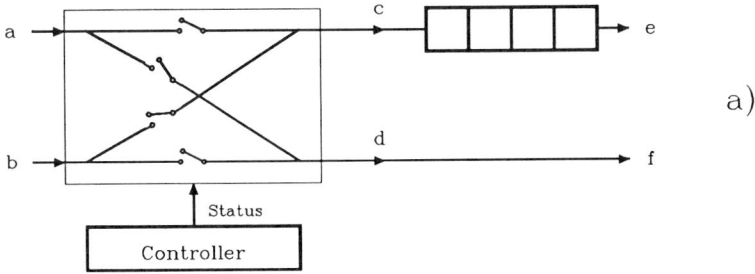

Figure 1.6
Commutator for FFT pipeline.

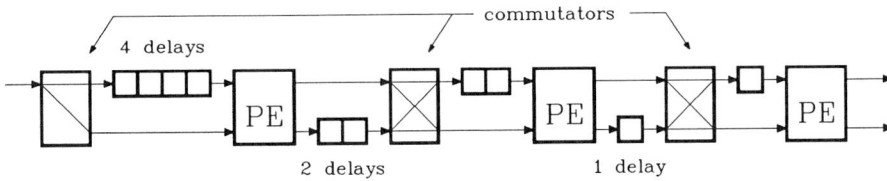

Figure 1.7
Pipeline for FFT implementation.

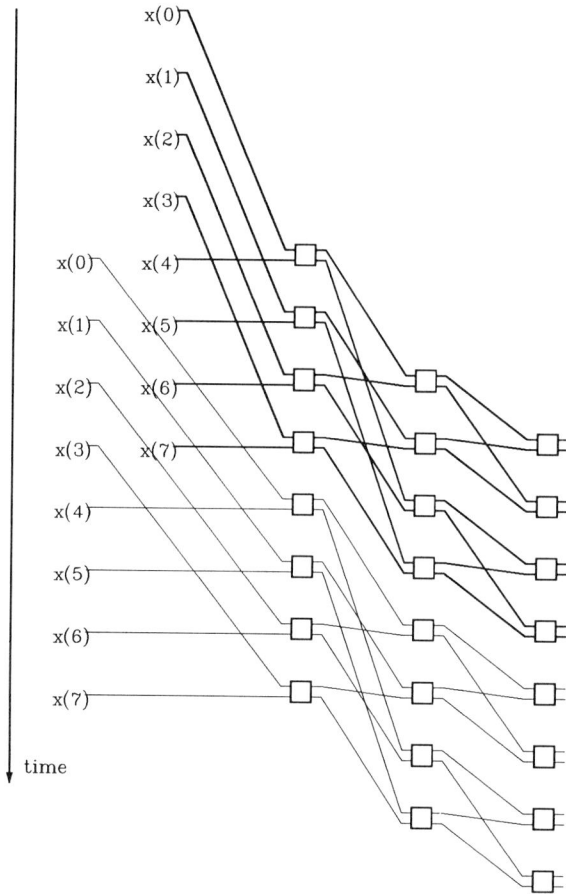

Figure 1.8
Timed flow graph of FFT: boxes are butterfly PEs.

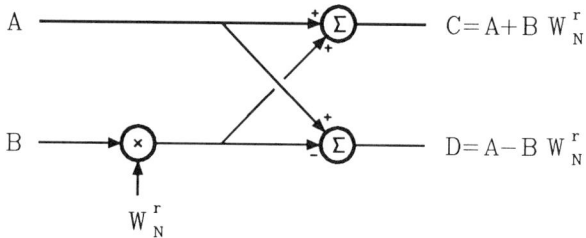

$$C = A + B\, W_N^r$$

$$D = A - B\, W_N^r$$

$$W_N^r$$

Figure 1.9
PE for FFT Pipeline (butterfly PE).

(2) a *row* of PEs (usually called a *pipeline*) that successively executes rows of operations in the graph. Timing of the pipe can be best understood by examining figure 1.8 where the same FFT flow graph is drawn differently, in order to show time steps on the vertical axis. Among PEs of the pipe lie the commutators necessary to store intermediate results and to change their order of flow. Note that the flow graph of figure 1.2 is also a two-dimensional timed flow graph, and this is the reason why it has been possible to map it so easily on a monodimensional pipeline, as seen. Horizontal lines in the graph correspond to data that flow with no delay, whereas downward directed lines correspond to data that must be temporarily stored inside FIFOs, waiting for their partner data. Of course, in this graph no line can be directed upward (no data can be used before having been produced!).

1.1.5. The Fast Fourier Transform (FFT)

Consider now the *FFT algorithm*. It can be shown that equation [1.1] can be modified by exploiting in many ways the intrinsic periodicity of its weights, due to equation [1.2] (see, e.g., [COO85] or [OPP75]). As a result of this manipulation, the number of complex multiplications and additions that must be computed significantly decreases and, consequently, the number of processing elements in the corresponding array also decreases sharply.

For example, the flow graph of figure 1.5 is obtained by one of these modifications, and corresponds to a *radix-2* FFT algorithm, where the number of multiplications has decreased from $N \cdot N$ to $N \cdot \log_2 N$ because the number of stages (columns) decreases from N to $\log_2 N$.

As already said, this flow graph can be drawn as in figure 1.8 so that the vertical axis shows time (i.e., clock pulses) and consequently it corresponds exactly to the pipeline of figure 1.7. The processing cells of this pipeline implement the step operation (known as *radix-2 butterfly*) shown in figure 1.9: this step operation constitutes the elementary flow graph repeated in the stages of figure 1.5. These processing cells are alternated with commutators, such as those described in §1.1.4., that simply change the ordering of data, and — by means of their FIFO memories — delay the data flowing along *sloping* lines of the timed flow graph.

Various other structures have been defined for FFT (see, e.g., [ANT88a, MCC78, SMI85]). *Multiple-radix* butterflies have also been proposed, for enhancing performances. For example, a radix-4 butterfly receives 4 input data at each clock pulse, and computes 4 intermediate results. If this butterfly is adopted, the pipeline can work at a frequency that is one fourth of the input sample frequency, thus multiplying by four the overall throughput. Of course, radix-4 butterfly PEs are rather larger than radix-2 circuits.

The FFT pipeline clearly shows how a DSP algorithm (corresponding to a two-dimensional flow graph) operating on a monodimensional signal can be implemented by a linear array.

Data sequence   Transform sequence

x( )   X_h( )

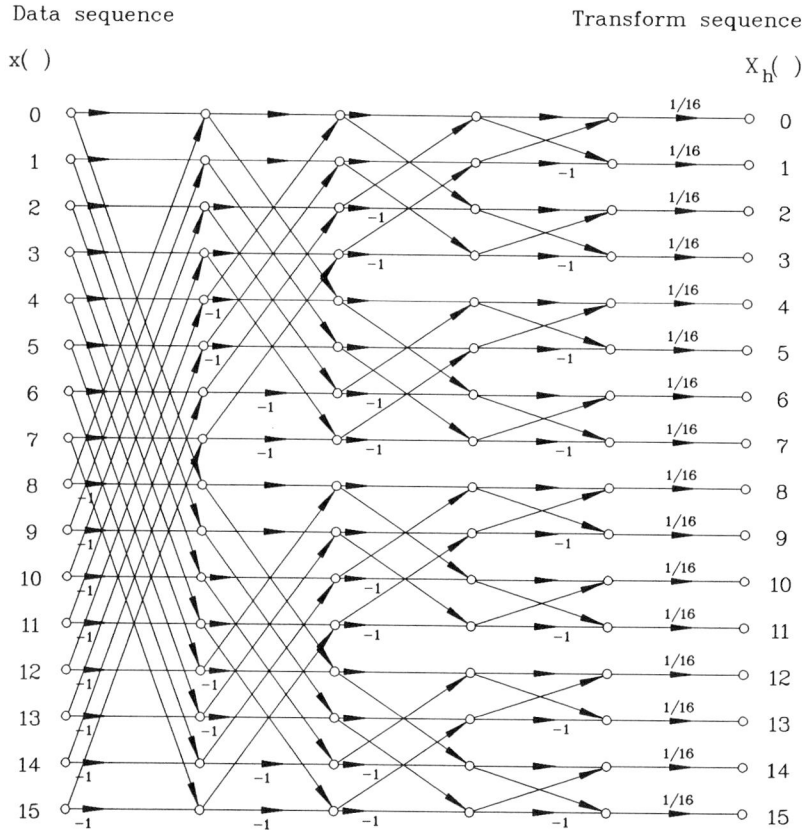

Figure 1.10
Flow graph for FWHT.

1.1.6. The Walsh-Hadamard Transform (FWHT)

Another transform useful in DSP is the *Walsh-Hadamard Transform* (WHT), whose fast version (FWHT) is depicted by the flow graph of figure 1.10 for the case of monodimensional, 16 samples signals. This transform decomposes the input signal in a sum of the 16 squared signals of figure 1.11, instead of in a sum of sinusoids like the Fourier transform. These square waves form an orthonormal base, as do the sinusoids in the Fourier analysis, and consequently they can be adopted instead of sinusoids for decomposing periodic signals. A signal can be decomposed in a summatory of these square waves multiplied by appropriate (unique) weights; the transform is the set of weights.

The corresponding pipeline of figure 1.12 (from [ANT88b]) can be understood by means of considerations similar to the ones previously seen for the pipeline for FFT.

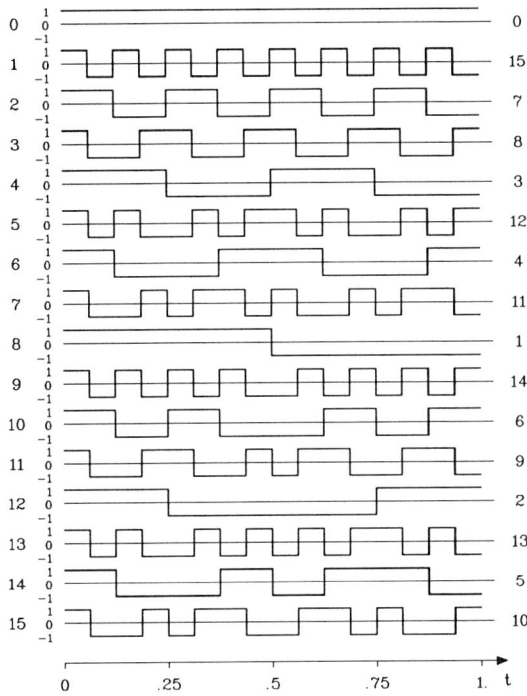

Figure 1.11
FWHT base signals.

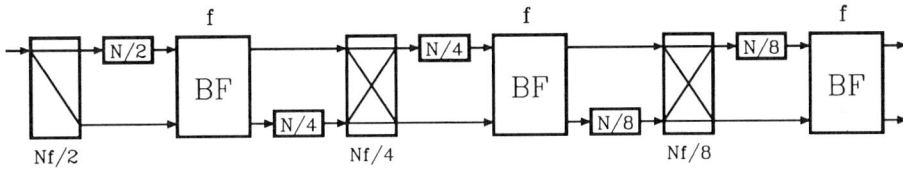

Figure 1.12
Pipeline for FWHT implementation.

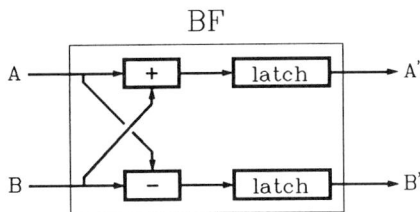

Figure 1.13
PE for FWHT pipeline.

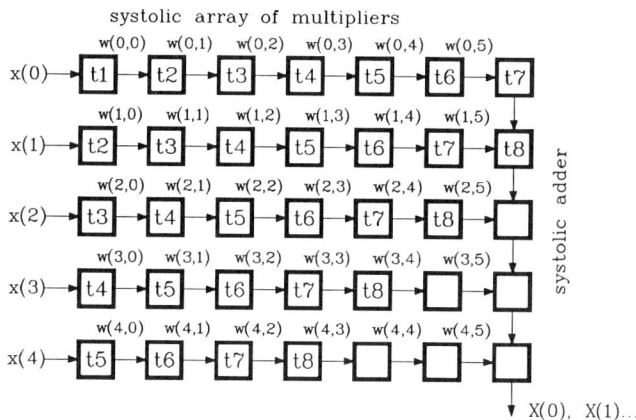

Figure 1.14
Multiple pipeline array for serial DFT.

It should be noted that processing cells (shown in figure 1.13) are simpler than FFT butterfly cells because the corresponding operations do not require true multiplications.

1.1.7. A bit-serial multipipeline DSP array

Another typical interconnection structure found in DSP arrays is that of the so-called multipipeline (i.e., multi-row) arrays, shown in figure 1.14.

In some instances, all pipelines are identical while individual stages of a pipeline may be different. In other instances, as can now be seen in greater detail, all the PEs of the structure are homogeneous. A typical example of circuit showing exactly this property is the convolver, that can also be used as a Discrete Fourier Transformer, presented in [BRU86].

This synchronous, truly systolic circuit performs the operation defined by

$$X = \sum_{j=0}^{N-1} x(j) \cdot w(j) \qquad [1.3]$$

where $w(j)$ are fixed weights. Every pipeline j contains the corresponding fixed multiplicand $w(j)$ (the weight) and it multiplies this weight by its input term $x(j)$.

The individual PE r of pipeline j contains the (fixed) r-th bit of weight $w(j)$, i.e., $w(j,r)$: cell r simply computes the partial product

$$x(j) \cdot w(j,r) \qquad [1.4]$$

and adds it to the partial products generated by the other PEs of the pipeline; thus, the whole pipeline constitutes a serial multiplier implemented in a systolic way.

At the pipeline end, the product bits are passed serially to the corresponding adder stage $A(j)$, that accumulates the product to the partial sum of products arriving from the adder stages of previous pipelines. In order to obtain the correct results, suitable delays must be introduced between operands $x(j)$ fed to the different pipelines: $x(j)$ at generic pipeline j must be input one clock period after $x(j-1)$ at pipeline $j-1$ and so on, so that the two products timely meet at the adder cell of pipeline j.

Successive waves of products can be computed by feeding the pipelines with successive waves of terms: each wave produces a single result, i.e., the corresponding convolved sample. It is sufficient to separate successive waves of numbers by means of filling zero bits, so as to provide for the increased bit length of results produced by multiplication and addition.

A simple external FIFO memory can be used to guarantee that the same input sequence of data (as, for example, produced by an analog-to-digital converter) is repeatedly input to the structure, shifted by one place as required by the convolution operation.

In the case of DFT, following equation [1.1], the operation can be implemented by first bit-serially loading into the structure the data samples (one sample for each pipeline) instead of the weights; then, waves of Fourier coefficients (i.e., the rows of the Fourier matrix defined by equation [1.2]) available in a fixed memory can be input to the array, that computes one transformed sample for each row of coefficients.

As shown in [BRU86], the array can be designed to implement complex additions and multiplications, using two's complement or sign-magnitude serial number formats. Figure 1.15 shows one actual implementation of a convolver operating on sign-magnitude bit-serial data. It is, in fact, only the upper part of the array: pipelines can be added at the bottom arbitrarily, depending on the number of samples. Input signal x flows through the tapped delay line, and from this it feeds the pipelines.

In particular, consider one pipeline, i.e., one serial multiplier. The input stages Ix's separate the sign bits from the magnitude bits of the data samples, and the input EXOR gates compute the sign of the multiplication result: this sign flows through the pipeline and arrives to the final adder, where it is used to convert the result to two's complement format, for the final addition. The control signal, CT, identifies the first bit of each data sample.

The length of the pipelines, i.e., the number of bits of reference data, can be adjusted by means of the vertical SET control signals (one for each column), broadcast to all PEs of a column. The only column where this signal is active works as the complete final adder stage of the array. Of course, all PEs are identical and the SET signal conditions the PE to operate either as a cell of the pipeline multiplier or as a cell of the final adder. Figure 1.16, from [BRU86], shows the internal details of the PE, built around the full adder FA that computes either the sum of the partial products, if acting inside a pipeline, or the final sum if acting as the adder stage at the end of the pipeline (small D boxes are flip-flops).

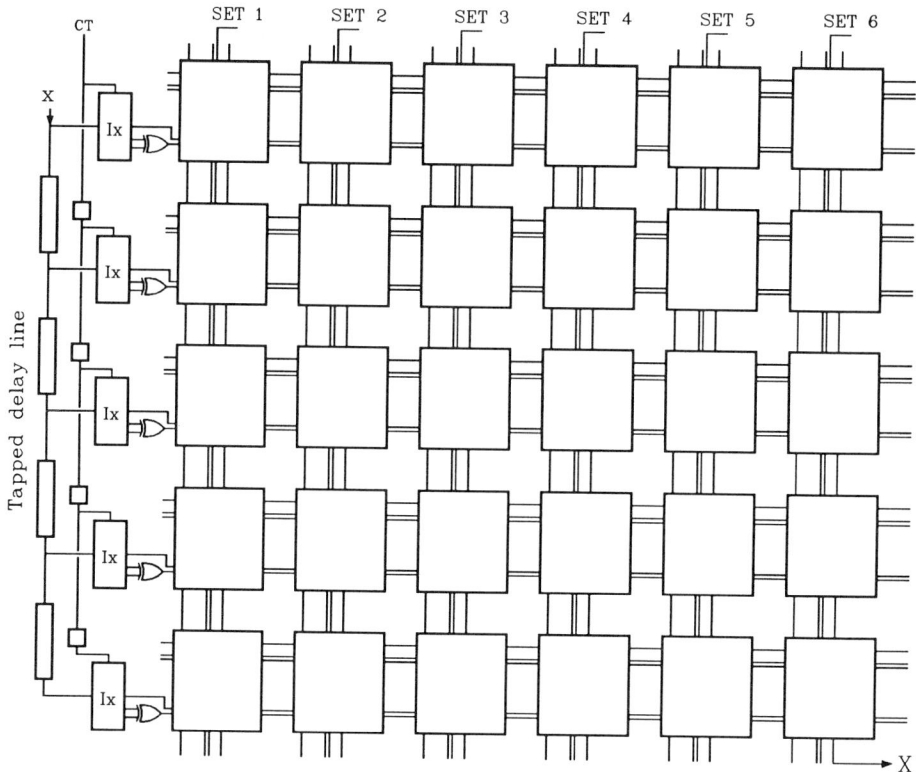

Figure 1.15

Array for serial convolution, based on the array of figure 1.14.

The two lines *SIGN* and *MAGNITUDE* carry the sign bit of the result, and the magnitude bit of the data samples; the bit of the weight is carried, before starting the operation, through the line *WEIGHT* in the flip-flop on this line.

The two vertical lines *SET* and *SUM* are activated only in the last stage of the pipeline (as already seen when discussing *SET*): along the *SUM* line, flow the bits of the final result, represented in two's complement. The other control signals (*C*1, *PERO*, *CNT*, *C*2) are used to synchronize the cell and to reset it at the end of each product phase. For details — here useless — on their functions and on functions of multiplexers *MUX* see [BRU86].

Note that the individual PE is very simple, and that interconnection paths, being serial, also require very little silicon area. As a consequence, a large number of PEs can be accommodated on a single chip and PE production yield can be very high. These high yields keep the costs of building an adequate number of on-chip spare cells very low.

These advantages of serial arithmetics are not unique to this structure: in fact, all the previously seen DSP arrays can adopt serial arithmetics, trading the

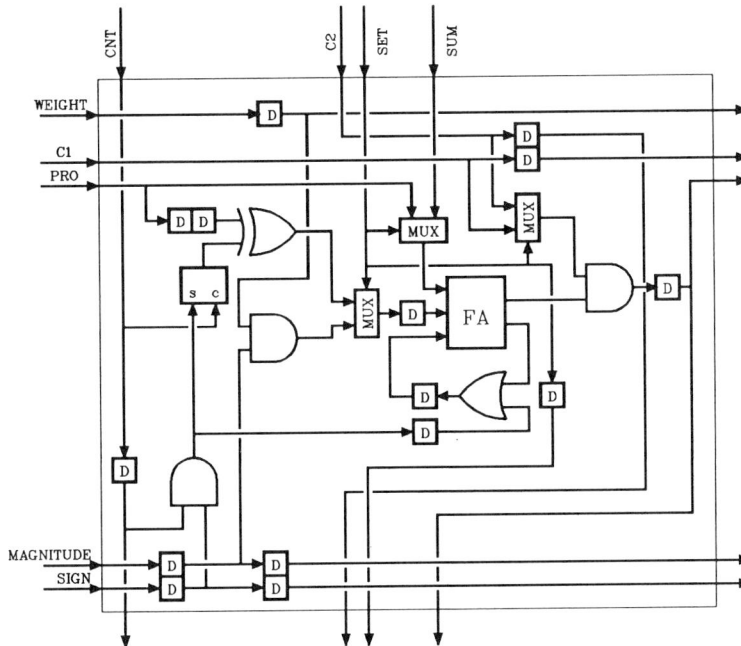

Figure 1.16
PE of the array for serial DFT.

advantages of higher integration against decreased performances.

1.2. Image Processing (IP) Arrays

Another field of applications well suited to arrays is that of Image Processing (IP). Different from DSP arrays, where floating point number formats can be required, IP arrays often adopt simple 8-bit fixed point arithmetics to implement algorithms on grey-scale and color images.

Most usual IP arrays (see, e.g., the survey paper [UHR86] by L. Uhr in [CAN86]) are:

(a) monodimensional arrays, where images are streamed one row at a time;

(b) two-dimensional arrays, that can directly process two-dimensional images by storing small subarrays of pixel data (e.g., 4×4 pixels) in each processing cell;

(c) pyramids, i.e., stacks of successively smaller two-dimensional arrays that are linked together by a tree structure. Thus, a PE in the two-dimensional array of level k is linked only to a PE (father) at level $k + 1$, to four neighbors at the same level, and to four PEs (called children, or siblings, or sons) at level $k - 1$. By introducing the image at the base of the pyramid, and moving it up

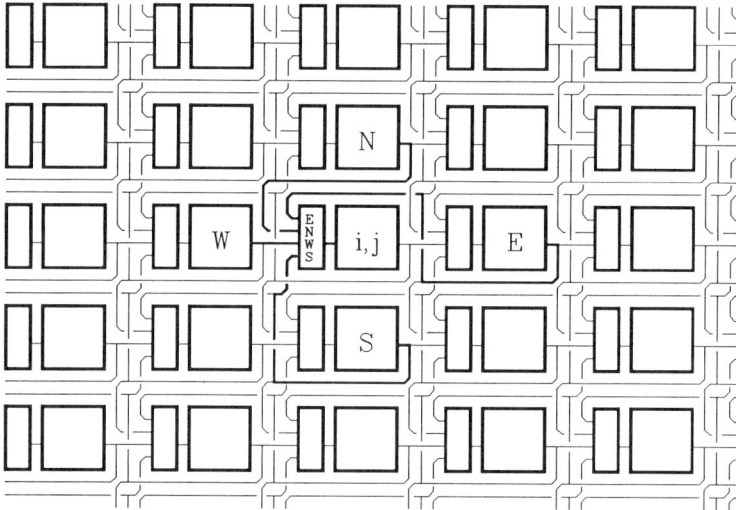

Figure 1.17
2-Dimensional mesh for image processing.

and down the planes, the image can be transformed and processed in many ways (it can be filtered, its contrast can be enhanced, or information can be extracted or averaged, and so on).

When mono- and two-dimensional IP arrays are specialized for implementing just a few low-level image processing algorithms (filtering and contrast enhancement, for example) the final structures are not very different from DSP arrays. The main difference usually lies in the fact that two-way communication links are needed instead of unidirectional links.

Figure 1.17 shows an example of structure for two-dimensional IP arrays. At the PE input, a multiplexer can choose from which of the interconnected neighboring PEs the information must be drawn. In this way, every PE can immediately exchange data with all its neighbors.

When (as is usual in IP) different processing algorithms are to be implemented by the same array, a small arithmetic and logic unit can be introduced in every cell. Each single operation is coded by an operation code, and algorithms can be programmed by synchronously broadcasting sequences of operation codes to all the cells. Each operation is executed synchronously by each cell on its private data. These arrays are thus single-instruction multiple-data (SIMD) computers.

A well-known example of these structures is the MPP architecture. As described in [BAT80] the MPP has a square mesh of 16384 PEs. The PEs are bit-serial processors with 6 one-bit registers, a shift-register of programmable length, and a 1Kbit RAM module.

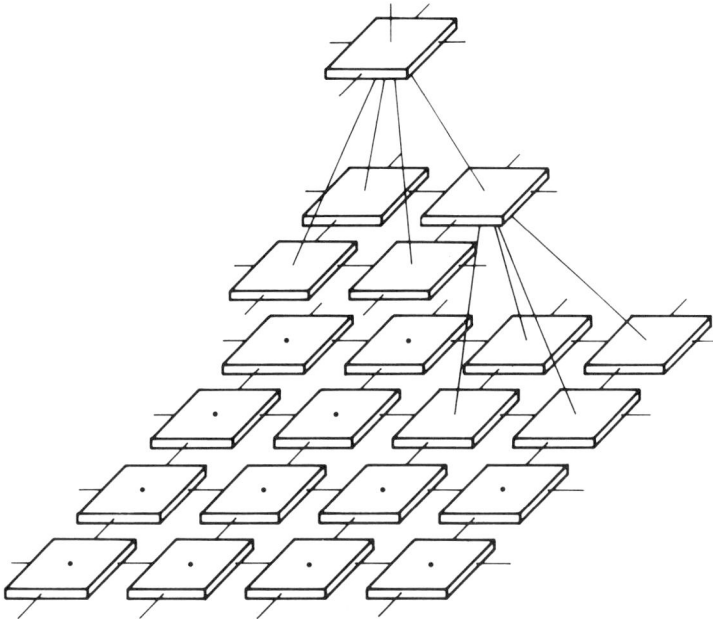

Figure 1.18
Pyramidal array for image processing.

High level IP algorithms (for extracting features like angles, complex curves or regions, and for understanding images and for computer vision) require highly complex data structures and flexible processing, that can best be implemented by more powerful arrays built by connecting programmable computers, i.e., by adopting tightly connected multiprocessors. Such systems, however, are outside the scope of this book.

Intermediate level algorithms found in texture analysis, edge detection, and local motion estimation have an intrinsic pyramidal structure that can well be mapped on a physical pyramid. Note anyway that it is also easy to map pyramidal algorithms onto (flat) two-dimensional arrays, with a not too severe performance degradation.

An example of homogeneous pyramidal array is PAPIA (acronym for Pyramidal Architecture for Parallel Image Analysis and also the Latin name Pavia, the Italian town where this architecture was conceived; see [CAN86]). The elementary PE is a one-bit processor, so that it is suitable for boolean operations and for implementing bit-serial arithmetic. In its first version, the PE contains a set of single-bit registers, and two variable length shift registers, acting as accumulators. A local, 256x1 bit memory is also added to each PE.

This structure can be considered as a classic example of homogeneous multi-SIMD pyramid (see figure 1.18) in which the PEs of every plane can execute contemporarily an instruction which is different from those of the other planes, but all the PEs of the pyramid have the same instruction set.

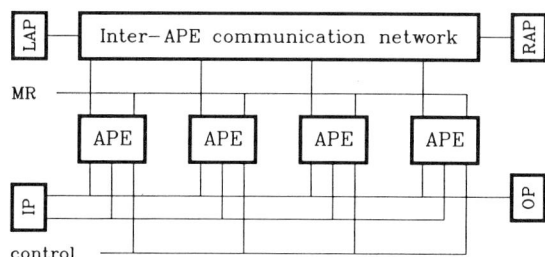

Figure 1.19

An ASP structure, containing APEs.

A notably different case is that of non homogeneous pyramids; where different planes are specialized for different SIMD operations; lower planes can implement low level algorithms like filtering, whereas intermediate planes can implement intermediate algorithms like edge detection and motion estimation, and upper planes can compact image information and attempt image understanding. Each plane can be reconfigured separately from all the others, and then links between planes can be formed, this approach being practically imposed by the dissimilarities among planes.

1.3. Arrays for non numeric processing

A good example of arrays for processing highly structured data is given by the WASP (WSI Associative String Processor) architecture; one variant targeted to real-time low to medium resolution image processing (e.g., for high speed graphics and industrial robotics) is described by Lea in [LEA86].

The basic PE of this structure is the APE (Associative PE): a large number of APEs are interconnected as shown in figure 1.19 to constitute an ASP (Associative String Processors). The ASP provides programmable support of abstract data structures (e.g., sets, strings, queues and stacks, trees and directed graphs). ASPs are chained by their Left and Right Activity Ports (LAP-RAP), and are controlled by a data-activity control bus.

Each APE contains a data register, an activity register, a comparator, and logic for local processing and communications with other APEs. In operation, the ASP supports a form of set processing, in which the subset of active APEs (i.e., those that match the broadcast data and activity values) support scalar-vector and vector-vector operations. The match reply line MR indicates whether none or some APES match.

Figure 1.20 shows how ASP modules can find a place on a wafer, and how they can be linked together by means of a hierarchically tree-structured interconnection. Each ASP substring is implemented as a chain of simply linked segments of *branches*. Each branch incorporates at least one RAM module, for local data, and a communication and control module for local control and for communication

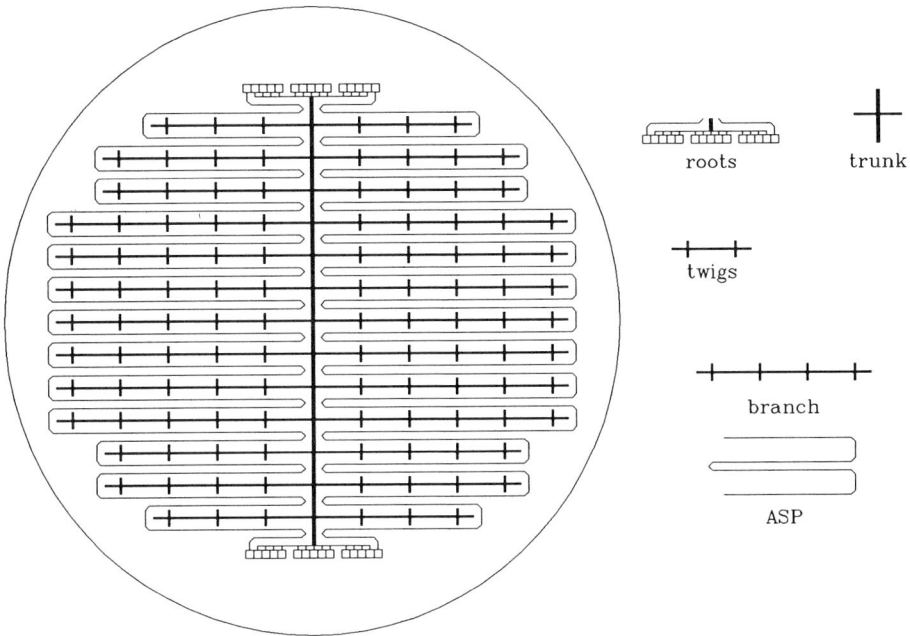

Figure 1.20
A WASP tree.

with other ASP substrings and global input-output data within the *trunk* of the *tree*. Wafer interface circuits constitute the *root* of the tree.

Because of the circuit complexity of the basic modules, the WASP approach explicitly foresees a Wafer Scale implementation as the only means for obtaining the necessary integration level.

The complex interconnection structure and the presence of different basic modules tend to limit the number of defect-tolerance techniques that can be adopted. A first approach suggested in [LEA86] is that of adding spare APEs to an ASP and of cutting interconnections to faulty APEs, taking advantage of the Inter-Ape interconnection network, that selects APEs in a functionally transparent way.

1.4. Iterative cellular arithmetic arrays

Well known even in the past, arrays of very small combinational cells have been variously proposed and implemented to build highly parallel arithmetic devices, in particular, very fast and easily expandable multipliers and dividers.

The schematic diagram of figure 1.21 shows the structure of a 5 × 5 unsigned array multiplier. Every cell is constituted by a simple full-adder; the array works

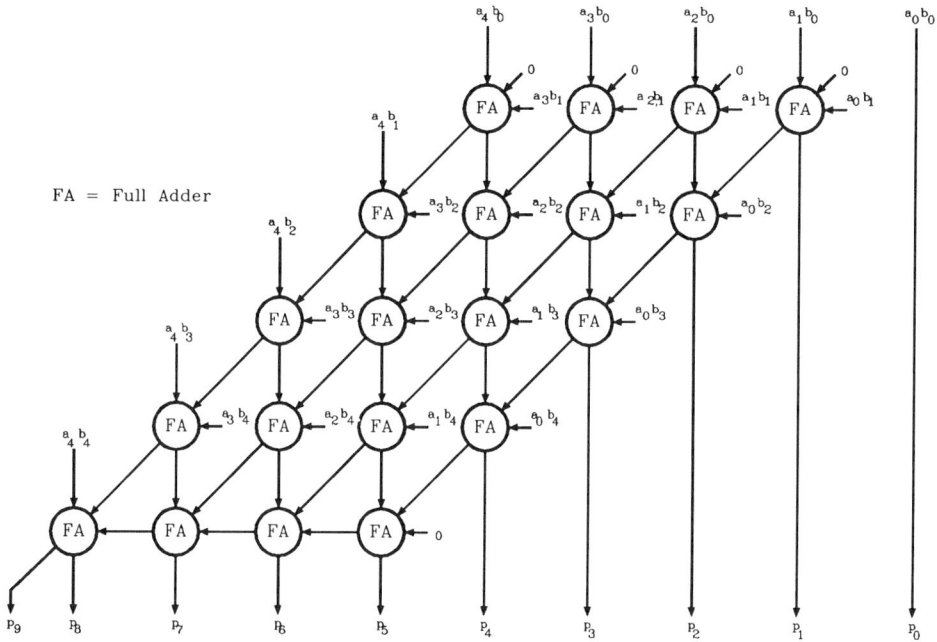

Figure 1.21

A 5×5 unsigned array multiplier.

directly by implementing the multiplication algorithm commonly used in by-hand multiplication.

Input operands are $a_4 a_3 a_2 a_1 a_0$ and $b_4 b_3 b_2 b_1 b_0$; the product is (in binary natural format):

$$P_9 P_8 P_7 P_6 P_5 P_4 P_3 P_2 P_1 P_0$$

The i-th row of full adders adds the i-th partial product

$$(a_4 a_3 a_2 a_1 a_0) \cdot b_i$$

to the similar partial products that are previously added in the upper part of the array and propagates the partial result towards the bottom of the array, as in the computation carried out by hand.

Other structures characterized by cells of similar complexity have been introduced for various kinds of multipliers and dividers (see [HWA79]). These cells are at least two orders of magnitude smaller than the processing elements of the previously considered arrays. They thus have very high production yields and a very low necessity of reconfiguration. At the same time, for such structures, the overhead introduced by spare paths and reconfiguration circuits would be too costly in comparison to cells.

1.5. Characteristics of PEs

The arrays previously examined constitute a sufficiently large and significant set of implementation cases. We now identify the characteristic structural parameters of these arrays that can influence performances and choice of reconfiguration algorithms.

Our focus will be primarily upon the requirements imposed by reconfiguration (and restructuration) against failures (and defects). Reconfiguration to alternatively implement different functions on the same circuit, chosen among a small set of functions, or to change performances (e.g., precision of number formats) will also be considered.

It is important to first consider the nature of the individual PE: *combinational* PEs show little difficulties in comparison to sequential PEs. Combinational PEs are usually small (see, e.g., the case of §1.4.) and are thus characterized by very high production yield and high mean time between failures. Also, they obviously do not require rollback to previous correct states and allow less complicated testing or self-testing procedures.

Small PEs suggest adoption of very small reconfiguration or restructuration circuits, based upon very few spare paths and links. Clusters of PEs can, in any event, be assumed as the smallest replaceable units (e.g., rows in the array of figure 1.21), thus allowing for more complex reconfiguration circuits at the price of lower yields.

In the case of *sequential* PEs, different requirements arise depending on whether memory and registers of a PE can or cannot be accessed by nearby PEs, and on whether it is reasonable to assume that the processor of a cell fails independently of the failure of its memory and registers.

In this case, it is possible to adopt methods similar to those introduced for microprogrammed control units (see e.g., [ANT85]). It is possible, for example, to introduce backup memory and registers where, from time to time, the status of the PE is saved synchronously for the whole array. After reconfiguration, the last saved status can be restored and processing rolled back to that point. The usual problems of rollback methods (correct identification of the moment when the status must be saved, effects of faults hitting backup memories, etc.) are present here as well.

If a PE memory can be isolated from its PE and linked to a nearby PE, so as to substitute the memory of this last one, it is possible to see the array as the superposition of two different arrays: the first one being composed of the memories and the second one of the remaining combinational PEs. The two arrays can be reconfigured with a certain degree of independence if every working combinational PE can be connected to a small set of nearby memory cells.

In fact, some applications do not require such fine-grained rollback techniques. For example, in case of on-line reconfiguration for fault-tolerance, an array can be simply reset in the event of a new fault arising during normal processing; the same happens in a case of power-up testing and reconfiguration only.

Sequential PEs can also be very large, then showing correspondingly low manufacturing yields. In the case of ULSI or WSI implementation, it can be useful to adopt a finer-grain approach to reconfiguration: by identifying PE internal subparts as interchangeable modules separately tested and reconfigured, the overall PE yield can be enhanced so that low-cost reconfiguration algorithms (adapted to high yield PEs only) can be used at array level.

Sequential PEs usually introduce other difficulties related to the presence of clocks. Apart from the cases of purely local clocks at each PE, or local clocks derived in simple, homogeneous ways from array-level clock in the case of globally synchronous arrays, the presence of clocks limits reconfiguration freedom.

Another extremely important character of PEs is the adoption of *serial or parallel* arithmetic; this could in fact be a main criterion for analyzing the processing arrays. Serial arithmetics cause a loss of inherent processing speed of the single PE, but they allow the packing of a higher number of PEs onto each chip. This choice is determined mainly by technological constraints.

Another common advantage of serial implementations is the decrease of length and number of interconnection paths in comparison to parallel arithmetic versions. This is fundamental in decreasing the complexity and cost of the reconfiguration circuits.

Furthermore, serial arithmetics allow the design of smaller PEs, which then leads to higher PEs production yields; this also permits the adoption of simpler fault-tolerance techniques.

Even resulting global throughput loss can be quite low if array sizes (number of PEs) can be increased sufficiently to compensate the single PE throughput decreases, because of the better integrability recalled in §1.1.7.

In conclusion, table 1.1 reports for some of the structures previously discussed, the number of two input NAND gates approximately necessary to build a processing element (*equivalent gates*) and the number of required Input/Output connections, both being key parameters for deciding what reconfiguration algorithm is most suited to a given array. The last column shows mean figures for PEs production yield with standard industrial processes, following figure 1.22 and assuming that a 2-input NAND gate requires from 50 to 100 square microns.

1.6. Other characteristics of arrays

A very important characteristic of the array, which conditions the choice of reconfiguration algorithms, is the presence of a clock that synchronously drives information exchanges inside the array. Synchronism explicitly refers here to communication among PEs, not to adoption of logic synchronous circuits for building PEs. Note that, in principle, even arrays thus classified as *synchronous* can have PEs that — being synchronous circuits — are driven by local asynchronous clocks.

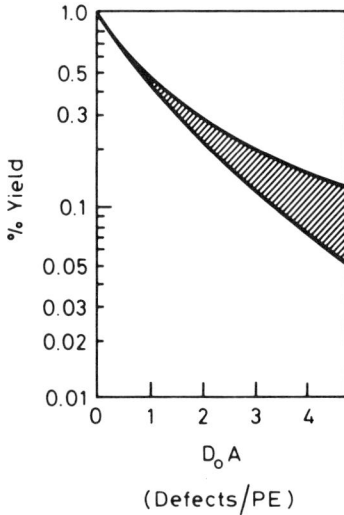

Figure 1.22
PE production yield

Synchronous arrays show clock skew problems; clock frequency can be imposed by the length of the longest link between two logically adjacent PEs, or by the presence of relay PEs that do not operate upon data passing through them but simply act as communication relay elements. In other words, the throughput can be conditioned by the properties of the reconfiguration algorithm.

Relay PEs in synchronous arrays can also introduce clock step delays along communication paths (see, e.g., [KUN84a]: this problem will also be considered in chapters 7 and 11) forcing identification of array cut-sets of PEs to which are attributed the same delay steps to maintain perfect synchronism in information flow.

In *asynchronous arrays* (e.g., *wavefront propagation arrays*, see [KUN84b]) communication among PEs is driven by asynchronous handshaking protocols; absence of clocks guarantees higher freedom in the choice of reconfiguration algorithms but, of course, maximum interconnection lengths continue to condition the global throughput.

In practice, a choice between synchronous or asynchronous implementation of inter-PEs communication is forced by technological aspects. Synchronous arrays tend to prevail, and dominate in the case of linear (one-dimension) arrays, where it is quite easy to eliminate all clock skew problems.

Another important parameter is the *geometric structure* of the array. Linear arrays can be quite simply reconfigured by approaches that guarantee optimality of harvesting or of interconnection lengths (i.e., throughput).

Table 1.1

Order of magnitude of complexity of examined arrays

ARRAY	PE complexity (PE equiv. gates)	Connection complexity (number of PE I/O lines)	Approx. expect. product. yield (% of good PE)
for signal processing:			
Pipe DFT (parallel arithm.)	6000-12000	50-60	<10
multi-pipe DFT (serial arithm.)	100	10	>95
Pipe FFT (parallel arithm.)	8000-15000	80-100	<10
for image processing/coding:			
Pipe FAST WALSH-HADAMARD (parallel arithm.)	4000-8000	60-80	<10
MESH (4 neighb., serial arithm.)	1000-2000	20	30-40
PYRAMID (serial arithm., 4 sons)	1000-2000	20	30-40
for "non numeric" processing:			
TREE (parallel arithm.)	2000-4000	80-100	10-30

Two-dimensional arrays, like trees or meshes, are complicated by the necessity of maintaining precise two-dimensional topologies and constraints. In many cases, this fact keeps the reconfiguration algorithms far from optimality, and not all the working PEs can be used.

Three-dimensional arrays show many difficulties; they can be mapped upon two-dimensional meshes and sometimes this can be done with only a very few performance losses. Practically all the structures seen here have to be actually mapped upon *planar* semiconductor wafer surfaces.

It can be concluded that linear arrays and meshes allow for the highest practical efficiency. This chapter's analysis confirms that, as far as structure is concerned, we can reasonably limit our attention to four basic structures (as shown in figure 1.1) that are most frequently used:

(1) monodimensional (linear) arrays

(2) two-dimensional meshes

(3) trees

(4) multiple linear arrays (multipipelines)

Note that multiple pipelines can also be treated by means of reconfiguration algorithms introduced for restructuring linear arrays, and that trees may be obtained by standard topologic mapping over reconfigured two-dimensional meshes. This helps to explain why linear arrays and meshes will receive detailed treatment in subsequent chapters.

1.7. References

[ANT85]   A.Antola, R.Negrini, M.G.Sami, N.Scarabottolo: *Transient Fault - Management in Microprogrammed Units: A Software Recovery Approach*, Proc. EUROMICRO 85, Bruxelles, Sept. 1985, North-Holland.

[ANT86]   A.Antola, R.Negrini, N.Scarabottolo: *An Approach to Fault Tolerance in Architectures for Fourier Transform*, Microprocessing and Microprogramming, Vol. 18, Dec. 1986, North-Holland.

[ANT88a] A.Antola, R.Negrini, N.Scarabottolo: *Arrays for Discrete Fourier Transform*, Proc. EUSIPCO 88, Grenoble, Sept. 1988.

[ANT88b] A.Antola:*Multiple Transform Pipelines for Image Coding*, Proc. EUSIPCO 88, Grenoble, Sept. 1988.

[BAT80]   K.E.Batcher:*Architecture of a Massively Parallel Processor*, Proc. 7th Symp. on Computer Architecture, May 1980, IEEE.

[BRU86]   O.Bruschi, R.Negrini, S.Ravaglia:*Systolic Arrays for Serial Signal Processing*, Microprocessing and Microprogramming, Vol. 20, Apr. 1987.

[CAN86]   *Pyramidal Systems for Computer Vision*, (V.Cantoni, S.Levialdi eds.), Proc. NATO ARW, Maratea, May 5-9 1986, Springer-Verlag.

[CAP84]   *VLSI Signal Processing* (P.R.Cappello et al. eds.), 1984, IEEE books, IEEE.

[COO65]   J.W.Cooley, J.W.Tukey:*An Algorithm for the Machine Calculation of Complex Fourier Series*, Math. Comput., Vol. 19, Apr. 1965.

[COS86]   *Parallel Algorithms and Architectures*, (M.Cosnard, Y.Robert, P.Quinton, M.Tchuente eds.), Proc. Int'l Workshop on Parallel Algorithms and Architectures, Luminy, Apr. 14-18 1986, North-Holland.

[HWA79]   K.Hwang: *Computer Arithmetic. Principles, Architecture and Design*, 1979, John Wiley and Sons.

[KUN80]   H.T.Kung, C.Leiserson: section 3 of chapter 8 in: C.A.Mead, L.A.Conway: *Introduction to VLSI Systems*, 1980, Addison Wesley.

[KUN82]   S.Y.Kung, K.S.Arun, R.J.Gal-Ezer and D.V.Bhaskar Rao: *Wavefront array processor: language, architecture and applications*, IEEE-TC, Vol. C-31, N. 11, Nov. 1982, 1054-1066.

[KUN84a] H.T.Kung, M.S.Lam: *Fault Tolerance and Two Level Pipelining in VLSI Systolic Arrays*, Proc. MIT Conf. on Adv. Res. in VLSI, Jan. 1984.

[KUN84b] S.Y.Kung:*On Supercomputing with Systolic/ Wavefront Array Processors*, Proceedings of the IEEE, Vol. 72, N. 7, July 1984.

[KUN88] S.Y.Kung: *VLSI Array Processors*, Prentice Hall, Englewood Cliffs, 1988.

[LEA86] R.M.Lea: *A WSI image processing module* in *Wafer Scale Integration*, (G.Saucier, J.Trilhe eds.), Proc. IFIP WG 10.5 Workshop, Grenoble, Mar. 17-19 1986, North-Holland.

[LEG86] *Parallel Processing by Cellular Automata and Arrays*, (T.Legendi, D.Parkinson, R.Vollmar, G.Wolf eds.), Proc. Third Int'l Workshop on Parallel Processing by Cellular Automata and Arrays, Berlin, Sept. 9-11 1986, North-Holland.

[MCC78] J.H.McClellan, R.J.Purdy:*Applications of Digital Signal Processing to Radar*, in: *Applications of Digital Signal Processing*, (A.V.Oppenheim ed.), 1978, Prentice Hall.

[MOO85] *Wafer Scale Integration*, (W.Moore, C.Jesshope eds.), Proc. Int'l Workshop, Southampton, July 10-12 1985, Adam Hilger.

[MOO86] *Systolic Arrays*, (W.Moore, A.McCabe, R.Urquhart eds.), Proc. First Int'l Workshop on Systolic Arrays, Oxford, July 2-4 1986, Adam Hilger.

[OPP75] A.V.Oppenheim, R.W.Shafer:*Digital Signal Processing*, 1975, Prentice Hall.

[POT85] J.L.Potter ed.:*The Massively Parallel Processor*, 1985, MIT Press.

[REI86] *Highly Parallel Computers*, (G.L.Reijns, M.H.Barton eds.), Proc. IFIP WG 10.3 Working Conf. Sophia-Antipolis, Mar. 24-26 1986, North-Holland.

[SAU86] *Wafer Scale Integration*, (G.Saucier, J.Trilhe eds.), Proc. IFIP WG 10.5 Workshop, Grenoble, Mar. 17-19 1986, North-Holland.

[SMI85] S.G.Smith:*Fourier Transform Machines*, in: P.Denyer, D.Renshaw: *VLSI Signal Processing, a Bit-Serial Approach*, 1985, Addison-Wesley.

[ULL84] J.D.Ullman:*Computational Aspects of VLSI*, 1984, Computer Science Press.

[SWA86] E.E.Swartzlander:*VLSI Signal Processing Systems*, 1986, Kluwer Academic Press.

[UHR86] L.Uhr: *Parallel, hierarchical software/ hardware pyramid architectures* in *Pyramidal Systems for Computer Vision*, (V.Cantoni, S.Levialdi eds.), Proc. NATO ARW, Maratea, May 5-9 1986, Springer-Verlag.

Fault models are the basis of any fault-tolerance technique. To evaluate the efficiency of such techniques it is mandatory to identify the set of faults that have to be detected and against which a system must be hardened. Most fault-tolerance techniques deal with *permanent faults*: this is true, in particular, for the case of processing arrays. This restriction will be adopted here also.

Although proposals for fault-tolerance of digital systems refer either to *logical faults*, i.e., faults that manifest themselves as unexpected values of logic functions, or to *functional errors*, that manifest themselves as unexpected results of high-level functions in response to known input data, it is necessary to consider, even quite briefly, some characteristics of the physical defects or failures causing the faults. A number of failure causes for highly integrated circuits have been listed by various authors. Our interest is not so much in the specific defects and in the mechanisms causing them as in their distribution (both onto a chip/wafer and in time) and in their relative relevance with respect to the total amount of failure causes. We will therefore consider such characteristics of fault types and fault distributions that are most likely to affect subsequent reconfiguration policies aiming to achieve fault-tolerance.

A basic preliminary classification has been proposed in [STA83]. It refers primarily to failures appearing during the production process — and, as such, affecting *yield* rather than *reliability* — and it distinguishes between *gross failures* and *random failures* (leading, in Stapper terminology, to *gross yield* and *random yield*, respectively). The authors of [STA83] examine fault distributions in *wafers*, each consisting of a regular geometry of *chips* whose dimensions and complexity are obviously dependent on the specific case. Stapper's first works actually referred to memories.

Gross yields are caused by manufacturing errors as a consequence of which wafers or at least major parts of wafers have no functioning chips. Such errors do not cause the chips to fail in random patterns on the wafer: losses are associated with the fact that resulting device parameters are completely out of acceptable bounds, and they cannot be recovered from through fault-tolerance techniques. In a similar way, Mangir and Raghavendra refer to *performance failures* [MAN84a] that affect speed and relative timing of signals, and that prevail over the whole wafer or — if the individual VLSI system is considered — over the whole chip. It is self evident that faults of this class can be overcome only at the *process* level, by adopting the most careful process control techniques in what is basically a *fault avoidance* approach. Therefore, in this study we will not consider either their causes or their effects. Statistics ascribe only 16.5 percent of yield losses to gross defects — and, obviously, such defects have no influence on run-time reliability — while random defects account for 83.5 percent of yield losses and are responsible also for most of those run-time failures that cannot be ascribed to environment causes. Thus, random defects will be considered here in far greater detail.

Some of the random defects affect relatively large areas and can be identified by visual inspection: these are typically area and line defects [MAN84a]. Other defects are actually due to errors that occur during processing: missing contact windows, misaligned gates, and missing device regions are characterized as *spot* defects and are considered to be localized and randomly distributed over the whole wafer and, therefore, over large chips as well. Spot defects are responsible for dependence of yield on chip area: in [MAN84a] their impact on yield loss has been evaluated as approximately 60 percent of total yield losses.

Photo defects, i.e., varying sensitivity of photolithographic patterns to defects of different size, can be classed among spot defects. It must be emphasized that such defects do not necessarily cause failures, and hence errors, since probability of failure is a function of defect size and it ranges from zero, for very small defects, to one for very large ones. Failures caused by such defects are identified by probe testing and, again, as in the case of gross defects, they affect *yield*, rather than *reliability*.

A large class of defects is defined as the *particulates* class [SCH86]. Impurity particles may be present either in the environment or in gases used during the production process, and they may lead to electrical failures either directly or through interfering with subsequent process steps. Defects thus arising are small in size and affect only a localized area of the circuit.

Oxide defects are also responsible for many failures in VLSI circuits. In contrast to defects previously examined, they affect not only yield but also reliability. Some oxide defects appear as infant failures due to electrical overstress or static discharge that result in short circuits. While in principle these should be considered as *run-time defects* rather than *production-time* ones, usually infant mortality is accounted for during burn-in, so that resulting failures may be classed together with *production-time* ones as far as subsequent treatment with fault-tolerant techniques is concerned.

A second main class of oxide defects that cause both production-time and run-time failures are so-called *pinholes*. A pinhole is a microscopic defect in insulators, the most important occurring, in the oxide layers. As a consequence of a pinhole, a short circuit between conductors on different levels of a MOS chip may be created, so that the physical *defect* (pinhole) will result in an electrical *failure* (the short circuit) and this, in turn, will originate a logic fault.

Not all pinholes generate failures and, more precisely, not all areas of a chip are as *sensitive* to pinholes, meaning that equal distributions of such defects in different sections of the chip will not create equal distributions of errors. Stapper identifies as the *critical area* for pinholes an area in which conductors on different levels cross each other. Other authors also relate pinhole density with factors such as wire length and to the number of crossovers or contacts, in order to deduce the probability of pinhole-induced faults. Since a pinhole results from a weakness in the oxide layer, it may also appear during device operation; thus, by inducing run-time faults it affects reliability besides yield.

There are also other defects appearing in the semiconductor (e.g., junction leakage) that affect yield and reliability. Many derive from scaling-down of devices when higher and higher integration density is achieved. Scaled-down devices are in fact more process-sensitive for a number of reasons:

- a thinner dielectric, that is both more difficult to manufacture and that may suffer from reliability problems;

- higher current density in power supply scaled-down lines, that may again lead to reliability problems;

- a less sharp distinction between *on* and *off* states of devices, impacting on circuit performance [MAN84b].

An important point stressed by most authors is that in the case of almost all random defects the probability of subsequent electrical failure depends on circuit topology, layout, circuit complexity, and regularity. All prior considerations lead to a classification of defects. For our purposes it is necessary to see how these defects are translated into logic faults.

2.1. Fault models

Traditionally, faults in digital circuits have been modeled on the so-called *stuck-at* model. As a consequence of a failure, a line (or *net*) is considered to be *stuck* at the logical value 0 (*stuck-at 0*) or 1 (*stuck-at 1*). Additional assumptions are that of the mutual independence of faults, that of the identical probability for all possible faults, and — at least, as far as testing is concerned — that of single-fault appearance in a given circuit. When higher integration density is reached all such assumptions become far from acceptable. All faults are not equally probable (this derives from process characteristics), they are technology- and layout-dependent, and all physical defects cannot be represented by the simple stuck-at model. A single defect at the physical level may result in multiple faults throughout the circuit; in particular, this happens when it affects a signal line feeding into multiple points. Faults of structural origin may be better modeled as *shorts* or *opens* (the electrical meaning is obvious). In programmable logic structures, such as PLAs, a further fault source comes from so-called *cross-point defects*, in which missing connections between signals that should be coupled, or alternatively couplings between lines that should not be interacting, modify circuit functions. Moreover, C-MOS technology — the one most largely used today for VLSI and for experimental Wafer-scale integrated systems — introduces further classes of failures, defined in current literature as *stuck-at opens* and *stuck-at ons*, and which cannot be represented by the conventional models since they introduce a sequential behavior even in the simplest combinatorial circuits.

The complexity deriving from all these diverse fault classes (as well as their being caused by their possible relationships with physical defects) is such that relating fault-tolerance techniques to logic-level faults would not be practicable for complex systems such as processing arrays. Moreover, it would make the techniques themselves too strongly process-dependent, which would be unacceptable in a rapidly changing technology. Thus, proposals for fault-tolerance of VLSI devices through reconfiguration make use of *functional error models*: i.e., models corresponding to an abstraction level higher than the logic one and dealing with the *functions* expected from a given *block* of a circuit. If the *block* is considered as the *minimum replaceable unit*, it is not necessary to locate faults within it with greater detail. A simple information *good* or *faulty* may be associated with it, and this is sufficient for subsequent reconfiguration strategies. This type of fault (or, more precisely, *error*) model will be adopted throughout this book: the *block* is usually made here to correspond to the individual processing element in the array.

Having identified the *abstraction level* at which errors are defined, it is also necessary to see how — in correspondence with defect analysis — errors are distributed both in space, i.e., on the chip, and in time. To this end, we must make a careful distinction depending on whether the reconfiguration is performed statically at the end of production, so as to enhance yield, or dynamically at run-time so as to enhance reliability. This distinction is also related to the physical defect causes involved in the two different instances.

2.2. Reconfiguration for yield

It is obvious that in this first case only distribution in space is of interest. Reconfiguration to increase yield has an interest for very large chips on which a reasonably high number of blocks can be integrated. In reality, it becomes viable for the highest-density VLSI devices and for Wafer-Scale integrated systems. Gross defects lead to immediately discarding the wafer or the affected chips. Even for *random defects*, however, an immediate assumption of a random (e.g., Poisson) distribution is not an a-priori acceptability. A number of important points have been observed in the literature:

- it has been experimentally found that about 30 to 40 percent of defects are located in the interlogic connections, while the remaining ones are in the logic/device area. This leads to the conclusion that techniques for static reconfiguration cannot refer to the assumption of defect-free interconnections, and that large defective *areas* (comprising interconnections as well as processing elements) will have to be taken into account;
- defects in the logic subareas depend on the complexity of the logic itself;
- defects tend to cluster — this is particularly true for defects of particulate origin, since particles are more liable to deposit where there are already other particles. Clustering is more evident along the border and in the center of a wafer.

The following assumptions can thus be made concerning distribution of *faults* and *errors* on a wafer, in view of production-time configuration:

(1) errors can be both random and clustered;

(2) random faults in interconnections can — if the reconfiguration strategies provide for this — be assimilated to errors in some connected block. As for clus-

tered errors, it must be assumed that interlogic connections running among faulty blocks are in turn faulty and cannot be used for the reconfiguration process;

(3) I/O logic along the borders must provide for faults in that area also, unless it is foreseen that the border region of a wafer is apriori discarded from the working system.

Since end of production reconfiguration (or, as some authors prefer to define it, *configuration* or *restructuring*) is controlled by a host computer external to the wafer/chip, complexity of reconfiguration-controlling circuits is not a subject of discussion. Nevertheless, the problem of final *yield* has to be carefully considered, since yield is a function of area and also of circuit complexity. While this point will be discussed in detail in the following chapter, it must be noted here that the same factors guiding the definition of an error model have to be taken into account when the yield model is formulated. Specifically, the fact that a simple Poisson distribution for production-time faults is not wholly satisfactory has been recognized by most authors dealing with such problem.

2.3. Reconfiguration for reliability

The alternative problem will now be considered: run-time reconfiguration aiming at higher reliability, i.e., aiming to achieve higher lifetime for a device with the given nominal functions. In this case, both space- and time-distributions of errors must be taken into account; the following factors can be accepted when an error model is defined:

(1) Errors that appear at run time are a consequence either of environment factors or of operation stress. Although pinholes may become manifest some time after operation has begun, it may be assumed that most of them have appeared during burn-in and therefore have been taken care of, if necessary, by static reconfiguration techniques.

(2) Assumptions usually adopted when reliability of ICs is considered may be adopted. In particular, it can be assumed that logic areas are more subject to faults than interlogic connections, and that logic complexity is a factor to be taken into account to evaluate reliability. A first-approximation criterion often used relates failure rate to transistor count. As a consequence, final evaluation of different reconfiguration techniques should also take into account the added complexity due not only to spares but also to reconfiguration-controlling circuits (if on-chip control for reconfiguration is envisioned).

(3) Points (1) and (2) lead to the assumption that errors are:

- confined to the logic blocks (in this case, processing elements) and to circuits of greater complexity, without affecting interlogic connections;

- mutually independent, excluding the possibility that failure of a circuit component causes undue electrical stress in other components;

- occurring randomly through the circuit and with a random distribution in time (distribution of failures in time can be reasonably assumed to be a Poisson one).

The above considerations have allowed many authors to adopt classical Markov models for reliability evaluation, with possible modifications introduced only to account for reconfiguration and fault-tolerance. Some details on this subject are given in the Appendix.

As will be seen in the following chapters, most reconfiguration approaches actually make use of this second fault model even for the case of end of production restructuring; the assumption most relevant is, in fact, that of confining faults to PEs only. We will explicitly point out the few instances in which the first error model, in its completeness, is adopted, and the main characteristics of the corresponding approaches will be highlighted as well.

2.4. References

[MAN84a] T.E.Mangir, C.S.Raghavendra:*Issues in the implementation of fault-tolerant VLSI and Wafer-scale integrated systems*, Proc. ICCD 84, 1984, IEEE.

[MAN84b] T.E.Mangir:*Sources of failures and yield improvement for VLSI and restructurable interconnect for R-VLSI and WSI: Part I: Sources of failures and yield improvement for VLSI*, Proceedings of the IEEE, June 1984, 690-708.

[SCH86] P.Schvan, R.Hadaway, M.King: *Defectivity and yield analysis for VLSI and WSI*, Proc. ICCD 86, 1986, IEEE.

[STA83] C.H.Stapper, F.A.Armstrong, K.Saji: *Integrated circuits yield statistics*, Proceedings of the IEEE, vol.71, N.4, Apr. 1983, 453-470.

3 BASIC PROBLEMS OF FAULT-TOLERANCE THROUGH ARRAY RECONFIGURATION

A basic choice can be made between two alternative approaches to the problem of fault-tolerance through reconfiguration of processing arrays:

(a) It is possible to achieve fault-tolerance by adapting the *algorithm* implemented by the processing array introducing very low circuit/time overheads for the array itself. In this case, modifications of the computing structure are introduced only to grant testability and error confinement: we could speak of a *fault tolerant algorithm mapped onto a fault-intolerant array*. Obviously, this approach is inherently application-dependent: it can be adopted only insofar as the algorithm lends itself to such *software reconfiguration*. If this type of approach is chosen, additional structure costs are quite low. Degradation will appear in the performances of the reconfigured mapping of algorithm onto array, typically through decrease of processing speed. Consider, for example, mapping of a DFT algorithm (see §1.1.1) onto a linear array. If the number N of PEs present is lower than the number of steps in which the algorithm is subdivided, it will always (i.e., even initially, on a fault-free structure) be necessary to *fold* the algorithm onto the array, achieving a given initial nominal speed. Presence of F faults (if suitable bypass links are provided to exclude the faulty PEs) simply appears as availability of a linear array of $N' = N - F$ PEs, and reconfiguration of the algorithm is obtained by a new folding of the algorithm onto this smaller structure. The choice of the folding policy will tend to limit the speed decrease caused by the decreased dimensions of the target array onto which the algorithm is mapped. Some solutions in this line can be found in [HUA84, JOU86, RED84].

(b) The alternative approach is to keep the algorithm unmodified, as regards its mapping onto the architecture as well, and to design an array capable of reconfiguration. In this second case, the algorithm (or set of algorithms) for which the array is designed is originally mapped onto a *logical* or *target architecture*; the *physical array* structure supporting this architecture must then be capable of undergoing reconfiguration, in presence of faults, so as to keep unchanged the logical array. This goal is achieved by augmenting the interconnection network in a suitable way (data flow between cells must be modified when faults are present), and by making use of some form of redundancy either in the structure (through introduction of spare PEs) or in the processing time (through introduction of spare processing phases).

In this book, only solutions based on this alternative approach are analyzed in detail. Reference is made to some *basic classes of architectures* rather than to specific algorithms or implementations, and only approaches of the second type can be examined from such a general point of view.

3.1. Reconfigurable arrays: the rationales

Two different goals may be defined for designing arrays capable of fault tolerance through reconfiguration:

(1) for *static reconfiguration*, performed at end of production (also defined as *configuration* or *restructuring*), the aim is to achieve *higher production yield*. The goal is achieved by accepting the presence of a (limited) number of faults on a chip or wafer, and by introducing a reconfiguration technique to circumvent such faults and obtain a correctly operating device. Since yield for very large chips (if no measure of fault tolerance is provided) can be as low as 10 percent or even lower, increasing such figures is a requirement of obvious importance for making technologies and also processing architectures economically viable.

(2) for *dynamic reconfiguration*, performed at run-time, the aim is to achieve higher *availability* and *reliability* of the device. Besides the intrinsic necessity of adding to the lifetime of very complex devices, it should be noted that many application areas of processing arrays are typically mission-critical. Consider the largest application area, i.e., real-time signal processing. Whether the final user environment be telecommunications, radar imaging or spaceborne signal processing, it is self-evident that reliability and availability are prerequisites.

Techniques to achieve these two goals are at least partially different: variations arise from the *fault models* adopted (see chapter 2) and they are also due to the *technologies* used to reach the reconfiguration proposed and even to the different *goals* of the reconfiguration procedures.

In the case of end of production configuration (or static reconfiguration), there are also two possible goals:

(A) to obtain the largest possible array from a chip (wafer) given the existing fault distribution. In this case, the chip can be considered *good* until an array of *any* dimensions (consistent with the target architecture envisioned) can be extracted from it;

(B) to obtain a nominal array of predefined dimensions even in the presence of faults. In this case, obviously, a percentage of chips — with fault distributions that do not allow the goal to be achieved — will be discarded.

In the case of run-time reconfiguration, the most reasonable aim is to keep available a logical array of predefined dimensions until the fault distribution exceeds limits leading to *fatal failure*, i.e., until a fault configuration is reached such that recovery becomes impossible, with the given reconfiguration algorithm. Apart from technologies used to implement reconfiguration, this second problem can be reduced to (B).

Referring then to the *redundancy* required to achieve reconfiguration, a number of different instances can be discussed:

- In case (A) above it is not precisely correct to take into account a *redundancy* as far as the number of PEs is concerned. If N is the total number of PEs on the chip, and X faults are present, the configuration algorithm will at most allow the obtaining of a working array with $M = (N - X)$ working interconnected PEs. None of the PEs is considered as a *spare* one. The number of PEs effectively interconnected in the working array will usually be lower than M, which constitutes an upper limit for the configuration algorithm. The performance of the configuration algorithm, measured in relation with the percentage of fault-free PEs actually inserted in the configured array, is usually called *harvesting*. The only form of redundancy that can be evaluated refers to the interconnection structure, since interlogic connections must provide for alternative paths to allow configuration around faults.

- Case (B) above — from the point of view of processor redundancy — presents no difference from the case of dynamic reconfiguration: distinctions appear in the technology by which reconfigured interconnections are activated (and therefore, ultimately, in the area taken by the interconnection subsystem). Basically, we need to introduce a number R of *redundant* PEs (*spares*) capable of performing the functions of faulty ones. With the best-performing algorithms, these spares will ideally be able to support up to R faults. Two alternative techniques can again be distinguished:

(B.1) *Space* or *structure redundancy.* R spares are physically added to the *nominal* array of N PEs: unused spares are not active, i.e., they do not participate in array operations. When a working PE is found to be faulty it is excluded from operation and a spare one becomes active. The upper limit for reconfiguration is clearly R faults. Practically, for most algorithms reconfiguration will become impossible for some distributions of $X < R$ faults as well, due to limitations on complexity of both algorithm and interconnection network. Processing speed of the array is not affected by reconfiguration (the only speed degradation factor is due to additional delays introduced by longer interconnection links). Structure (or *area*) redundancy results from the R spare PEs and to the augmented interconnection network.

(B.2) *Time redundancy.* No *physical* spare PEs are added to the array in this case. When a PE is found to be faulty, it is excluded from operation and one of the the fault-free PEs will perform its functions during an added operation phase. Thus, in a way, all nominal PEs act as *time spares* during such added operation phases. Here also, obviously, the interconnection network must be augmented to allow this type of operation: structural redundancy is limited to such added links and to the controlling circuits. On the other hand, as soon as there are faults the processing speed decreases, even in a relevant way: redundancy can be said to appear in the number of operation phases required to perform the processing functions.

The *costs* of the two alternatives (B.1) and (B.2) will now be analyzed in detail.

3.2. Case (B.1): structure redundancy

Reconfiguration through structure redundancy involves three main cost factors that should be examined when evaluating an algorithm:

- loss of structure regularity through addition of the augmented interconnection network and of the related control circuits;

- decrease of interconnection locality to allow connection of PEs that are not physically neighbors;

- increase of area due to spare PEs to augmented interconnection network and to control circuits;

- increase of connectivity for the individual PEs to allow alternative connections.

The first two factors also lead to further area increases, since nonregular structures and nonlocal interconnections lead to less compact layout. The discussion may be simplified by assuming that they affect primarily layout complexity and communication delays. While the first point is too dependent on design rules and implementation technology to be discussed in general terms, the second point is specifically taken as a factor of cost for many reconfiguration algorithms and, as such, will later be taken into account from case to case. Accordingly, the present focus will be on the effects of area increase. The most trivial effect is that redundancy requires larger chips or, alternatively, it leads to lower processing power per given chip. Even more interesting are the effects on yield and reliability.

Yield of VLSI circuits has been extensively studied. While relatively simple models have also been proposed, it has been found that they hardly fit experimental data. No generally accepted model has in fact been presented, although solutions proposed by Stapper and by Mangir and Avizienis have achieved a consensus. The large variety of defects, as well as the different characteristics of the circuits and consequently of their sensitivity to defects, easily explain why no single yield projection model exists. Stapper in [STA83] analyzes various types of statistics for definition of a yield projection model, concluding that mixed and compound Poisson statistics lead to a formula providing a good fit to experimental data and enabling the clustering of defects to also be taken into account. While Stapper's first work [STA76] concentrated mainly on large RAM chips, later papers by other authors have dealt with other architectures, and in particular with processing arrays. This is very important for the analysis of yield in presence of redundancy, since redundancy and reconfiguration policies are quite different in the case of memories and in the case of processing arrays.

A basic paper on yield and its relation with redundancy in fault-tolerant arrays was published in 1982 by Mangir and Avizienis [MAN82a, MAN82b]. To derive yield expressions, they preferred to adopt Einstein-Bose statistics, as this approach keeps its validity in the presence of redundancy as well. Moreover, derived expressions referred to the *cumulative* effect of defects, rather than to the individual defect types. Again, the statistics employed are valid in general as long as defect distribution can be described with an average defect density. The chip architecture envisioned is that of a two-dimensional array of identical modules (the processing elements), each pair of rows and of columns being separated by a *wiring channel* in which intermodule connection lines run in a multiple-bus organization. Of the N modules present in the array, R are spares; $N - R$ are required for operation of the target architecture. It is assumed that redundancy is used to recover from end-of-production defects only, and no specific assumptions on the technology adopted to *implement* reconfiguration are made. No redundancy is provided for the intermodule connections (an assumption consistent with the reconfiguration techniques proposed in the literature), so that defects in intermodule connections are considered catastrophic.

Given all these initial assumptions, Mangir and Avizienis derived yield equations depending on a number of factors, in particular:

- redundancy (evaluated with reference to area);

- total area required;

- percentage of total area reserved for intermodule connections;

- conditional probability that a faulty module is bypassed and succesfully substituted in performance of its functions by a fault-free spare. (This factor is strongly algorithm-dependent, and its value is 1 only if the reconfiguration algorithm is such that for any distribution of up to R faults the reconfiguration algorithm will be successfully applied);

- probability of distributing k defects into R out of N modules, i.e., a factor translating the *density of physical defects* into a *density of functional errors* of the modules.

Rather than a simple area-to-yield plot, a figure of merit FM is defined as

$$(A_0/A_R) \cdot (Y_R/Y_N)$$

where:

- A_0 is the *nominal* area (in the absence of redundancy);

- A_R is the area of the redundant chip;

- Y_R is the yield with redundancy;

- Y_0 is the basic yield of the nonredundant chip.

This figure of merit can be evaluated against a number of factors: redundancy, the area reserved for intermodule connections, relative importance (dimensions) of modules with respect to interconnections, and internal module complexity. Developments presented by Mangir and Avizienis led to plotting a number of curves (see figure 3.1) whose relevance can be summarized as:

- *self-contained* modules (i.e., modules whose internal functional complexity is very relevant with respect to interconnections with, and therefore to information transfers to and from, other modules) allow higher yield improvement when redundancy is added to the basic array. This is consistent with solutions

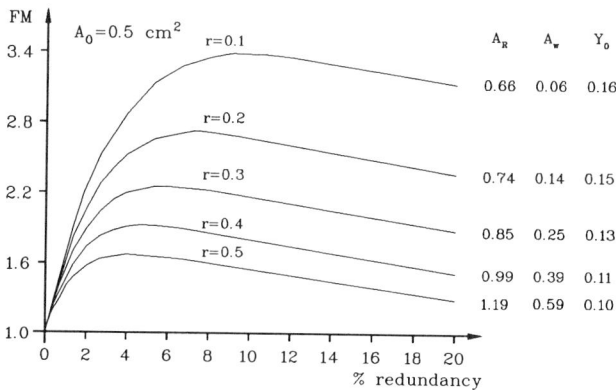

Figure 3.1.a

Figure of merit (FM) vs. percent redundancy for varying interconnect area ratios $r = A_W/A_R$ (A_W is the intermodule interconnect area). Values derived by Mangir and Avizienis for a sample chip definition).

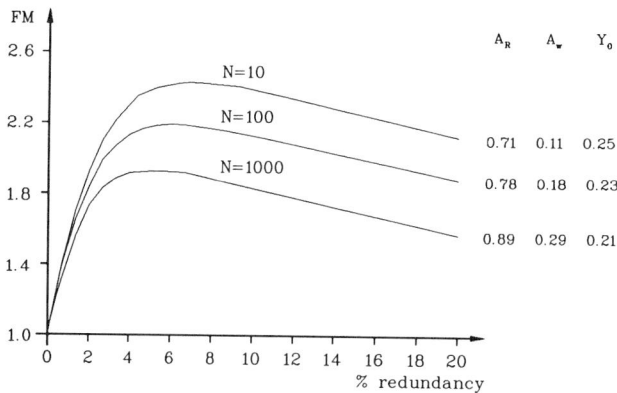

Figure 3.1.b.

Effect of number N of blocks on FM (values derived as for figure 3.1.a).

proposed for systolic arrays made of relatively complex PEs, possibly of parallel type (such was the case, for example, of the Systolic Processor Module originally proposed by H.T.Kung).

- For modules with simpler interconnect requirements, the number of modules becomes the dominant factor for determining yield improvement, even in the case of relatively large modules. This is consistent with solutions proposed for arrays made of PEs interconnected by serial communication links (see, e.g., the machine of [EVA85]).

- For designs with high degree of connectivity between modules, module complexity becomes the limiting factor to yield improvement achieved through use of spares.

The same figure of merit was later adopted also by Franzon [FRA86] who concluded that to maximize yield from a wafer the size of the single processing element must be chosen with fault tolerance in mind, rather than with the single purpose of maximizing the processing power of the nonfault tolerant array. Franzon points out that increasing granularity, i.e., increasing *detail* of the processing element with respect to the whole system, leads to increasing values of the figure of merit already recalled, even though it involves some overhead with regard to required connections.

The preceding discussion highlights the necessity of finding a good balance between added complexity (deriving from redundant PEs as well as from interconnection network) and reconfiguration efficiency. Most reconfiguration algorithms introduce a trade-off between complexity of the reconfiguration algorithm — which usually also involves complexity of the augmented interconnection network — and probability of surviving to increasing number of faults. This factor is accounted for, e.g., by the *conditional probability that a faulty module is bypassed* in Mangir-Avizienis formulas.

The second point, i.e., *run-time availability and reliability*, can be taken into account only for dynamically reconfigurable systems. The definition of such factors for reconfigurable arrays has not received the same attention in the literature that has been given to yield for comparable architectures. This is probably a result of the fact that general (possibly simplified) rules could be applied here (see the Appendix). A well-known rule gives a direct relation between failure rate and transistor count: Koren and Breuer [KOR84] were the first to have discussed the problem of reliability for dynamically reconfigurable arrays in a detailed way. In their study a *degradable array* is considered, i.e., it is assumed that no spares are available on the chip and that reconfiguration leads to creation of smaller working arrays with lower processing power: faults are assumed possible both in processing elements and in interconnection links. A Markov model philosophy is adopted for evaluating the reliability of such array. This model can in fact be extended to different reconfiguration strategies so as to give a measure of their capabilities. A factor termed *area utilization measure* is introduced. Koren and Breuer define it as the ratio between the equivalent array computational availability after a given number of faults and the total chip area. It also gives a reasonably good idea of one other main factor concerning reconfiguration algorithms, i.e., the measure to which redundancy can be *exploited* at run time as related to the total redundancy introduced. While it can be deduced intuitively that a relation also exists between area utilization and reliability (through the percentage of unused spares that will actually only contribute to a *decrease* of reliability), the complete dependency still needs to be formalized. A more extensive treatment of this subject is given in the Appendix.

Some common requirements for reconfiguration strategies based on structure redundancy can now be identified deriving from yield and reliability considerations:

(1) Reconfiguration algorithms must lead to reasonably low increment of area

reserved for intermodule connections. Interconnections are critical for overall reliability and yield and, in particular, *global* interconnections may lead to total system failure.

(2) Reconfiguration must create links of limited length between processing elements that become *logical neighbors*, i.e., that must directly exchange information.

(3) Area increment due to the various sources of redundancy must be kept within bounds; otherwise, the apparent *gain* promised by addition of spares and possibility of inserting them to replace faulty elements will be offset by the actual loss in yield and reliability.

As a consequence, reconfiguration algorithms will not necessarily aim for one hundred percent reconfiguration efficiency — that is, requiring that any distribution of faults be recovered from as long as there are spares available — but they will rather aim for a compromise between reconfiguration efficiency, interconnection locality and regularity, and area increment.

3.3. Case (B.2): time redundancy

In the case of time redundancy, the obvious cost accrues from speed loss. In this type of approach additional processing phases are added to each single operation step of the array so as to allow some of the working processing elements to perform the functions of faulty ones. Though no spare processing elements are added to the basic array, the interconnection network must be augmented to support the modified distribution of functions after reconfiguration has been performed. Moreover, the processing elements themselves must be augmented by the addition of registers storing information in a suitable way during the various phases. Array control also becomes more complex to account for phase sequencing and interleaving. A more complex control and, in particular, a multiple-phase clock may become critical for system reliability. Clock distribution is one of the most delicate points particularly where wafer-scale integration is concerned. Thus, an *area* cost must also be evaluated, and for this the figures of merit and the problem factors already analyzed in (B.1) must be taken into account.

3.4. References

[EVA85]   R.A.Evans: *A self-organizing fault-tolerant 2-d array*, VLSI-85, Tokyo 1985, 233-242, IEEE.

[FRA86]   P.D.Franzon: *Yield modeling for fault-tolerant arrays*, in: *Systolic Arrays*, (W.Moore, A.McCabe, R.Urquhart eds.), Proc. First Int'l Workshop on Systolic Arrays, Oxford, July 2-4 1986, 207-216, Adam Hilger.

[HUA84]   K.H.Huang, J.A.Abraham: *Algorithm-based fault-tolerance for matrix operations*, IEEE-TC, Vol. C-33, N. 6, June 1984, 518-528.

[KOR84]   I.Koren, M.A.Breuer:*On area and yield considerations for fault tolerant VLSI processors arrays*, IEEE-TC, Vol. C-33, N. 1, Jan. 1984.

[JOU86]   J.Y.Jou, J.A.Abraham: *Fault-tolerant matrix arithmetic and signal processing on highly concurrent computing structures*, Proceedings of the IEEE, Vol. 74, N. 5, May 1986, 732-741.

[MAN82a] T.E.Mangir: *Use of on-chip redundancy for fault-tolerant VLSI design*, PhD dissertation, UCLA 1982.

[MAN82b] T.E.Mangir, A.Avizienis: *Fault-tolerant design for VLSI: effect of interconnect requirements on yield improvement of VLSI design*, IEEE-TC, Vol. C-31, N. 7, July 1982, 609-615.

[RED84]   G.R.Redimbo: *Fault-tolerant digital filtering architectures using finite-field transforms*, Signal Processing, N. 9, 1985, 35-50.

[STA76]   C.M.Stapper: *LSI yield modeling and process monitoring*, IBM Journal of Research and Development, 1976.

[STA83]   C.M.Stapper, F.A.Armstrong, K.Saji: *Integrated circuits yield statistics*, Proceedings of the IEEE, vol. 71, N.4, April 1983, 453-470.

3.5. Further readings

V.K.Agarwal et al.: *Techniques for implementing Wafer-scale devices*, Proc. ICCD 86, Rye, Oct. 1986, IEEE. A detailed analysis is made of problems for a specific architecture, with particular reference to yield considerations, and to robust design aspects.

A.Antola, R.Negrini, M.G.Sami, N.Scarabottolo: *Time folding: a solution for functional and fault-tolerance reconfiguration of systolic FFTs*, Proc. 3rd ISCS, Boston, May 1988.

A.Antola, R.Negrini, M.G.Sami, N.Scarabottolo: *Policies for fault tolerance through mixed space- and time-redundancy in semi-systolic FFT arrays*, Proc. Int'l Conference on Systolic Arrays, San Diego, May 1988.

In these two papers, reconfiguration policies relating to FFT arrays are presented that make use of mixed-mode approaches involving both space redundancy and time redundancy philosophies.

A.V.Ferris-Prabhu: *Modeling the critical area in yield forecasts*, IEEE JSSC, Vol. SC-20, Aug. 1985, 878-882. The effect of photolitographic defect size distribution critical area of a device is analysed, critical area being defined as that in which occurrence of a defect results in yield loss.

I.Koren, D.K.Pradhan: *Introducing redundancy into VLSI design for yield and performance enhancement* Proc. FTCS 85, Ann Arbor, June 1985, 330-335, IEEE. Analytical models are presented, evaluating how yield enhancement and performance improvement may be achieved by introduction of redundancy.

S.R.Jones, R.M.Lea: *Interconnection strategies for the WASP device* in: *Wafer*

Scale Integration, (W.Moore, C.Jesshope eds.), Proc. Int'l Workshop, Southampton, July 10-12 1985, Adam Hilger. An Analysis of the basic cell yields required to configure array- and string-cells, referring to four-way and eight-way connected arrays.

G.Saucier, J.Trilhe: *Introduction*, in: *Wafer Scale Integration*, (G.Saucier, J.Trilhe eds.), Proc. IFIP WG 10.5 Workshop, Grenoble, Mar. 17-19 1986, North-Holland. A comprehensive overview of problems encountered in wafer-scale integration.

On-chip redundancy providing for array reconfiguration must be supported by a suitable technology that allows the actual performance of such reconfiguration. It is necessary to *route* in the correct way information between the various active processing elements and then — if reconfiguration is self-driven by on-chip circuits — to store in some way (either permanently or dynamically) the information on fault status of the elements on the chip.

Several alternatives can be considered with regard to the physical support of reconfigurable information routing and of fault information storage:

(a) for information routing (i.e., with regard to technologies for the implementation of "programmable" interconnections) interconnection paths can be modified by means of:

   (a.1) electrically programmable links;

   (a.2) electron-beam programmable fuses;

   (a.3) laser cutting/welding.

(b) for fault information storage:

   (b.1) latches storing information on good-bad devices and controlling the switching on information paths;

   (b.2) electrically programmable storage elements (acting as latches, but with higher persistency);

Some of these solutions (a.2, a.3) lead to permanent array configuration and are therefore suited for end-of-production, static configuration; others (a.1, b.1, b.2) allow dynamic modification and are therefore useful for dynamic run-time reconfiguration. We shall briefly consider some of the most widely used and discussed alternatives [MAN84].

4.1. Static reconfiguration

In the case of static configuration, various policies are possible:

- modules are unconnected until testing is performed and subsequent configuration decided upon; only at that moment are the interconnections completed;

- a *primary* set of modules are connected a-priori while spares remain unconnected until testing has been completed (connection of spares is then performed as above);

- all modules are already connected following a basic scheme prior to testing and failed ones are removed after testing by *cutting* a suitable subset of links.

The first alternative can be implemented by means of so-called *discretionary wiring* [PET67]. Some authors claim that such a technique is overly liable to faults and that it presupposes reconfiguration approaches that take into account only the failures of processing elements, not those of interconnecting links. Implicit in the discretionary wiring philosophy, in fact, is the assumption that *wiring*, i.e., the system of intermodule connection paths, is defect-free. Moreover, discretionary wiring requires the availability of sophisticated CAD and production tools for completing the wiring. A solution of this type has been adopted in at least one case [DON85]. This philosophy of building up a large system on a wafer starting from small dies proved to work correctly. The functional cells were then wired together by means of additional layers of metal interconnect. As evaluated by the authors of [DON85], the technique offered good yield. Dedicated CAD tools are adopted, so as to create a wiring pattern that is essentially guided by the testing probe. An obvious drawback of this approach, however, is the limited speed attained when manufacturing the final interconnection pattern, since this pattern is individually designed for the single wafer; a factor that will necessarily influence final costs.

Other authors have proposed techniques based on the assumption that some sort of *basic* interconnection structure — more or less uncommitted — has been initially provided (and is, or can be, subject to testing just as functional cells are). The interconnection pattern is then *customized* either by creating a conductive link between suitable (and working) segments, or by interrupting unwanted paths.

In this context, laser technologies have been widely advocated. Several alternatives are also possible with regard to laser personalization: it is possible to use it for blowing fuses that connect spare elements to the nominal array, or for disconnecting the contacts to faulty elements, or for joining together interconnect layers so as to create suitable links. As in the case of discretionary wiring, sophisticated systems are obviously required. On the other hand, a larger set of defects can be taken into account, since failures in interconnection links can be considered and overcome as well. A notable example of the use of laser technologies is presented in the RVLSI project of Lincoln Laboratories [CHA85, CHA87, RAF84, RAF85, RHO85]. There, laser is used not only to cut links but also to *complete*, so to speak, the interconnection patterns by means of a *microwelding* technique.

The basic idea in the RVLSI approach is to create an orthogonal matrix of two-level metal upon the wafer; at the crossings between the two levels, links are placed to provide possible connection points. In correspondence of the possible connection points foreseen, the two metals are separated by a sandwich of insulating materials. During the linking process, a laser beam is focused onto the spot where a link must be created, and — by melting the upper metal and the material underneath — an alloy is created that provides the required linking. The technique proved to be very reliable and has shown good electrical performances; moreover, use of laser pulses can also be extended to segment existing interconnections as well as to *stitch* together segments of signal nets.

Actions performed with E-beam technologies [SHA84] can be fairly similar to the ones seen in the case of laser technology, i.e., E-beam can be used to program fuses and antifuses and to join or disconnect metal lines. In such cases, E-beam technology can be said to be *destructive* just as are laser technologies, since the device is permanently modified by use of the technology. Neverthe-less, E-beam can also be used to implement a rerouting of interconnections in such a way that, while being nonvolatile, it is also reprogrammable. To this end, suitable semiconductor technologies are also required: floating-gate FETs can be used as electron-beam programmable switches [GIR86]. The programming crite-ria for these switches can avoid permanent damage to the semiconductor, thus guaranteeing integrity of access paths during the programming actions. Moreover, relevant integration density can be reached. A further advantage of electron-beam programmable switches is that it guarantees design propriety (switches can be embedded deep in the semiconductor). In our case, this may be of interest with regard to the reconfiguration algorithm implemented. The propriety factor is of obviously greater interest when this technology is used to program macrocell ar-rays or similar devices. Programmed switches can be reprogrammed, albeit not in a dynamic way, since they have to be driven by ad hoc systems: for example, a laser beam is necessary to turn a switch on again that has been turned off, and the switch can be switched on and off only a limited number of times. Interesting features of E-beam approaches include their application in the testing phase as well, so that basically one system is adopted for all phases of the fault-tolerance-oriented procedure, and their speed: E-beam systems can be capable of accessing over one million coordinates per second.

4.2. Dynamic reconfiguration

Technologies are necessarily quite different if *dynamic* reconfiguration, performed in the field, is envisioned. The parameters that characterize a reconfigurable in-terconnection network can then be summarized as follows:

- interconnection reconfiguration must not modify in a permanent way the sta-tus of the device or of any of its components;

- circuit components introduced to allow reconfiguration should not modify in a significant way signal propagation delays along the interconnection paths;

- complexity of intermodule links and related control circuits accounting for reconfiguration must be kept reasonably limited (these factors will be partially responsible for overall area redundancy, already seen to be a critical factor).

The specific criteria adopted in various instances will be described later together with the reconfiguration techniques to which they apply; outlined here are common factors and some basic approaches to solutions.

It is self-evident that electrically programmable connections between PEs are necessary: the basic, global interconnect network must contain — as if they were embedded into it — all alternative actual interconnection structures that can be made active at any given moment as a consequence of dynamic reconfiguration. So-lutions proposed in current literature have been based either on networks of buses, endowed with suitable switches, or on direct (alternative) inter-cell links, terminat-ing upon multiplexers (and, in some instances, originating from de-multiplexers). Switches, multiplexers and de-multiplexers act then as *control* devices, effectively creating the active interconnect as it is needed. Such devices should therefore be subject to checks as regards their correct operation. Most authors either assume switches and similar devices to be *a priori* working correctly (an assumption justified by their relative simplicity as compared with the processing elements, or else made acceptable by a presumed *robust design*) or, in some instances, they assimilate failure of such controlling devices to failure of some interconnected processing element. Failure on a switched bus is sometimes taken as a *fatal failure* effectively forbidding any subsequent reconfiguration. This is implicitly true also for those approaches that require correct operation of the interconnection network as a starting assumption.

Whatever the nature of the devices activating the interconnection links, a further problem must be considered in the case of dynamic reconfiguration: that of identifying the *agent* controlling reconfiguration itself on the basis of fault information. Two alternatives can be examined here as well:

(1) Fault information is individually associated with each processing element and stored in a local elementary *fault memory*. This technique is particularly suitable when processing elements are individually self-testing; it can be noted that it excludes faults in interconnection links (unless failure of a link is considered *as if* it were the fault of an attached PE).

(2) A *fault map* is stored in an external memory; this solution is particularly attractive when testing is performed — periodically, or even, possibly, concurrently — by a *host* device. In this case, the map can well include faults in the interconnection network and in related controlling devices.

When fault information is directly stored on chip, it is possible to envision also on-chip circuits that — based on such information — immediately control the reconfiguration. In this case, the *latency* of reconfiguration, i.e., the delay occurring between detection of a fault, confinement of the faulty device, and effective reconfiguration, can be kept to a minimum (reconfiguration-controlling circuits should obviously be designed so as to guarantee high operation speed). On the other hand, since such circuits (as well as fault memories) must also be taken into account when redundancy is evaluated, the constraints discussed in chapter 3 will lead to an acceptance of this solution only insofar as the reconfiguration algorithm can be implemented by relatively simple circuits.

Alternatively, a *fault map* available in external storage is best used by a reconfiguration-controlling host device (possibly the same machine that controls the testing operations). Control circuits on the chip are then limited to the links that carry configuration signals to switches or multiplexers. In this case, complexity of the algorithm is no longer a problem. Latency of reconfiguration, however, increases, since the procedure will be performed periodically (as can happen with testing).

The problem of latency in error confinement has obvious relevance with reference to dynamic reconfiguration. In this context, optimum performances would require that *information* produced by a faulty element be immediately confined and discarded, so as not to affect the results of other, fault-free processing elements.

Such optimum error confinement can be reached by suitable communication protocols and software modules in multiprocessors whose PEs are actual, complete microcomputers provided with acceptable amounts of working memory. In the processing arrays here considered, on the contrary, individual PEs are fairly simple devices and to reach the same measure of error confinement a relevant amount of structure redundancy should be added to each individual PE. Such redundant hardware should allow to:

- check the status of the associated PE;

- perform confinement of the faulty PE and (if necessary) drive on-line reconfiguration;

- repeat the last previous processing step on the reconfigured array (to this end, all PEs should be endowed with spare I/O and status registers, unless system operation be purely combinatorial);

- grant correct succession of processing steps even in presence of faults and of reconfiguration.

This would involve not only relevant costs in terms of memory, but also in terms of external system control (in order to obtain correct repetition of the phase) and, ultimately, in terms of processing time. Note that all would be totally independent of the specific implementation technology adopted. None of the authors cited take into account this particular point: rather, reconfiguration is performed by confining the faulty element, as soon as it is detected, but without discarding the information that it has produced and that is thus processed by interacting PEs. There will therefore be a *transient* during which array output results will actually be incorrect. If we consider signal-processing applications, it is quite acceptable to have a reduced amount of *noisy* results, provided the maximum amount of noise induced by faults is identified in advance and bound, as is true in our case.

4.3. References

[CHA85]   G.H.Chapman: *Laser-linking technology for RVLSI*, in: *Wafer Scale Integration*, (W.Moore, C.Jesshope eds.), Proc. Int'l Workshop, Southampton, July 10-12 1985, Adam Hilger.

[CHA87]   G.H.Chapman, J.I.Raffel, J.M.Canter, F.M.Rhodes: *Advances in laser link technology for wafer-scale circuits*, Proc. IFIP Workshop on WSI, Brunel, Sept. 1987.

[DON85]   B.J.Donlan et al.: *Computer-aided design and fabrication for wafer-scale integration*, VLSI Design, April 1985, 34-42.

[GIR86]   P.Girard, F.M.Roche, B.Pistoulet: *Electron beam effects on VLSI MOS, conditions for testing and reconfiguration*, in: *Wafer Scale Integration*, (G.Saucier, J.Trilhe eds.), Proc. IFIP WG 10.5 Workshop, Grenoble, Mar. 17-19 1986, North-Holland.

[MAN84]   T.E.Mangir: *Sources of failures and yield improvement for VLSI and restructurable interconnect for R-VLSI and WSI: Part II: Restructurable interconnects for RVLSI and WSI*, Proceedings of the IEEE, Dec. 1984, 1687-1694.

[PET67]   R.L.Petritz: *Current state of large-scale integration technologies*, IEEE JSSC, Vol. SC-2, N. 6, Dec. 1967, 130-147.

[RAF84]   J.J.Raffel, A.H.Anderson, G.H.Chapman, K.H.Konkle, B.Mathur, A. M.Soares, P.W.Wyatt: *A wafer-scale digital integrator*, Proc. ICCD-84, Rye, Oct. 1984, 121-126, IEEE.

[RAF85]   J.J.Raffel, A.H.Anderson, G.H.Chapman, K.H.Konkle, B.Mathur, A. M.Soares, P.W.Wyatt: *A wafer-scale digital integrator using restructurable VLSI*, IEEE TED, Vol. ED-32, 1985, 479-486.

[RHO85]   F.M.Rhodes: *Applications of RVLSI to signal processing*, in: *Wafer Scale Integration*, (W.Moore, C.Jesshope eds.), Proc. Int'l Workshop, Southampton, July 10-12 1985, Adam Hilger.

[SHA84]   D.C.Shaver: *Electron-beam customization, repair and testing of wafer-scale circuits*, Solid-State Technology, Feb. 1984, 135-139.

5 FAULT DETECTION

The preliminary step for any reconfiguration policy aiming at fault-tolerance is obviously the testing and diagnosis of the system itself. Management of test information will of course be quite different depending on whether reconfiguration is performed at production time or at run time. In the first instance, fault information is used immediately by configuration-controlling instruments and can afterwards be discarded; in the second instance, fault information is, in turn, a *dynamic* information, that can be modified at any iteration of the testing procedures, so that it must be stored and updated as necessary. While *static*, end of production, testing can be as accurate as allowed by testability characteristics of both interconnection network and individual PEs, and can therefore be performed by specific tools such as SEM (electron-beam technologies can well be adopted for both stages, testing and configuration) during *run-time* fault detection lower accuracy in fault definition may be acceptable since other problems such as concurrency, speed and error latency have to be taken into account. A totally different outlook is therefore required.

This chapter will deal with techniques aimed at *run-time fault detection*; the problem of initial, end of production testing will be considered only insofar as some approach is useful for both. This limitation is justified when it is noted that approaches suited for run-time fault detection (by involving increased array testability) implicitly also make initial testing easier, while *conventional* testing techniques that can be exploited for end of production testing are certainly out of the question at run time. On the other hand, the general problem of end of production testing, even for complex VLSI devices, is a much-researched subject and outside the scope of this book. Some recent papers dealing with particular aspects of *testability* for VLSI array devices are listed in the reference section of this chapter.

The fault sets against which array testing is performed must be defined as a preliminary step. Three different fault assumptions have been considered in the literature:

(1) Faults are assumed to be located only in the processing elements, not in the interconnecting links or in the interconnection-controlling devices. This assumption is usually justified by the consideration that PEs are much more complex than the interconnection structure and therefore much more liable to run-time failures. While dependence of failure rate on transistor count is widely accepted, several authors have also stressed that complexity of logic design influences liability to failures [MAN84a, MAN84b]. Moreover, experimental data confirm that defect density is strongly dependent on layout rules as well, and this fact holds also for those defects that become apparent as faults, as a consequence of stress, during the device lifetime. All these factors support the assumption that probability of run-time failures is much higher

for PEs than for interlogic connections and switches, at least in the majority of cases, where individual PEs are by far more complex and area consuming than the interconnecting structures, and require design rules much tighter than those used for buses and switches.

When this assumption is made, testing involves only the processing elements. Since most reconfiguration strategies are independent of the specific functions performed by the PEs of the individual arrays, testing of each PE will have an outcome of the *go-no go* type, and no further detail on fault origin will be required once the failed PE has been located.

(2) Besides PEs, interconnection-controlling devices (e.g., switches) are also assumed to be possibly faulty. In this case, detailed information on settings and (possibly) structure of the switches (or of equivalent devices) is usually required, and related testing actions are based upon it, independent of the testing criteria adopted for PEs.

(3) All components of the array, including also interconnection links, are taken into account as possible fault locations. When discussing reconfiguration policies, few authors consider this last instance separately. Even when the possibility of faults in interconnection links is accepted, either it is considered as leading to immediate *fatal failure*, i.e., impossibility of reconfiguration, or else a fault in any given segment of the interconnection network is considered equivalent to the fault of a suitably chosen adjacent PE. Faults in the interconnection network are specifically considered (and overcome) only in some approaches actually oriented to end of production configuration [AND85].

A further distinction derives from whether testing is performed on-chip by means of some kind of fault detection technique, or whether it is externally controlled by a *host* or *monitor*; the corresponding approaches are not wholly independent of fault assumptions. The various alternatives will now be listed, and some corresponding solutions will then be analyzed.

(a) *Autonomous fault detection.* Two different modes of operation can be adopted. In the first mode, each individual element (typically, a PE) is designed so as to be self-checking. In the case of arithmetic PEs, for example, function-specific criteria such as the use of residue arithmetics can be employed to achieve concurrent fault detection capacity, at least up to a certain number of faults and a given fault model [ANN83]. Error latency is then minimal, since a fault is detected as soon as it affects the correctness of a result. Consequently, the spread of errors through the array can also be avoided, since faulty results can be immediately discarded instead of passing them on as inputs to other PEs or as final results to external devices. (Note that this solution does not in any way depend on the array architecture.) This approach can also be used for switches (by adopting suitable self-checking design techniques) and even for the interconnection links, assuming suitable coding for information transmitted.

An obvious drawback of local, function-specific, concurrent self-checking is inherent in the redundancy required and in the area increase therefore involved (with all the negative aspects already underlined for such factor). In fact, such redundancy depends on local, functional characteristics of the individual device rather than on global, array-level architecture, and thus it does not exploit regularity and repetitivity of the architecture: it must be inserted at the level of the individual component, and globally, therefore, it will be quite relevant. On the other hand, operation speed is kept nominal or as near to nominal as modified design of the individual units permits.

In the second case, concurrent fault detection is performed at system (array) level, and is therefore strongly dependent on the specific architecture of the array. Several techniques have been proposed for this type of fault detection. All involve a measure of structure redundancy — although global area increase is usually kept lower than in the previous case, since architecture regularity is fully exploited — and also, quite often, a measure of time redundancy. Two different criteria can in fact be adopted:

(a.1) all components of the array (or, better, all components in which faults are assumed to be possibly located) are simultaneously subject to testing. Unless massive array-level redundancy is accepted, a time redundancy factor must be envisioned. Some authors suggest a triple time redundancy technique, allowing to achieve simultaneously testing, fault confinement and error masking (under suitable fault assumptions); nominal operation is then always continued, but a relevant speed decrease must be accepted. Alternatively, test phases are inserted between nominal operation phases; again, operation speed decreases while error latency is inversely proportional to test frequency, thence also to speed deterioration. Clearly, in both cases a (limited) measure of structure redundancy must be taken into account as well, to allow modified information transfers, storage and checking of results obtained in the various phases.

(a.2) A subset of the devices in the array are subject to testing during any given phase of array operation, while nominal operation of the array itself is continued. This requires a suitably augmented interconnection network as well as some further structure redundancy; error latency is related to the number of devices that can be tested in parallel, which again depends on structure redundancy introduced.

(b) *Externally driven testing.* In this second case, a host or monitor accesses the array in order to control its testing and observe related results. The first problem here consists in granting to the monitor access to all devices subject to testing: such capacity is in turn directly related to the fault coverage that can be reached and — in order to increase the possibility of observation and control — redundancy may be added, at least as far as the interconnection network is concerned.

Depending on fault assumptions and on functional characteristics of the PEs as well, testing may be performed by using nominal data (in which case no speed decrease is caused) or by forcing suitable test vectors (whereby, periodicity of tests is related to speed decrease). Error latency again depends on

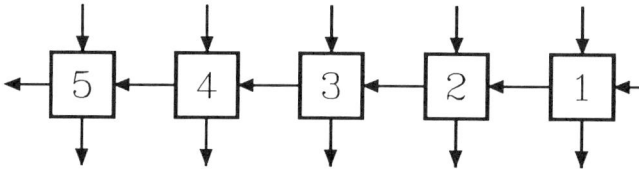

Figure 5.1

Repeated Execution with Shifted Operands (RESO). Example of a linear array

the periodicity with which the individual components of the array are subject to testing.

While fault information derived from autonomous fault detection (whatever the particular technique adopted) can well be stored into individual *fault memories*, connected with each corresponding device, in the case of externally monitored testing fault information will be stored into a *fault map* that is in turn part of the external monitor. With regard to the subsequent reconfiguration actions, while local, individual fault memories can be used by on-chip circuits controlling dynamic reconfiguration, the external *fault map* is better used in the case of *host-driven reconfiguration*.

This chapter will present a detailed analysis of the instances of array-level concurrent fault detection and of externally monitored testing; device-level self-checking is better examined with regard to the specific device functions and is therefore outside the scope of this book.

5.1. Case (a): Array-level autonomous fault detection

Three main types of approaches can be distinguished: the first makes use of some type of *time redundancy*, or rather of a *repeated operation* philosophy; the second exploits some measure of structure redundancy to create pairs (or subsets) of devices simultaneously performing the same operation, so as to allow diagnosis through comparison; and the third derives from classic *t-testability* techniques.

5.1.1. Time-redundancy techniques

Repeated Execution with Shifted Operands (RESO) was introduced initially for linear modular structures dedicated to arithmetic functions [LAH83]. As shown in figure 5.1, all modules in the linear array are identical and all perform identical operations. Each module or cell receives a *primary* input and a *carry-in*, and produces a primary output and a carry-out. Carries propagate from right (cell 1) to left (cell N); a ripple-carry adder is a typical example of such structure. Testing proceeds in three phases during which the same data are fed to the modules — but a different ordering is adopted each time. A *rotation* of the operands rather than a simple shifting is in reality performed. The fault assumptions underlying this technique are summarized as follows:

(a) if module i is faulty, both primary output and carry out are faulty (i.e., different from the expected ones, whatever the inputs);

(b) if module i is fault-free, but at least one of its inputs (primary or carry-in) is faulty (i.e., different from the expected one) both outputs are *faulty*, i.e., different from the expected ones;

(c) results produced by a faulty cell as response to different inputs, are also different, albeit faulty. Moreover, different but equally faulty inputs to a fault-free module produce different — albeit faulty — outputs;

(d) inputs, and therefore also outputs, of different modules are all different;

(e) faults are located in modules — interconnections and related control circuits are fault-free;

(f) at most one module is faulty.

Given such assumptions, if module i is faulty all outputs (primary and carry) from module i to module N will be faulty. Let us now denote as :

- *physical* index that identifying the module in the array

- *logical* index that identifying the operand element (or *segment*) related to the function performed.

In a first phase, a given set of operands is fed to the array following the direct correspondence of physical to logical indices and the results produced by the individual modules are stored in suitable memories. In a second phase, a circular shift of operands is performed leftwards so as to obtain the following correspondence:

physical indices	N	$N-1$	$N-2$	$\cdots$	3	2	1
logical indices	$N-1$	$N-2$	$N-3$	$\cdots$	2	1	N

and, again, results obtained are stored.

In a third phase, finally, a rightward circular shift is performed so that the following correspondence is obtained:

physical indices	N	$N-1$	$N-2$	$\cdots$	3	2	1
logical indices	1	N	$N-1$	$\cdots$	4	3	2

If module i is faulty, erroneous outputs will be obtained during the three phases as follows:

- phase 1: from module N to module i

- phase 2: from module N to module $(i-1)$ for $i \neq 1$;

    in module N for $i = 1$

- phase 3: from module N to module $(i+1)$ for $i \neq N$

    all faulty for $i = N$.

Under the fault assumptions already given, a comparison between two outputs with the same logical index will give an Agreement (A) if and only if both outputs are correct.

Figures 5.2.a, 5.2.b, 5.2.c show logical index mapping onto the physical modules in the three phases with $N = 5$. The same figures outline the structure and organization of signals in the three phases. Thick lines denote objects and data links active in each phase.

Figure 5.2.d shows the array structure comprising the augmented interconnection network (buses and switches) that support communications as denoted in the previous figures.

Figure 5.2.e outlines the structure of a module; latches L1, L2, L3 store the outputs pertaining to the three phases while C(1,2) and C(1,3) compare the results of the first phase respectively to the ones computed during the second and third phase.

Table 5.1 gives logical outputs (C=correct, F=faulty) during the three phases, as well as outputs of comparator circuits between outputs of phases 1, 2 and 1, 3, for the array shown in figure 5.2, given the basic assumption that only one module is faulty. Obviously, in the absence of any fault all inputs and outputs are correct and all comparators give an Agreement.

Assume now that k is the first comparator giving a Disagreement (D) between phases 1 and 2. The module with physical index $k + 1$ appears to be faulty and a conclusive outcome of testing is already available after the first two phases. If, however, all comparators give a Disagreement (D), there is ambiguity as to fault status between the first and second module; the third phase must then be completed before fault location is achieved. Comparison on the first logical output between first and third phase gives a Disagreement if the first module is faulty and an Agreement if the second one is faulty.

The example in table 5.1 proves that comparisons between pairs of outputs of phase 1 and 2 in the first $N - 2$ modules and of outputs from module 1 in phases 1 and 3 not only enable the detection of overall faulty behaviour, but also allow the finding of faulty modules in the assumption of single fault.

In [LAH83] a particular application of the above general instance is considered with reference to the case of $N = 3$. In this case, some of the restrictions noted previously with regard to fault assumptions can be relaxed; specifically, some instances of error masking that result from a faulty module producing correct outputs for some specific inputs can be accepted.

The technique is obviously simple and the structure redundancy required is low; limited basically to the logic controlling the rotation of the operands and to the comparators. The single fault assumption is quite reasonable at run time for moderately complex systems. On the other hand, if we relax fault assumption (b), erroneous fault location may be generated whenever a given set of inputs to a faulty module is such as to mask the fault itself. The main drawback of the approach is due to the increase in operation time that doubles the nominal time.

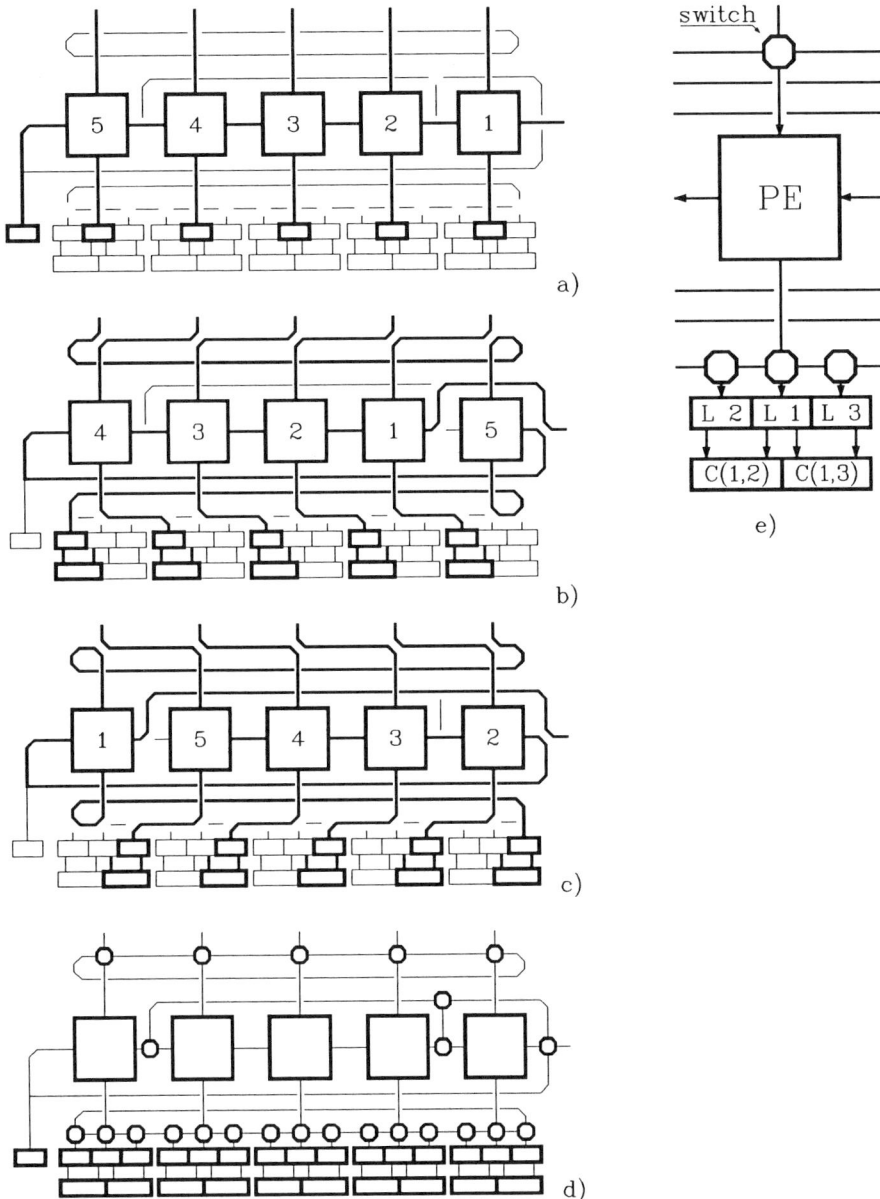

Figure 5.2

Repeated Execution with Shifted Operands (RESO). General scheme of a linear modular structure.
a, b, c: Organization of signals during the three phases. d: Augmented interconnection network for
RESO. e: Architecture of a single module.

Table 5.1:

Output from modules and comparators for different positions of the faulty module. Symbols used have the following meaning: **E, C** (faulty and correct data); **A, D** (agreement and disagreement); **C(i,j)** (comparison between phase i and phase j); **O(i)** (logical output in phase i)

Faulty module n. 1					
O(1)	E	E	E	E	E
O(2)	E	C	C	C	C
O(3)	E	E	E	E	C
C(1,2)	D	D	D	D	D
C(1,3)	D	D	D	D	D
Faulty module n. 2					
O(1)	E	E	E	E	C
O(2)	E	E	E	E	E
O(3)	E	E	E	C	C
C(1,2)	D	D	D	D	D
C(1,3)	D	D	D	D	A
Faulty module n. 3					
O(1)	E	E	E	C	C
O(2)	E	E	E	E	C
O(3)	E	D	C	C	C
C(1,2)	D	D	D	D	A
C(1,3)	D	D	D	A	A
Faulty module n. 4					
O(1)	E	E	C	C	C
O(2)	E	E	E	C	C
O(3)	E	C	C	C	C
C(1,2)	D	D	D	A	A
C(1,3)	D	D	A	A	A
Faulty module n. 5					
O(1)	E	C	C	C	C
O(2)	E	E	C	C	C
O(3)	E	E	E	E	E
C(1,2)	D	D	A	A	A
C(1,3)	D	D	D	D	D

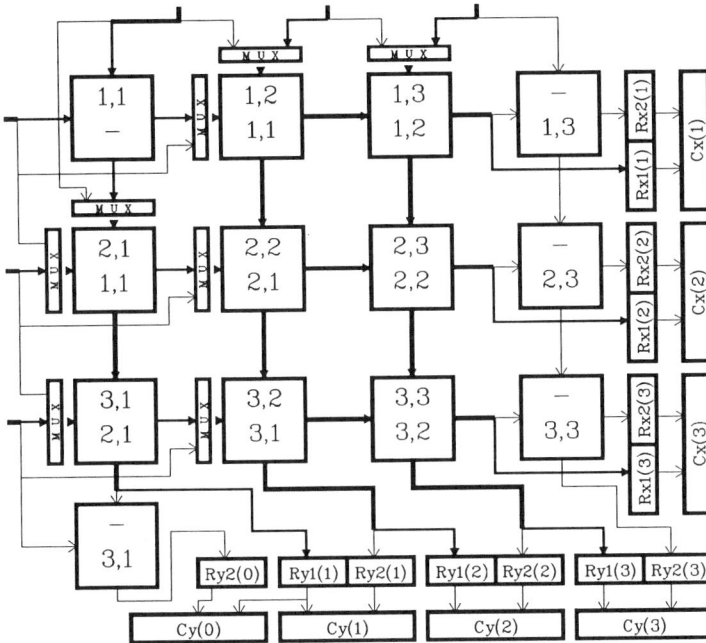

Figure 5.3

Testing of a 3 × 3 nominal array through time redundancy (two phases). R_x and R_y are registers; C_x and C_y are comparators. Thick interconnection lines are active in both phases. Medium lines and thin lines are active respectively in the first and second phase. Upper logical indices: first phase. Lower logical indices: second phase.

This criterion can be extended to the case of a rectangular array and, by immediate extension, to that of any array in which each device is connected to a regular set of n neighbors. Given the requirement of locality, it is now necessary to consider *shifted operands* not *rotated operands* (the latter case would involve very long interconnection links between extreme rows and columns) so that a number of redundant cells must inevitably be added. We refer here to the solution presented in [SAM84].

The basic structure is given in figure 5.3 while the operating principle is outlined in figure 5.4. Consider first the simpler fault model, identical to the previous one and represented by the *cell model* in figure 5.5.a. Links and related control circuits are assumed to be fault-free.

The following two-phase fault detection algorithm can then be applied to a rectangular $N \times N$ array operating in a *wavefront computation* way, with a spare column added in position $N + 1$:

Phase a: logical and physical indices of all cells coincide, i.e., cell in position (i,j) performs operations corresponding to logical indices $(i',j') = (i,j)$. Results produced in each row by cell (i,N) are stored in a corresponding

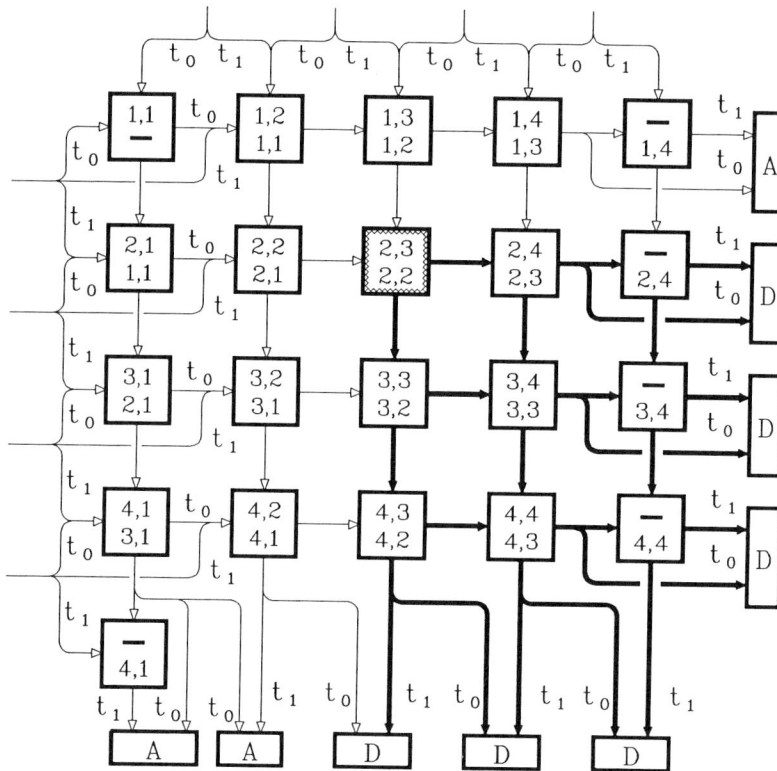

Figure 5.4

Testing of a 4 × 1 nominal array. t_0 and t_1 correspond in order to first and second phase. Cell (2,3) is faulty. Thin lines denote correct data, thick ones denote faulty data. A= agreement, D= disagreement. Logical indices for the two phases are inserted in each cell.

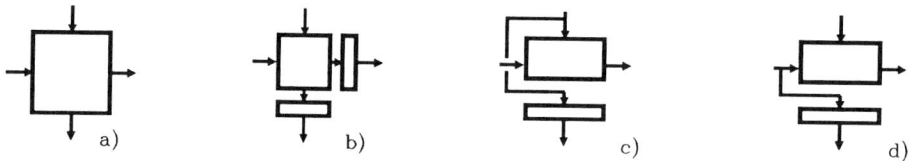

Figure 5.5

Fault models: a) all outputs of a cell are affected equally by a cell fault; b) functional logic of the cell is distinct from logic controlling output communications; therefore, independent faults may occur in any of the three sections; c), d) independent paths directly connecting an input of a cell to the logic controlling one of the outputs exist. Independent faults may occur in the distinct sections.

register $R_{x1}(i)$, and results produced in each column by cell (N, j) are stored in a register $R_{y1}(j)$;

Phase b: logical indices are reassigned as $i' = i$, $j' = j-1$. The same operands used in phase a are fed again to the logical array; thus, no cell will process twice the same inputs. Results of each cell $(i, N+1)$ are stored in a register $R(i)$, and results of each working cell (N, j') in a register $R(j')$.

Comparisons between each pair of registers with the same row (column) index will — under the given fault assumptions — give Agreement whenever both cells operating upon the same data in the two phases receive correct inputs and give correct results; while Disagreement appears as soon as one of the pair is faulty or when one of them receives incorrect inputs. Thus, a very simple algorithm allows location of the faulty cell by checking the first row and the first column in which disagreement is detected. Even failures in the interconnect links can be provided for, the only limitation being that they will be attributed by the diagnostic algorithm to the receiving cell.

Here, area redundancy is low, increasing with N both as regards added cells and as regards supporting devices (albeit quite simple) such as multiplexers, registers, and even augmented interconnection links. Operation speed is obviously halved — if diagnostic phases are introduced at each processing phase, so as to minimize error latency — and error latency is proportional to N; still, the fault model may often be considered rather restrictive. An extension to far more general fault models can be made if we assume that a faulty cell can produce an unexpected value on one output only; this allows for separated output sections on the two output lines (see figure 5.5.b). This problem, which clearly does not arise in pipelines or in strictly systolic linear arrays (where each cell has one input and one output), is reasonable for rectangular arrays, since quite often in a PE of relative complexity the purely processing part will be separated from the output sections. It can be assumed, moreover, that whenever a cell receives faulty (unexpected) data the results on both outputs will be *faulty*. While the previous technique would give erroneous fault location, a three-phase testing action provides the solution. To this end, both a spare row and a spare column must be introduced and mapping of logical indices onto physical indices in the three phases must be such that no single cell will communicate with the same physical neighbors in all three phases corresponding to each single processing step. Figures 5.6.a, 5.6.b and 5.6.c show for each of the three phases, respectively, interconnections among and logical indices associated with the various cells. While presence of a fault is again very easily diagnosed, location becomes more complex than in the previous case, depending as it is on the actual type of fault in the cell (global, horizontal output, vertical output) and on the indices themselves. Table 5.2 pinpoints the faulty device and the kind of fault associated with it. Such greater detail is obviously paid for with greater area redundancy (there are now $2 \times N$ spare cells) and, even worse, with far greater time redundancy, since operation time is tripled if testing is repeated at each processing step. Error latency is identical with that previously evaluated.

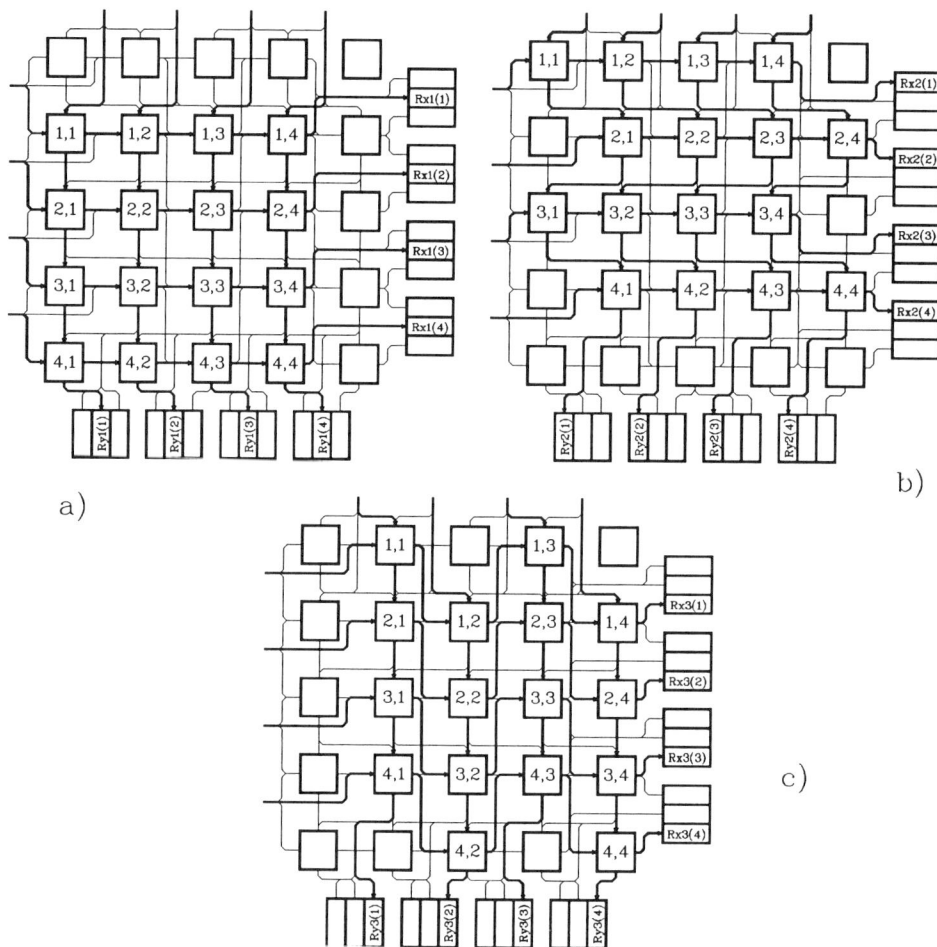

Figure 5.6

Testing of a 4×4 nominal array through time redundancy: three-phase algorithm (fault model of figure 5.5.b). a) first phase; b) second phase; c) third phase.

Table 5.2
Symbols used in the table have the following meaning: **A, D** (agreement and disagreement); **E, O** (even and odd); **n** is the lowest index associated with a horizontal comparator Cx23 or Cy13 giving disagreement; **m** is the corresponding index for vertical comparators; **i,j** indices of the recognized faulty cell; **hor, ver** and **both** denote respectively a fault on horizontal or vertical output section or on both output sections.

Cx23	Cy13	n	m	i	j	fault type
D	A	E	E	n+1	m+1	both
D	A	E	O	n+1	m	hor
D	A	O	E	n	m+1	vert
A	A	O	O	n+1	m+1	both
A	A	O	E	n+1	m	hor
A	A	E	O	n	m+1	vert
D	D	O	E	n+1	m+1	both
D	D	O	O	n+1	m	hor
D	D	E	E	n	m+1	vert
A	D	O	O	n+1	m+1	both
A	D	O	E	n+1	m	hor
A	D	E	O	n	m+1	vert

Both of the techniques described above permit the identification of a single fault. Should several cells be faulty at the same time, then:

- if there is a faulty cell (i, j) such that for any other faulty cell (h, k) it is $i \leq h$ and $j \leq k$, then only (i, j) is detected (and correctly located)

- if the above condition is not satisfied, erroneous fault location will occur to a cell characterized by

  i'=min(row indices of all faulty cells)

  j'=min(column indices of all faulty cells)

These criteria can be adopted for self-reconfiguring arrays, where detection of a fault is immediately followed by fault confinement and array reconfiguration. In that case, multiple faults can be provided for, with the only restriction that they appear singly in time (i.e., after a fault has been detected, no other fault will occur until reconfiguration has been performed).

The previous approaches assume that individual processing cells are fairly simple and not capable of general-purpose functions. An alternative solution still

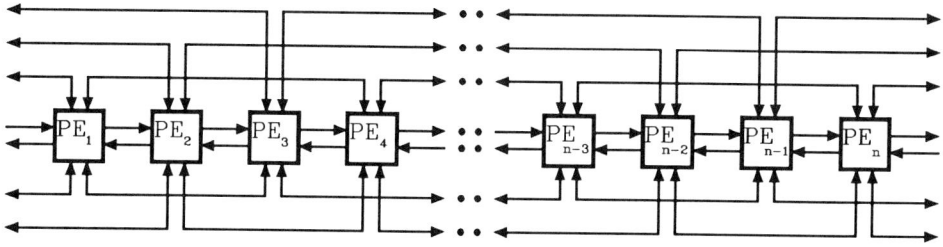

Figure 5.7
Linear systolic array with modified interconnection structure.

exploiting time redundancy but assuming rather complex, "intelligent" PEs is discussed in [MAJ88]. The class of processors taken into account is exemplified by the WARP machine, [ANN86].

In this solution, actual triple modular redundancy is used for each computation step on a *linear systolic array*; thus, besides fault detection and confinement, error masking is also concurrently achieved. While in the *nominal* array each i-th PE would be connected only to the $(i-1)$-th and to the $(i+1)$-th PEs, to allow fault detection the interconnection network is modified so that each PE i has bidirectional links with all PEs in positions from $(i-3)$ to $(i-1)$ and from $(i+1)$ to $(i+3)$ (see figure 5.7).

Denoting by $Y_i(l)$ the computation that must be performed by PE l during the i-th computational cycle, the cycle itself is split into three subsequent ones and $Y_i(l)$ is computed independently by PE_l, PE_{l+1}, PE_{l+2}, as follows:

- Cycle 1:

  $Y_1(1)$ is computed by PE_1, PE_2, PE_3;
  - PE_1 transmits its results to PE_2, PE_3, PE_4;
  - PE_2 transmits its results to PE_3, PE_4;
  - PE_3 transmits its results to PE_2, PE_4.

- Cycle 2:

  PE_2, PE_3, PE_4 vote on the three independent copies of $Y_1(1)$ they have, and compute $Y_1(2)$. As before, results are propagated to PE_3, PE_4, PE_5.

- Cycle 3:

  PE_3, PE_4, PE_5 vote on the three independent copies of $Y_1(2)$, and compute $Y_1(3)$.

- Cycle 4:

  PE_4, PE_5, PE_6 vote on $Y_1(3)$ and compute $Y_1(4)$. At the same time, PE_1, PE_2, PE_3 compute $Y_2(1)$, and so on.

A result is thus available from PE_n every third cycle, and the final result $Y_m(n)$ is available after $3m + n - 3$ cycles.

A functional error model is used here also; PEs are assumed to fail independently, while interconnections and related logic are assumed to be fault-free. While the *fault detection* itself is performed in a distributed way by all the PEs, its *management* is the task of a central processor CP. More precisely, whenever PE_i finds a mismatch between the three copies of an operand, it sends a message to the CP; the CP updates the array-level fault information, possibly checking also transient *vs.* permanent faults. Fault information on neighboring PEs is stored also inside each PE; no reconfiguration has to be introduced unless the fault distribution is such that either

(1) two adjacent PEs are faulty, or else

(2) there is exactly one non-faulty PE between two faulty adjacent ones.

Error latency here is minimal (a faulty PE is confined as soon as it provides unacceptable results); structure redundancy is limited to the interconnection network and to the related control logic. On the other hand, processing speed reduces to one-third the nominal one; moreover, the technique can be adopted only when the PEs are fairly complex ones.

5.1.2. "Roving-monitor"-like approaches

An example of the second class of approaches will now be examined, by which the presence of a number of redundant devices is exploited to obtain replicated performance of operation steps and subsequent comparisons and voting. Again, the specific characteristics of a very regular architecture permits the utilization, in a particular way, of an otherwise classical technique so that reduced redundancy can be obtained (instead of the classical duplication). Criteria used are typically related to the *roving monitor* approach, by which a few redundant elements are exploited to cyclically check a large number of different devices, without decreasing the nominal processing speed.

An example of this approach has been presented in [CHO84]. The discussion concerns a linear array but extension to rectangular arrays does not present any difficulty. The basic criterion can be defined as follows: Let a linear array of n *nominal* PEs be endowed with k redundant ones; these k redundant elements can duplicate operations performed by k nominal ones and a comparison can then be performed on the respective results. As long as the comparison is performed upon only two elements fault location is not possible. However, assuming permanent faults, by repeating the same operation upon different sets of PEs location can also be achieved.

Figure 5.8 represents one particular phase of this technique concerning diagnosis of one specific cell. Cell $(i + 1)$ is bypassed by means of a bypass link (provided for subsequent reconfiguration), so as to be excluded from nominal array operation, and it receives the same data fed to cell (i). Both perform the same operation step and results are compared locally, so as to check agreement or disagreement of results. Fault assumptions adopted are, again, functional. Moreover, it is assumed that:

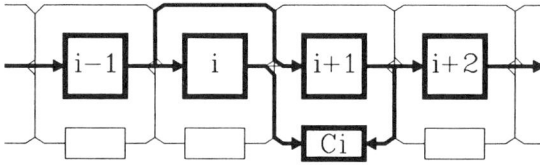

Figure 5.8

Testing of a linear systolic array through roving monitor philosophy. Modules i and $i + 1$ receive the same input data: circuit C_i compares outputs of the two modules.

- a faulty cell will always produce faulty results;
- two faulty cells, given the same inputs, will produce different results;
- comparators and links are fault-free.

Disagreement then immediately signals presence of a fault. Consider now the possibility that in two following operation steps the duplicated operation is shifted, i.e., duplication and comparison is effected between cell $(i + 1)$ and cell $(i + 2)$. If results agree, fault is located in cell (i). If disagreement is again found, by repeating comparisons along the chain of processing cells the faulty device can be identified, the only restriction being that no fault-free cell can ever be located between two faulty ones. Thus, operation duplication is dynamically assigned to pairs of adjacent cells. Since each cell, in turn, must be isolated from the array and the related bypass link must be activated, a stream of tokens (proceeding along with the data stream in the linear array) is *passed along* in the processor chain. This obviously requires a slight increase in the complexity of each cell that must now be endowed with a local control unit checking whether nominal or duplicated operation must be performed and, in the second case, activating the comparator. If k redundant cells are present, k tokens may be inserted simultaneously and corresponding comparisons may be activated; error latency is thus immediately related to the number of redundant cells. Since reconfigurable arrays are envisioned, after faults have occurred and corresponding spare cells have been introduced in operation, the parallelism obtainable correspondingly decreases.

In this new case, no degradation in processing speed can be observed, assuming that the time required by comparisons is negligible with respect to nominal operations performed by the cells. On the other hand, structure redundancy is significantly higher than in the previous case; each single cell must be endowed with a comparator in addition to the added control logic discussed previously. Error latency depends only upon the degree of parallelism reached by testing operations, in the simplifying assumption that a faulty PE will always produce faulty results whatever the input data. Otherwise it is obviously related to the characteristics of input data distribution *vs.* fault model and to the relative frequency with which test tokens are passed along the PE chain, with respect to frequency of data passing along the data stream.

The technique has been extended in [CHO85] for the case of off-line — possibly, end of production — testing. The problem considered there relates in partic-

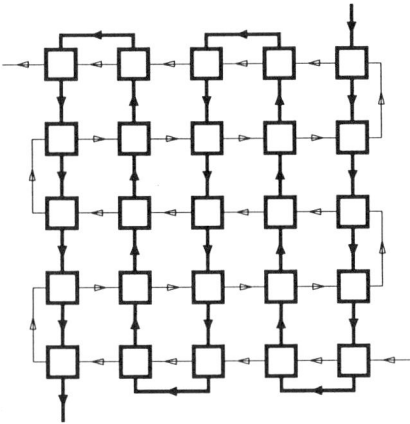

Figure 5.9
Orthogonal linear arrays configured for two-stage testing token algorithm.

ular to wafer-scale arrays: a case that presents unusual testing problems given the extremely unfavorable ratio of internal functional complexity *vs.* number of observation and control points. Autonomous fault detection capacity thus becomes mandatory even for off-line testing. The concept of a token-passing organization is exploited to select the PEs subject to the testing procedure, while complete test vectors are fed from the outside in order to reach acceptable fault coverage. Test tokens are now rather complex, since they must also signal when a test vector for an individual PE ends; the fault detection algorithm has been designed to be *safe*, in this case, meaning that while a fault-free cell might be declared faulty as a consequence of unforeseen fault patterns on adjacent PEs, a faulty cell cannot be declared fault-free.

Validity of the *test token* criterion is then very high as far as detection is concerned, but it may become lower when location of the faulty PEs is considered. The concurrent-test technique explicitly assumes some restrictions on fault distribution that are less satisfactory in the case of very large (possibly, wafer-scale) arrays. For this second case (referring to the off-line approach) it is suggested that multiple passes of the algorithm be repeated, involving pairs of cells at increasing distance (a solution whose simplicity is offset by the increasing amount of added interconnection links and the decrease in interconnection locality), or, alternatively, that the linear algorithm described be repeated along two dimensions. The interconnection network would still be augmented, but locality would be kept (see figure 5.9). Since this second approach is off-line — in the sense that nominal computation must be suspended — the concept of error latency is here meaningless; rather, if it is envisioned for periodical testing of an array, the amount of time redundancy involved may be considered. As in the strictly time-redundancy approach, such a factor is first of all dependent upon fault information that will guide the determination of the test periodicity. An additional factor is constituted by the parallelism that can be reached for the testing operation. Optimization of the parallelism is far from obvious since it also involves the effective possibility of feeding independent test vectors in parallel to a possibly high number of internal PEs.

5.1.3. System-level approaches

Off-line fault detection is also envisioned in system-level approaches referring, with suitable modifications, to the classical Preparata-Metze-Chien (PMC) technique [PRE67] and to the subsequent Hakimi-Amin extension [HAK74], usually known as *t-testability*. The PMC criterion, designed for multiprocessor systems, foresees that in a system consisting of $N > 2 \cdot t$ units, each unit is tested by t neighbors — without making any preliminary assumption on the fact that some of these unit also may be faulty: in fact, test of each unit leads to creation of a table, and only final analyisis of the complete test results will allow to identify faulty and fault-free units. The only restriction to fault distributions — necessary to grant validity of these final results — is that no more than t units be faulty at any given time. The original PMC proposal further required that no two units test each other — i.e., if unit u_i belongs to the set of neighbors testing u_j, u_j cannot belong to the set of neighbors testing u_i: Hakimi and Amin subsequently relaxed this restriction, otherwise fairly severe for systems of limited dimensions and connectivity. Still, straightforward application of this technique to a systolic array would not be very satisfactory, given the low connectivity of such arrays — each PE will usually have two or four neighbors, only seldom six, and the *rectangular* array is by far the most commonly envisioned. Therefore, the limit to the number of faulty PEs present in an array would reduce, respectively, to two, four or six — in any case a too stringent limitation given the chip dimensions and the fault-tolerance requirements. In [SOM84] a major generalization of this classical criterion was introduced, allowing to overcome such limits by directly exploiting the systolic nature of the array that — during the test phase — executes a suitable systolic algorithm.

The algorithm operates in two phases: the first one performs detection, the second one locates faults. Detection follows the classical *t*-testability pattern: each PE is tested by its t neighbors, using a predetermined test vector, and each PE i that performs a testing operation upon a PE j generates a binary result stating whether it declares j to be fault-free $(A_{ij} = 0)$ or faulty $(A_{ij} = 1)$. At the end of the testing phase, presence of faulty PEs is detected; whenever at least one A_{kh} is 1, it can be deduced that an error is present in the array, except in the extreme case of an array consisting only of faulty PEs that moreover all agree on the same erroneous test result. If presence of faults is detected, the phase dedicated to fault location is begun: the restriction introduced for allowable fault distributions is far less stringent then in the classical PMC or even in the Hakimi-Amin proposals, since it is stated as follows:

- define an *Uniform open system* as a subarray consisting of a center unit, of its t immediate neighbors (at distance 1) and of the t neighbors at distance 2 such that the total number of external edges is $t \cdot (t-1)$. An *acceptable set* of faulty PEs in the array is defined as a fault distribution such that no uniform open system in the array will contain more than t faulty units.

Given this definition, the Somani-Agarwal algorithm will detect any acceptable set of faults in the given array. Starting from the set of $A_{i,j}$ created in the detection phase, majority votes on all $A_{i,j}$ are computed by each unit, thus generating a set of *confidence level* values that identify whether a given PE i is *likely good* or *likely faulty*. All such values are communicated by each PE to all its neighbors, so that in a third step each PE will combine preliminary test results with confidence levels in order to obtain revised confidence levels. By repeating this step a number of times equal to the maximum distance between units in the systolic array, finally a syndrome allowing to locate all faulty units is obtained — provided no more than t faulty units are present in each of the open uniform systems of $2 \cdot t + 1$ units: this holds even in the worst case — i.e., when each uniform open system contains exactly t faulty units. An evaluation of algorithm performances can be deduced simply by noticing that — while the classical PMC method only allowed presence of 4 faults in any rectangular array — here the number of faults for which the algorithm may be capable of location is related to the number of PEs in the array and that, if a rectangular array of 10×10 PEs is envisioned, up to 44 faults can be located (always respecting, of course, the restriction on fault distribution already presented).

Clearly, error latency is once again uninteresting, since the approach is designed for off-line testing. Moreover, here no structure redundancy whatever is required — not even the modest one required by time-redundancy approaches — while performance degradation can only be defined in terms of the periodicity with which test actions are repeated. One singular point, that keeps this approach apart from other ones excepting that described in [MAJ88], is the fairly high level of *intelligence* — i.e., of general computation capacity — requested to the individual PEs: while most techniques previously considered only required that a PE be capable of its own nominal functions and, at most, of processing some very simple protocol (as in the test-token driven procedure), here PEs must be capable of feeding test vectors to neighbors, observing the results, communicating with other neighbors, voting, updating confidence levels etc. In other words, they are expected to be full computers — although possibly simple ones — rather than just processing units such as, e.g., ALUs (as most other testing criteria assume).

5.2. Case (b): Host-Driven (concurrent or semi concurrent) fault detection

The case of host-driven fault detection presents an actual interest if we consider that in many applications systolic arrays are effectively controlled (and even programmed) by an external *host* machine, characterized by less brilliant application-specific performances but of a general-purpose nature. Moreover, some reconfiguration algorithms for overcoming faults presume in turn (as it will be seen in later sections) availability of such host, that must be capable of obtaining and storing test data on all PEs in the array, as well as of controlling the array's internal interconnection network. A host with such capacities can well be in charge also of test actions, i.e., of activating test operations in correct sequence on the various PEs and of evaluating results of such actions.

Obviously, most end of production test procedures will fall in this cathegory: criteria such as the ones described in [SCA86], [CHA86] refer to optimization of test vectors in order to reach optimum test results notwithstanding the limited number of observation and control points available (presence of extra pads on the VLSI or on the wafer in order to increase the number of test points would again increase the chip area without increasing run-time performances). Still, as already said, we are not here interested in such type of testing, but rather on those approaches that — while host-driven — can be considered as *concurrent* or *semi-concurrent*, so as to be a viable alternative to autonomous fault detection.

A solution that can be defined as a *mixed-type* one is described by Evans and McWhirter [EVA86]: the purpose is to achieve a complete test of the whole array (both as regards PEs and as regards interconnection network and reconfiguration-controlling circuits) and to this end it is suggested that host-driven actions for the interconnection network (in particular, for the switches) and for the reconfiguration-controlling circuits be paired with self-testing solutions for the individual PEs. The technique is explicitly related to the particular reconfiguration philosophy adopted in the *WINNER* approach [EVA85]: still, the main criteria can be considered fairly general and therefore applicable also in a different context.

The test strategy suggested is a *hierarchical* one, based on the partitioning of test actions already outlined. Individual PEs are self-checking: the authors suggest that a signature-analysis type of strategy is adopted for this purpose. This, of course, could create a *hard-core* section consisting of signature analysis comparators, that would be required to be always correctly working — something that the authors want to avoid. First of all, then, host-driven testing is performed upon the reconfiguration-controlling circuitry (and, implicitly, upon the interconnection links) by using a scan-path technique: since control circuits associated with each PE are simple combinatorial circuits, simple test vectors can be used for them. Additional circuits are — obviously — required to create this scan-path organization at test time, but again they are reasonably simple, organized along well-known lines and add little to the overall redundancy (see figure 5.10): the scan register is suitably modified in order to tolerate faults in the control circuit.

Also as a host-driven action, testing of the signature comparators is performed: this requires to inject test patterns into the comparators and to observe test results. To this end, scan path are once more used — as a technique general enough to be applicable even for different internal organizations. Obviously, a criticism could be brought against this approach by noticing that the scan-path themselves could become the ultimate hard-core section of the system: although simple, they involve additional area and thus can lead once more to yield (or reliability) decrease. In [EVA86] it is suggested to duplicate the scan paths, in order to avoid such bottleneck: alternatively, a *fail-safe* philosophy could be adopted,

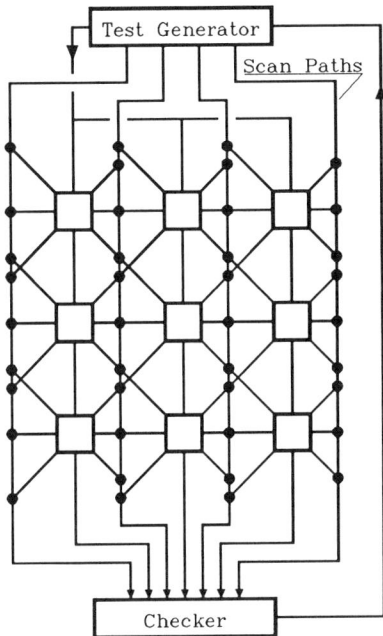

Figure 5.10
Schematic of the scan path testing approach.

by a proper design of test vectors.

Host-driven techniques — whether fully concurrent or, as the above one was, *semi-concurrent* — have been given particular attention with regards to the problem of testing the interconnection network and the related control circuits, i.e., a subject seldom taken into account otherwise. A number of reconfigurable arrays adopt an interconnection network consisting of switched buses running inside routing channels between pairs of rows or/and columns to achieve either functional or fault-tolerant reconfiguration: this class of arrays is considered for application of the technique described in [CHO86]. As in the previous case, the technique there described is specific for a given structure: yet, it has in fact a very wide field of applications.

The set of faults considered in [CHO86] is fairly comprehensive, since it involves not only switches but also buses and interconnection links: although faults are listed as if belonging to the *stuck-at* type, actually switch faults are seen as functional, since only faults that cause the switch to malfunction are taken into account. While no assumption is made on number or distribution of faults, it is considered as a reasonable condition that a high number of switches and connection lines will be active in a wafer (otherwise, even with reconfiguration, harvesting would be so poor as to make the wafer practically useless). The basic philosophy is to avoid augmenting the interconnection network for testing purposes (actually,

Figure 5.11

Triple-bus interconnection network supporting reconfiguration algorithms in rectangular arrays.

a *hidden redundancy* exists, since PEs are assumed to have also switching capabilities, in order to allow testing of communication links between cells and adjacent switches). Relatively simple test actions (*straight-line tests*) are repeated so as to identify *regions* inside which faulty switches are confined, and to achieve better and better definition of such regions (to limit, that is, their dimensions) until in practice fault location is reached. The algorithm is actually relatively complex (it requires $O(n)$ comparisons, in a $n \times n$ array, to find a fault-free routing path for test data to be applied); it is also *fault-secure*, so that presence of any fault will certainly be detected.

A fully host-driven approach is presented in [DIS88]: there, no self-checking capacity is required from the PEs — on the other hand, a measure of redundancy is again required, even though limited to the interconnection network. The solution is presented with relation to a class of reconfigurable structures, all making use of a switched-bus interconnection network: again, it can be immediately extended to any structure based upon the same type of interconnection networks.

The technique can be defined as a *semi-concurrent* one, in the sense that interconnection network and switches are tested by periodical actions, off-line, while PEs are tested concurrently with nominal operation. To achieve this second aim, it is necessary to augment the interconnection network (consisting, in the basic reconfigurable structures here envisioned, of triple buses suitably switched as in figure 5.11) by adding in each routing channel a *diagnostic bus*, related switches and additional *input observation* links connecting diagnostic buses to PE inputs (see figure 5.12). It is assumed that reconfiguration — performed by efficient, but fairly complex, algorithms — will be host-driven in order to limit on-chip complexity and area redundancy: therefore, testing is limited to the interconnection network (buses, links and circuits) and to the PEs.

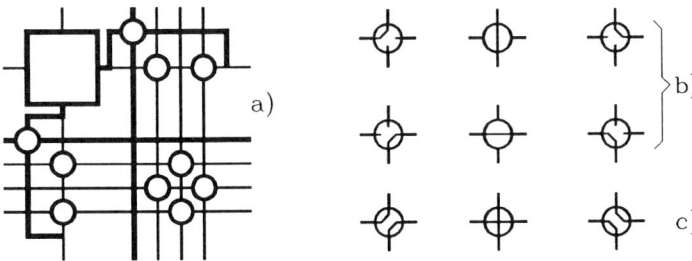

Figure 5.12

Triple-bus structure augmented with diagnostic bus. a: Elementary cell. b: allowed setting for switches on the horizontal and vertical diagnostic busses. c: allowed setting for switches on the reconfiguration busses.

Basically, the test organization assumes that faults in the interconnection network will render impossible any attempt to reconfiguration: thus, test of the interconnection structure is simply of the *go-no go* type, without requesting exact location of the faulty link or switch. Test of PEs, on the contrary, requires location of the faulty devices without any measure of ambiguity. Test of the network is, as already said, performed off-line, periodically, and test vectors are designed with reference to structure and technology of both buses and switches: the test procedure for the network consists of a set of configuration commands, issued by the host, that set all switches of the network in given positions. The underlying fault assumptions for switches can be summarized as follows:

- functional faults are considered: a switch may be open, shorted or *stuck-at state* (i.e., it will not react to setting commands);

- no two inputs can merge into a single output;

- no single input can split into two outputs.

With these limitations, a test procedure using the test patterns in figure 5.13 together with specular ones, allows to test for faults in all buses and switches: it is yet incapable of identifying faults in links connecting PEs to adjacent buses — faults that will then subsequently be detected, during the second phase, *as if* the interconnected PEs were faulty. While giving incorrect fault location, for reconfiguration purposes it can be noticed that a PE incapacitated from interconnecting itself with the adjacent buses is effectively isolated from array operation so that considering it as inactive is quite correct.

Refer now to testing of PEs: here, aims of concurrent fault detection can be summarized as follows:

- reasonable fault coverage must be reached;

- error latency should be kept as reduced as possible — this, in turn, leads to requiring that the largest possible subset of PEs be concurrently subject to testing at any given time.

In the approach here considered, nominal operation of the array is performed; concurrently, at each processing step, inputs and outputs of a suitably chosen

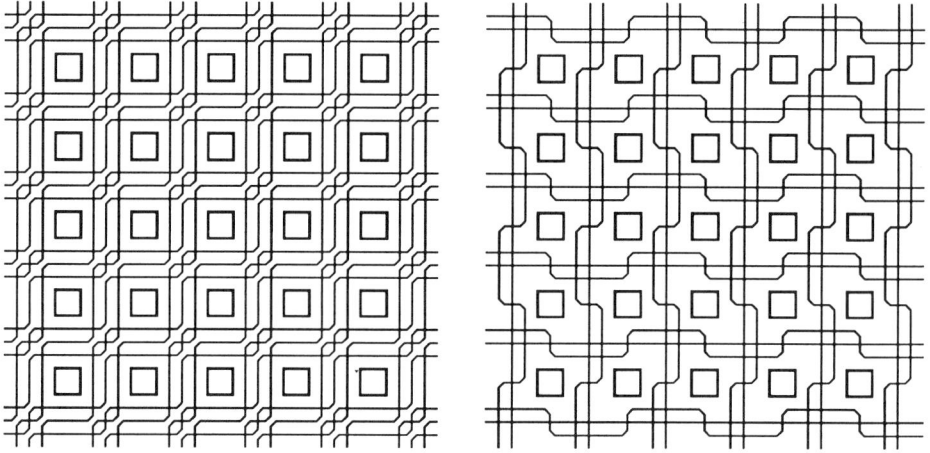

Figure 5.13
Test patterns for bus and switch testing.

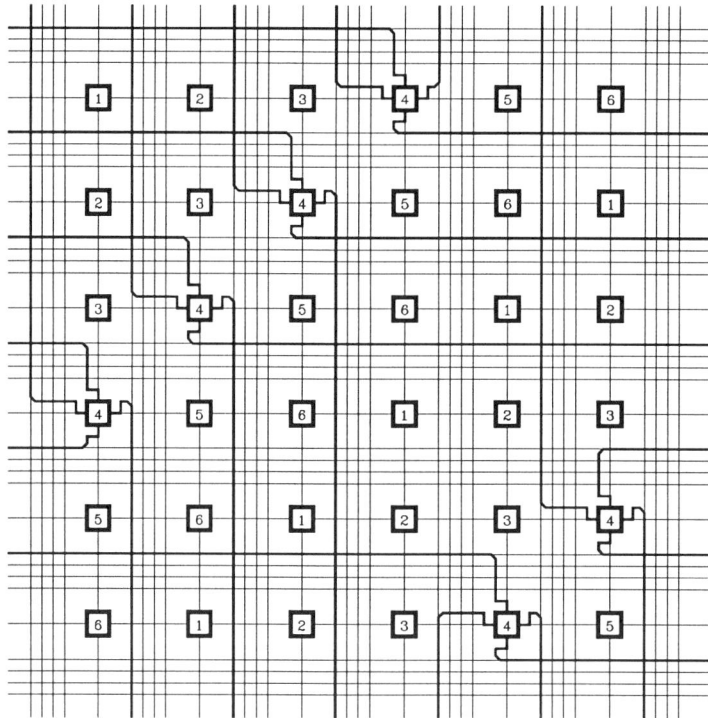

Figure 5.14
Triple-bus structure augmented with diagnostic bus: Diagnostic switch setting.

subset of PEs are observed (via diagnostic buses) and, for each PE, outputs are compared with expected results corresponding to the given inputs. This is a viable solution if PEs are relatively simple, as it might happen, e.g., with an array designed for signal-processing operations. Then, the procedure is not so far from the *token-triggered* criterion, if we consider that duplicate operation is actually performed by the host acting as a gold unit (or, better, read by the host in a protected memory). Array structure actually allows to observe $(N-1)$ PEs simultaneously at each step, organized along diagonals as in figure 5.14: thus, if the subset of PEs under test were renewed at each processing step, the number of processing steps required to *go through* the whole array would be proportional to N if we accepted the most elementary fault assumption that a faulty PE will always produce faulty results (the same ratio would then also define error latency). If, on the other hand, we accept the more realistic assumption that some input data may effectively *mask* a fault in a PE, evaluation of error latency will very much depend on the distribution of input data: assuming that such *invalid* (from the testing point of view) data are known a-priori, a solution can be to keep a subset of PEs under test until all of them have received at least one set of valid input data.

5.3. References

[AND85]   A.H.Anderson: *Computer-aided design and testing for RVLSI*, in: *Wafer Scale Integration*, (W.Moore, C.Jesshope eds.), Proc. Int'l Workshop, Southampton, July 10-12 1985, Adam Hilger.

[ANN83]   M. Annaratone, R.Stefanelli: *A multiplier with multiple error correction capabilities*, Proc. ARITH-6, 1983, IEEE.

[ANN86]   M.Annaratone, E.Arnould, T.Gross, H.T.Kung, M.Lam, O.Menzilcioglu, J.A.Webb: *WARP architecture and implementation* Proc. 13th Annual Symp. on Computer Architecture, Tokyo, June 1986, 346-356, IEEE.

[CHA86]   A.Chatterjee, J.A.Abraham: *C-Testability for generalized tree structures with applications to Wallace Trees and other circuits*, Proc. ICCAD 86, 1986, 288-291, IEEE.

[CHO84]   Y.W.Choi, S.H.Han, M.Malek: *Fault diagnosis of reconfigurable systolic arrays*, Proc. ICCD-84, New York, Oct. 1984, 451-455, IEEE.

[CHO85]   Y.H.Choi, D.S.Fussell, M.Malek: *Token-triggered systolic diagnosis of Wafer-scale arrays*, in: *Wafer Scale Integration*, (W.Moore, C.Jesshope eds.), Proc. Int'l Workshop, Southampton, July 10-12 1985, Adam Hilger.

[CHO86]   Y.H.Choi, D.S.Fussell, M.Malek: *Fault diagnosis of switches in Wafer-Scale arrays*, Proc. ICCAD 86, 1986, 292-297, IEEE.

[DIS88]   F.Distante, M.G.Sami, R.Stefanelli: *Testing techniques for complex VLSI/ WSI processing arrays* in *Supercomputing Systems*, (S.E.Kartashev, S.P.Kartashev eds.), Van Nostrand Reinhold, 1988.

[EVA85]   R.A.Evans, J.V.McCanny, K.W.Wood: *Wafer scale integration based on self organisation*, in: *Wafer Scale Integration*, (W.Moore, C.Jesshope eds.), Proc. Int'l Workshop, Southampton, July 10-12 1985, Adam Hilger.

[EVA86]   R.Evans, J.McWhirter: *A testing strategy for self-organising fault-tolerant arrays*, in: *Systolic Arrays*, (W.Moore, A.McCabe, R.Urquhart eds.), Proc. First Int'l Workshop on Systolic Arrays, Oxford, July 2-4, 1986, Adam Hilger.

[HAK74]   S.L.Hakimi, A.T.Amin: *Characterization of connection assignment of diagnosable systems*, IEEE-TC, Vol. C-23, 86-88, Jan. 1974.

[LAH83]   S.Laha, J.H.Patel: *Error correction in arithmetic operations using time redundancy*, Proc. FTCS-13, Milano, June 1983, 298-305, IEEE.

[MAJ88]   A.Majumdar, C.S.Raghavendra, M.A.Breuer: *Fault-tolerance in linear systolic arrays using time redundancy*, Proc. 21th Annual Hawaii Int'l Conference on System Sciences, Kailua-Kona, Jan. 1988, IEEE.

[MAN84a]   T.Mangir: *Sources of failures and yield improvement for VLSI and restructurable interconnect for R-VLSI and WSI: Part I, Sources of failure and yield improvement for VLSI*, Proceedings of the IEEE, June 1984, 690-708.

[MAN84b]   T.Mangir: *Sources of failures and yield improvement for VLSI and restructurable interconnect for R-VLSI and WSI: Part II: Restructurable interconnects for RVLSI and WSI*, Proceedings of the IEEE, Dec. 1984, 1687-1694.

[PRE67]   F.P.Preparata, G.Metze, R.T.Chien: *On the connection assignment problem of diagnosable systems*, IEEE Trans. Electronic Computers, Vol. EC-16, N. 6, Dec. 1967.

[SAM84]   M.G.Sami, R.Stefanelli: *Self-testing array structures*, Proc. ICCD-84, New York, Oct. 1984, 677-682, IEEE.

[SCA86]   J.T.Scanlon,W.K.Fuchs: *A testing strategy for bit-serial arrays*, Proc. ICCAD-86, 1986, 284-287, IEEE.

[SOM84]   A.K.Somani, V.K.Agarwal: *System-level diagnosis in systolic systems*, Proc. ICCD-84, New York, Oct. 1984, 445-450, IEEE.

5.4. Further readings

R.K.Gulati, S.M.Reddy: *Concurrent error detection in VLSI array structures*, Proc. ICCD-86, New York, Oct. 1986, 488-491, IEEE. A time-redundancy approach is presented for wavefront-computation arrays, limited to particular classes of algorithms.

S.Z.Hassan, N.Alten et al.: *Testing and diagnosis of wafer-scale integration logic*, Proc. ICCAD-84, 1984, 108-112, IEEE. A set of techniques adopted to test wafer-scale devices is described. While not dedicated to arrays, it exemplifies well the challenge provided by end of production testing of WSI devices.

C.Jay, G.P.Eynard, G.Saucier: *Test facilities and test plans in WSI 2-D arrays*, in: *Wafer Scale Integration*, (G.Saucier, J.Trilhe eds.), Proc. IFIP WG 10.5 Workshop, Grenoble, Mar. 17-19 1986, North-Holland. End-of production testing for a 2-D Wafer-Scale array is considered. A start-small strategy is suggested, involving in sequence PE accessibility, PE operating capacity, diagnosis of faulty nodes and identification of fault-free connection and links.

A.K.Somani, V.K.Agarwal, D.Avis: *A generalized theory for system-level diagnosis*, IEEE-TC, Vol. C-36, May 1987, 538-546. Extension and generalization of the work presented in [SOM85]. Results envisioned are particularly useful in diagnosis of WSI multiprocessor systems.

A.Vergis, K. Steiglitch: *Testability conditions for bilateral arrays of combinational cells*, IEEE-TC, Vol. C-35, N. 1, Jan. 1986. End-of production testing of systolic arrays is considered. Conditions making 2-D arrays of combinational cells testable for single faults are discussed.

A number of different classification criteria could be adopted for the analysis of the many reconfiguration approaches proposed in the current literature for VLSI/WSI processing arrays. Such criteria include *dynamic vs. static reconfiguration* philosophies, *graph-theoretic approaches vs. technology-related* proposals, or identification of some main factor of merit such as *algorithmic simplicity* vs. *full exploitation of available resources.* These factors will be discussed in the following chapters when analyzing the various reconfiguration techniques proposed in the literature, although they will not necessarily be used to classify the different approaches.

A basic consideration, independent of all such factors, must now be introduced that, although generally valid for processing architectures, is particularly useful in the case of processing arrays. It involves distinction of *physical architectures* from *target architectures.* A target architecture (also designated a *logical* or *virtual architecture* by some authors) may be said to constitute the functional representation of the problem (or class of problems) that the processing system is called upon to solve.

A physical architecture, on the other hand, constitutes a hardware implementation upon which one or more target architectures can be mapped.

Adoption of an array architecture to solve a specific problem involves, first of all, identification of a solution allowing partitioning of the overall computation process into a number of subprocesses (elementary processes). Information transfers among subprocesses are performed in a regular, uniform way and each involves only a limited number of strictly related subprocesses. In recent years, much attention has been paid to the problem of finding, for various algorithms, an organization into subprocesses following this type of philosophy. Identification of *systolic organizations* for different algorithms is typical of this line of research. In this instance, most relevant for the present study, all subprocesses are identical and no *global* information transfers that affect all processors at once through broadcast-type communications are allowed (see §1.1.3.). The *network* of subprocesses thus defined is then mapped onto a network of processors, all identical and interconnected in a structure characterized by high regularity and locality. The interconnection paths activated by the algorithm can be a subset of the ones actually present in the physical system: these active paths then define the architecture of the target array. In chapter 1 several examples were discussed with regard to applications: the commonest target architectures are *linear arrays, rectangular arrays (meshes), binary trees* and *multiple-pipeline arrays.*

The physical architecture supporting the target array is comprised of a sufficiently high number of identical PEs each capable of supporting a logical subprocess. Depending upon the application class envisioned, therefore, the PE can be a simple ALU, a combinational network, or a full computer endowed with a memory and characterized by an instruction set (although most authors assume PEs to be fairly simpler than complete microcomputers).

We are not concerned here with the specific processing functions of a PE: in some instances, reconfiguration techniques involve some restrictions, e.g., they can be applied only to arrays consisting of purely combinatorial PEs. *For our purposes, a PE is a black box*, and assumptions that concern it are usually made only with regard to faults and fault model. As a rule, however, it must be simple enough to allow implementation on a single VLSI chip of an array consisting of fairly large number of PEs together with the related interconnection network (in the case of WSI, the number of PEs per wafer obviously increases in proportion). It is quite probable, clearly, that the full processing array system will be made up of several chips (unless WSI implementation is envisioned).

Constraints — and figures of merit — concerning on-chip reconfiguration are quite different from those concerning inter-chip system-level organization; here, the on-chip case will be consistently discussed.

Mapping of a target architecture on silicon involves three subsequent problems:

(a) *Silicon implementation* of the physical array: the chip area usually allows creation of a rectangular *distribution* of PEs among which a suitable interconnection network is provided so as to allow mapping of target functions. Note that this distribution does not coincide with the *physical architecture* (this is determined by the interconnection network).

(b) Possible *functional configurability* of the interconnection network, allowing to map different target architectures onto the given physical array. To this end, the interconnection network will be suitably "programmed", i.e., links will be activated in a selective way.

(c) Possible *reconfigurability* of the interconnection network to achieve fault-tolerance, i.e., to allow creation of the requested target machine (possibly with degraded performances) upon the physical architecture even in the presence of faults. Programmability of the interconnection network is again required.

Physical array structures presented in the literature usually fall into two main categories: linear arrays and rectangular-mesh arrays. The basic physical architecture may be provided with an augmented interconnection network (augmented, that is, with respect to the immediately corresponding target architecture) capable of supporting a number of other different target architectures. Thus, an augmented interconnection network of sufficient complexity, superimposed upon a linear array, can allow creation of any target architecture. Other physical architectures — such as the multiple-pipeline one — are actually designed to support only one, corresponding class of target architectures, allowing at most some measure of reconfiguration within the same class induced by fault-tolerance (or, less often, by varying functional requirements within the same class).

When a number of different target architectures are to be mapped onto one physical architecture, we will speak of *functional (re)configuration*, configuration being a one-shot procedure performed at production time, while reconfiguration is a dynamic procedure allowing repeated modifications at run time. The most evident case of functional (re)configuration is, obviously, the mapping of different array structures such as rectangular mesh and binary tree onto the same physical array. Functional reconfiguration is also requested in simpler instances such as multiple-pipeline arrays with different pipe lengths being mapped onto a single multiple-pipeline structure.

With regard to reconfiguration induced by fault-tolerance, its meaning and aims can be summarized as consisting of a mapping of the target array functions onto the physical array organization so as to isolate, confine and overcome faults. Most solutions envisioned in the literature refer to reconfiguration induced by fault-tolerance for target arrays whose organization coincide with that of the physical arrays. Very few instances exist in which the two configuration problems are afforded concurrently (one of these, the "Diogenes" approach, will be analyzed in detail in chapter 7).

The so-called *Diogenes approach* is among the best known reconfiguration techniques proposed for processing arrays implemented by VLSI technology [CHU86, ROS83, ROS85], and it is of special interest for two reasons that make it unique among the techniques here examined.

- Its foremost aim is functional reconfiguration by means of a purely architectural technique that does not in any way involve the algorithm mapped onto the array. It constitutes in fact a general solution to the problem of *mapping general target architectures onto the simplest physical one: a linear physical array*.

- The philosophy adopted to achieve fault-tolerance is very simple: *straightforward bypassing of faulty PEs*. No additional components are introduced to achieve fault tolerance (with the exception of the required spares), neither new functions are added to the basic structure.

The physical architecture envisioned in the Diogenes approach is a linear array of identical PEs. PEs are not directly connected to each other (see figure 7.1), since a stack of *wires* (or buses) runs parallel to the PEs above the array and the interconnections between the bus segments and the various PEs, as well as between the individual bus segments, are controlled by simple switches (marked as SW in figure 7.1). The number of wires, i.e., the *depth* of the stack is directly related to the class of target architectures that are to be implemented. Thus, for example, a single wire allows the creation of a target linear array, while the number of wires requested to configure binary trees is equal to the depth of the tree. Interconnection between the PEs, in appropriate order, is effected by setting the switches on the buses. Switches are actually implemented by simple pass transistors inserted on the single wires, so that for each PE a set of such elementary switches will be present. Nevertheless, to better analyze the basic philosophy we prefer to introduce a functional representation of the complete switch set (the *complex* or *functional switch*, as it will be called here) associated with the individual PE in order to support a specific configuration capacity. The complex switch is considered to be directly associated with the PE whose connections it controls.

If the target array is a linear one, the functional switch used is the simplest one: it has one input x from the bus, one output y to the bus, and the related PE is connected with it via an input a and an output b (see symbols marked on the switch associated with PE n. 1 in figure 7.2). PEs and switches may actually be positioned on the chip in a rectangular pattern as in figure 7.2 (such physical distribution, however, is irrelevant to our purposes). Interconnection is straightforward, and reconfiguration as a consequence of faults is effected immediately by a simple bypass that connects the bus in a *through* fashion, excluding the faulty PE.

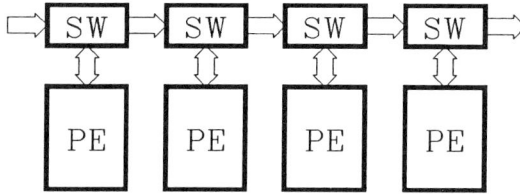

Figure 7.1

Physical architecture of a linear array in the Diogenes approach.

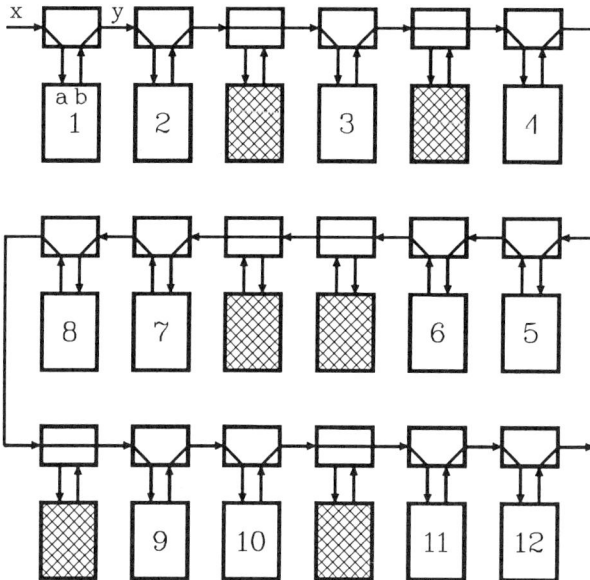

Figure 7.2

Linear target array. Example of reconfiguration after faults.

It is assumed implicitly that reconfiguration, either functional or due to fault-tolerance, is managed externally. It is quite obvious that for functional reconfiguration the host-driven alternative is the only acceptable one, since the final target architecture depends on overall user requirements and these, in turn, involve global host-controlled system configuration. As regards fault-tolerance, on the other hand, on-chip run-time reconfiguration is a viable alternative. Switch settings in fact are then related only to the correct behavior of the individual PE, and if PEs are provided with local self-testing capacity (an assumption accepted for most reconfiguration approaches) immediate setting of the corresponding switches to the *through* position or propagation of different settings to the switches associated with the nearest nonfaulty PEs can be achieved by means of simple reconfiguration-controlling circuits. Unless the simple case of the linear array is considered, this latter issue, i.e., propagation of functional configuration signals in the presence of faults must be carefully analyzed, since it becomes necessary to also modify the settings of switches associated with fault-free PEs. It is possible to design the network controlling switch settings so that, as soon as a PE is declared faulty, the associated switch will disregard the related control signals, *passing them on* to the nearest working device. In this way the twin goals of host-controlled functional reconfiguration and of on-chip fault-induced self reconfiguration would be achieved.

To exemplify how the two interacting philosophies work, consider the binary tree case. The simple binary tree that will be implemented is shown in figure 7.3.a. Since its depth is four, four wires are necessary to support functional configuration. The complex switch allows connection to the PE and to the four-wire stack. Any PE has one input a from the switch and two outputs b and c going to the switch: each switch, moreover, has four lateral inputs (x_1 to x_4) from the wire stack and four outputs (y_1 to y_4) to the bus stack (see figure 7.3.b). If the PE is to be a leaf of the tree, bus input x_4 will be connected to PE input a, the PE outputs will remain unconnected, x_1 will be connected to y_2, x_2 to y_3, x_3 to y_4, and output y_1 will be set to 0; if the PE is a nonleaf node, bus input x_4 will again be connected to PE input a, x_2 will go to y_1, x_3 to y_2, and PE outputs b and c will go to bus outputs y_3 and y_4 (see table 7.1 and figure 7.3.c).

Table 7.1

Faulty	Leaf-Node	Non-Leaf-Node
$x_1 \rightarrow y_1$	$x_1 \rightarrow y_2$	$x_2 \rightarrow y_1$
$x_2 \rightarrow y_2$	$x_2 \rightarrow y_3$	$x_3 \rightarrow y_2$
$x_3 \rightarrow y_3$	$x_3 \rightarrow y_4$	$x_4 \rightarrow a$
$x_4 \rightarrow y_4$	$x_4 \rightarrow a$	$b \rightarrow y_3$
	$0 \rightarrow y_1$	$c \rightarrow y_4$

The binary tree of figure 7.3.a is thus mapped onto the physical array (figure 7.3.d).

Accounting for a faulty node is not difficult; it requires only an extension to the connectivity rules (table 7.1 also foresees the faulty node case). It is sufficient to bypass the faulty PE, excluding its input and outputs from connection with the buses and setting the switches to a *through* position for all wires (see the example in figure 7.4). Such technique does not require any modification of the functional configuration procedure, since the presence of a faulty node simply results in a longer segment of wire until the next working PE is reached.

A slightly more complex architecture is required to implement a rectangular array. In this case, in fact, the stack of wires must be deep enough to account for row and column configuration: finally, the total number of wires will equal the sum of row and column cardinality. In order to make the architecture (and the

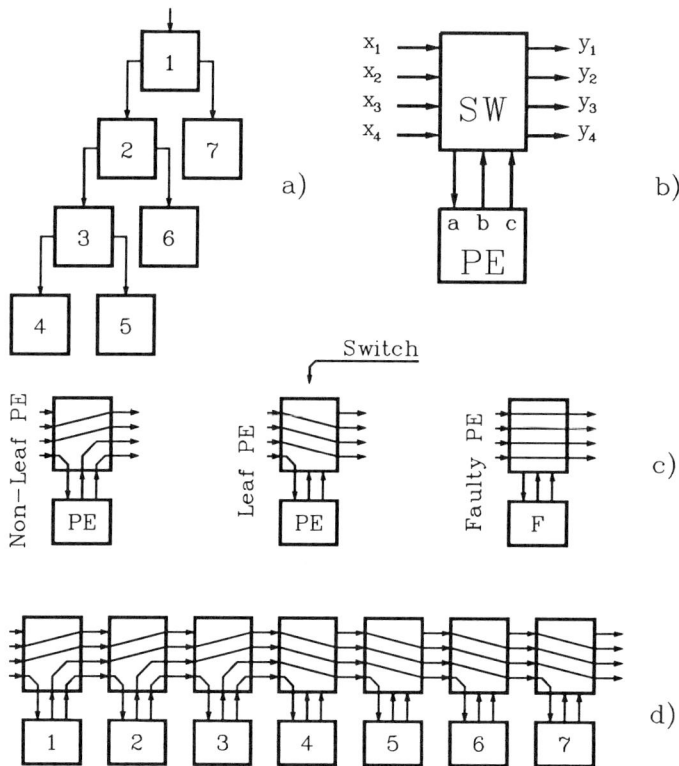

Figure 7.3

Binary tree target array. a) The target array. b) Structure and interconnections of the individual PE-switch pair. c) Switch settings for different PE positions. d) The configured Diogenes array.

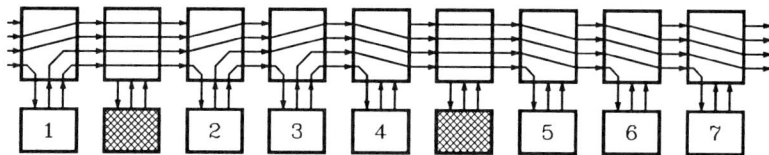

Figure 7.4

Reconfiguration of the binary tree case in presence of two faults and two spares.

reconfiguration technique) more easily understandable, it may be useful to split this huge set of wires and the complex associated switch into two separate stacks of wires associated respectively with two (less complex) switches. This does not actually modify the Diogenes philosophy, but it is a simpler representation of the global structure. Twin stacks of buses and related switches are then introduced (figure 7.5.c), a line of PEs being *inserted* between the two stacks. Each PE is connected to an *upper* switch SW1 and to a *lower* switch SW2; SW1 is dedicated to the creation of row interconnections, while SW2 manages column interconnections. The depth of wire stack is related to the dimensions of the rectangular array: the depth of the upper stack (passing through SW1) is thus identical to the number of rows, while that of the lower stack (passing through SW2) coincides with the number of columns.

Setting of switches for fault-free PEs is related to the mapping of the target onto the physical array. To this end, PEs in the target rectangular array are numbered in column-wise order: assuming an array or R rows and C columns, such ordering will proceed from $1, 2, \cdots, R$ to $(R+1)$, $(R+2), \cdots, (2R)$, etc. (See the sample 3×3 array in figure 7.5.a.) The same code numbers are then given, in order, to the PEs of the physical linear array, proceeding along the propagation direction (in the example of figure 7.5, from left to right). Switches will have the following input-output set-up:

$$\text{SW1 (\textit{row} \quad switch)} : \begin{cases} \text{bus} & \text{inputs } x_{11} \cdots x_{1R} \\ \text{bus} & \text{outputs } y_{11} \cdots y_{1R} \\ \text{one} & \text{PE input, one PE output} \end{cases}$$

$$\text{SW2 (\textit{column} switch)} : \begin{cases} \text{bus} & \text{inputs } x_{21} \cdots x_{2C} \\ \text{bus} & \text{outputs } y_{21} \cdots y_{2C} \\ \text{one} & \text{PE input, one PE output} \end{cases}$$

Setting of upper switches is identical for all fault-free PEs: $x_1(i)$ is connected to $y_1(i+1)$ for $1 \leq i \leq R-1$, $x_1(R)$ is connected to the PE input and y_{11} to the PE output. Lower switches have different settings depending on whether the corresponding PE belongs to the first (uppermost) row (the setting is identical to that of upper switches) in the array or to any other row (in which case x_{21} and y_{21} are connected, respectively, to PE input and output, while for all other values of i x_{2i} is connected to y_{2i}). Settings, for a 3×3 array, are exemplified in figure 7.5.c (the target architecture is shown in figure 7.5.a). Here also, faulty PEs are quite simply bypassed by setting both upper and lower switches to the *through* position. Thus, the fault-tolerance policy remains identical whatever the target architecture implemented by the linear array. Note that the physical architecture designed to support different target architectures has to undergo some modifications depending upon specific target architectures to be mapped onto it, since wire stacks and switch functions are different from case to case.

A basic fault assumption underlies the Diogenes approach, namely, that buses and switches are inherently fault-free. Otherwise, it is quite obvious that failure of just one switch would lead to collapse of the complete array. Nevertheless,

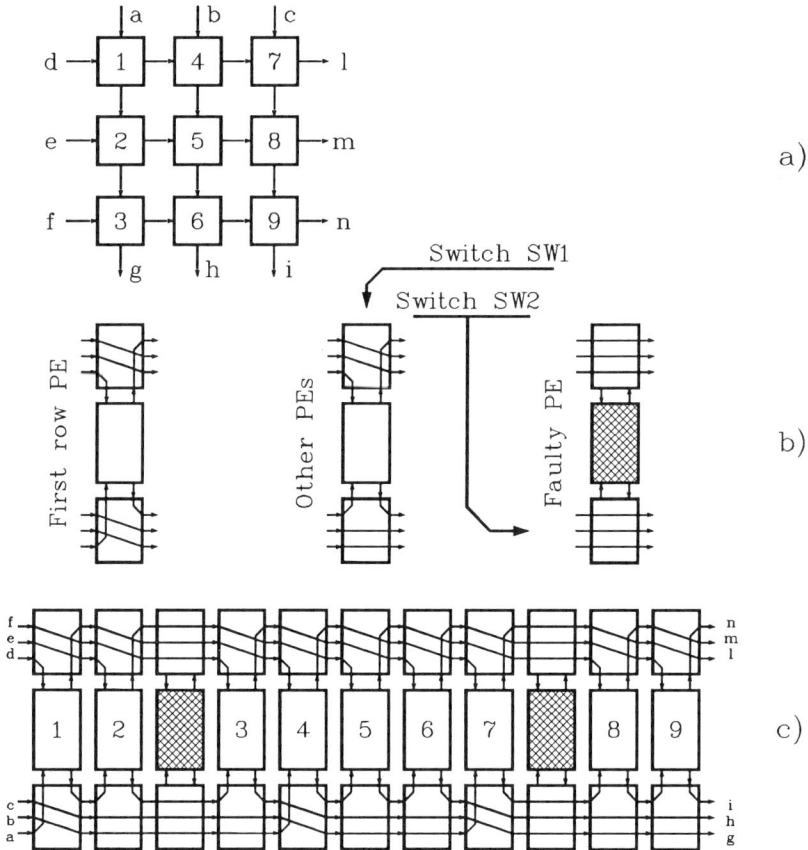

Figure 7.5

Rectangular target array. a) The target array with PE ordering codes. b) Structure and switch settings for different PE positions. c) Reconfiguration of the rectangular array in the presence of two faults and two spares.

this assumption is accepted in the greatest majority of reconfiguration strategies, so that it should not be considered as a peculiar limitation of this technique. Moreover, simplicity of the elementary switches and of the overall interconnection network makes this assumption acceptable, except for clusters of defects such as might appear on a wafer at production time. It is therefore reasonable to examine performances — as regards fault-tolerance — with relation to PEs only.

In the Diogenes approach, given R spares, the probability of survival is always 100 percent for any number of faults $F \leq R$: given the simple bypass philosophy, no limitations resulting from particular distributions of faulty units have to be considered as exceptions leading already to fatal failure for $F < R$. Thus, spares utilization is also 100 percent, a figure seldom reached by other reconfiguration techniques. The reconfiguration algorithm, moreover, is the simplest possible one: an immediate technique, in fact, rather than an algorithm proper.

Against all the above advantages, the lack of any bound on interconnection length — and, therefore, on propagation delay — as a consequence of reconfiguration must be considered as a negative feature. If k adjacent PEs are faulty, this will cause bypass interconnections through as many switches: delays, therefore, become directly related to the relative positions of the faulty elements. This could cause some problems if strict synchronization was required in the overall architecture (switches, obviously, add to the intrinsic delay of the interconnection bus, and there is one switch added to the *logical* connection between two PEs for each faulty PE found on the physical path). A limit to the length of bypass wires has been defined — in probabilistic terms — in [LEI86]: it has been proven that given an array of N cells, where each cell has a 50 percent probability of being faulty, the probability that the maximum wire length is $O(\lg N)$ is at least $1 - O(1/N)$. In particular, the probability that $k \lg N$ consecutive cells are faulty — and, therefore, that the bypass link will have length $k \log N$ — is less than one in N^k. By $O(f(N))$ the authors denote a function that is bounded above by $c \cdot f(N)$, for a fixed constant c and for all sufficiently large N. Chapter 9 will discuss in detail the theoretical bounds determined for this as well as for other instances.

Figures proposed by Leighton and Leiserson refer to a linear array extracted from a rectangular pattern of cells on a wafer: basically, the Diogenes structure once switches are ignored. For the most general Diogenes application, the problem of added delays due to the presence of faults must be compounded with that of delays introduced by functional configuration, since in any target architecture that is not a linear array the paths connecting PEs that are adjacent in the target architecture but not in the physical one may easily involve a relevant number of switches. Thus, in the implementation of a small rectangular array seen in figure 7.5.c, the horizontal link from 4 to 7 (PEs that are row-adjacent in the target array) passes through four switches. In such instances, composition of delays due to functional reconfiguration and to presence of faults may create some disadvantage with respect to other approaches, particularly in the case of large arrays, by introducing excessive delays.

Another figure of merit usually taken into account is area redundancy introduced as a consequence of fault-tolerance. Here, in fact, two causes of redundancy are present: one the result of functional reconfiguration — expressed by buses and switches — and a second caused by fault-tolerance, expressed by spare PEs and by the related bus segments and switches. While the first one does not directly concern us (even though it may become relevant for complex and large target architectures), the second one is directly evaluated as simply proportional to the number of spares and, therefore, to the number of faults that can be overcome, since no overhead to the structure of the interconnection network is introduced by fault-tolerance. In this respect too, the Diogenes approach is quite different from others that will be examined in the coming chapters, where the interconnection network (or, possibly, the individual PEs insofar as they act also as connecting elements) is modified — even in a relevant way — to support fault-tolerance.

The Diogenes approach for reconfiguration after fault is basically adopted also in [MAJ88], limited to linear arrays. The structure of switches, there, has greater intrinsic complexity because (as discussed in chapter 5), switches must provide alternative data paths for the fault-detection phase as well as for reconfiguration. Actually, with regard to fault-tolerance switches support only the bypass of faulty PEs, since subsequent reconfiguration (when necessary) involves a modified *mapping of the algorithm* onto the array, and is therefore outside our scopes.

Before leaving the Diogenes approach, it may be useful to compare it with another approach, dedicated only to fault-tolerance, that was introduced by Kung and Lam for systolic arrays [KUN84]. Systolic arrays discussed by Kung and Lam have a fixed structure, i.e., target architecture and physical architecture coincide, in the absence of faults. Our comparative analysis will therefore be confined to the fault-tolerance aspects alone.

A systolic array is a strictly synchronous structure: this mode of operation involves the presence of registers to support information transfers among the cells. As may be readily inferred, reconfiguration problems are quite different in the case of unidirectional information transfers as compared to bidirectional information transfers: in the first case, there is no *delay skew* problem between the two lines of information transfers, while in the second case the problem arises and is in fact rather delicate.

In a unidirectional linear systolic array (figure 7.6) data move through all the cells from left to right, controlled by a single clock. Introduction of a bypass around a faulty cell must of necessity keep the synchronization unchanged. To grant this requirement, not a simple *bypass link*, but rather a set of *bypass registers* (one register on each data stream) is then introduced (see figure 7.7). In this way, the computational rate of the overall array is kept unchanged but, of course, result latency is increased by one clock cycle for each bypass register introduced. Data travel systolically from cell to cell at the original clock rate: data along the two transmission lines represented in figure 7.7 keep their relative alignments, since both data streams are delayed by the same amount. Basically, correctness of the linear unidirectional array is kept by introducing uniform delays on all data streams between adjacent cells: the pipeline mode of functioning remains, and additional delays are just inserted in correspondence of faults.

For unidirectional linear arrays, correctness of the technique shown in figure 7.7 (discussed intuitively above) is formally proved by a simple but powerful theorem (the *cut theorem*) that proves the necessity — and the technique — for distributing delays (i.e., additional registers) along the data streams in order to achieve correct reconfiguration.

The cut theorem is based upon creation of an equivalent directed graph, representing a systolic array, in which nodes correspond to processing elements while edges represent the communication links and are characterized by a *weight* (equal to the number of registers along the corresponding link). Two graphs are said to be equivalent if, given an initial state for one of them and given one

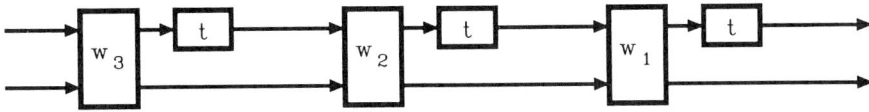

Figure 7.6
Basic unidirectional systolic linear array (non-fault tolerant): w_i are processing elements, t are latches, i.e., single clock step delays.

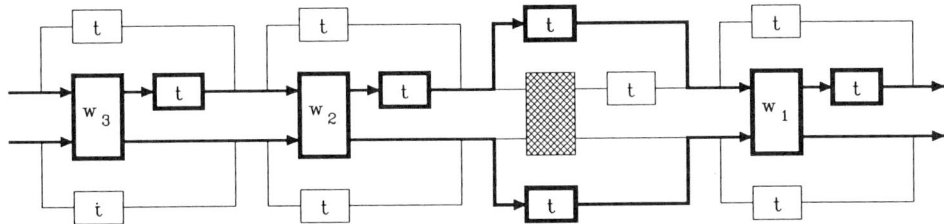

Figure 7.7
Example of reconfiguration of the augmented uni-directional linear systolic array in presence of faults: thick lines denote the active interconnections.

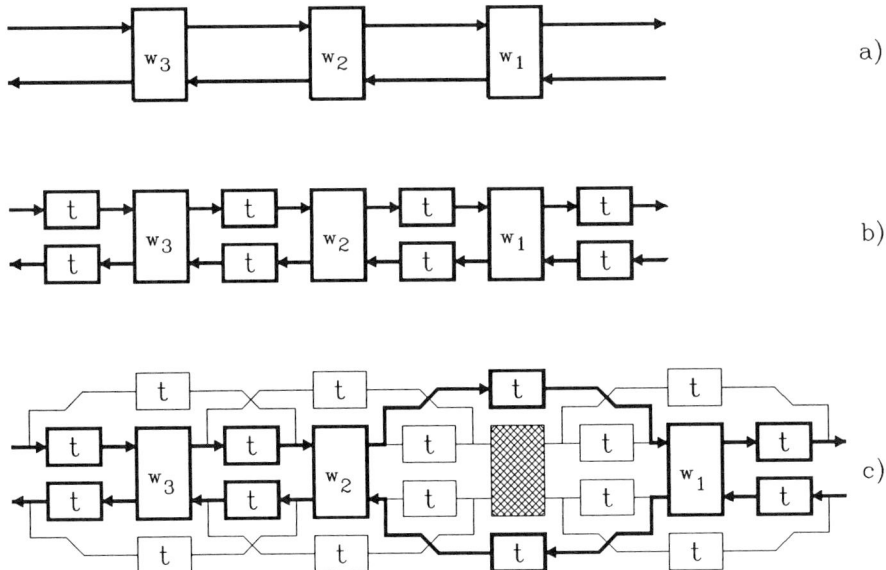

Figure 7.8
Fault-tolerance for bi-directional linear systolic arrays: a) basic (non-fault-tolerant) array; b) array with latches inserted; c) fault-tolerant solution (etched cell is faulty).

same set of inputs fed by the host machine, it is possible to find for the second machine an initial state such that the two designs will produce the same set of results — possibly with different delays. It can be noted that this definition of equivalence is based upon the well-known concept of equivalence between finite-state automata. As far as the host is concerned and provided the different timing can be accommodated, the two designs are then interchangeable.

Given this definition of equivalence, a *cut* is defined as a set of edges C that partitions the nodes of a graph representing a systolic array into two disjoint subsets, a subset S called the *source set*, and a subset D called the *destination set*; such that edges in C are the only ones crossing the boundary between S and D and that they are all directed from the nodes in S to the ones in D. Kung and Lam then prove the *cut theorem*, stating that: *for any design, adding the same delay to all edges in a cut and to those pointing from the host to the destination set D of the cut will result in an equivalent design.*

The problem becomes much more complex as soon as propagation is not as trivial as in unidirectional linear arrays. We will see in chapter 12 how Kung and Lam propose to solve the problem in the case of rectangular arrays. Nevertheless, when bidirectional linear arrays are considered (see the example in figure 7.8) the solution is far from straightforward. In fact, *feedback loops* are involved here and from the statement of the cut theorem it follows that no feedback loops should be cut. It may be noted that the Diogenes approach, basically asynchronous, could be applied without any difference in principle to bidirectional arrays but then the added delay problem would become even more relevant. A modified approach must then be adopted: a first solution suggested in [KUN84] involves the addition of a delay register on each edge (see figure 7.8.b). It can be easily verified that the structure in figure 7.8.b operates correctly, provided the data rate in both directions is exactly half that adopted in the nonfault tolerant structure (see figure 7.8.a). This happens because an *empty time slot* must be introduced after each data transfer, so as to allow transition of data through the bypass registers.

In examining figure 7.8.c, it will be noticed that the bypass of a faulty PE is effected by means of a register introducing a unit delay t, inserted between two fault-free PEs. This technique also allows reconfiguration in the presence of multiple faults, provided there are no adjacent faulty PEs in the array. To avoid k adjacent faults the structure in figure 7.8.b should be modified by insertion of delays corresponding to k time slots ($k \cdot t$ delays) for each cell. This would lead to decrease the data rate by a factor k through the array.

To reach a satisfactory solution for the problem of multiple faults without introducing such excessive speed degradation, Kung and Lam suggest substituting the bidirectional linear array with a *systolic ring*, in practice, a linear array with one single feedback edge leading from its last cell to the first one, rather than with a feedback path passing through all cells in the array. Proof that systolic rings can solve the same problems for which bidirectional linear arrays are designed is outside the scope of this study. With regard to fault-tolerance, of most interest is the fact that a single-bypass register can be inserted on each link — just as in the first structure proposed — and that the delays introduced are also identical to those evaluated for the unidirectional array.

In the Kung and Lam approach, as in the Diogenes one, the underlying fault assumption confines faults to the processing elements only. This is consistent with most proposals for fault-tolerance in systolic arrays, although it might be less justified here than in the Diogenes case since not only switches and links, but also registers are involved. In order to keep the evaluation of relative complexity to acceptable ratios, we should consider cells to be far more complex even than registers (and, in fact, Kung's Programmable Systolic Chip is a device of relevant complexity, as compared with array cells in other devices). Register failures might be accepted by inserting further redundancy schemes involving memories. Nevertheless, the basic requirement that delays along all data streams present in the architecture be preserved without modifications must at all costs be satisfied.

With respect to the Diogenes approach, performance analysis does not introduce any new facts beyond considerations of delays, at least as far as probability of survival and spares utilization are concerned. Area requirements are quite different, since redundancy is not limited any more to the k spare cells (and to related interconnection network segments), but it must involve also the k sets of bypass registers. On the other hand, this new factor cannot be evaluated in an abstract way, since it is obviously dependent on the parallelism of data in the specific implementation.

7.1. References

[CHU86]   Fan R.K.Chung, F.T.Leighton, A.L.Rosenberg: *Diogenes: a methodology for designing fault-tolerant VLSI processing arrays*, Proc. FTCS-13, 1983, Milano, 26-32, IEEE.

[KUN84]   H.T.Kung, M.S.Lam.: *Fault-tolerant VLSI systolic arrays and two-level pipelining*, Journ. Parall. and Distr. Proc., Aug. 1984, 32-63.

[LEI86]   F.T.Leighton, C.E.Leiserson: *A survey of algorithms for integrating wafer-scale systolic arrays*, in: *Wafer Scale Integration*, (G.Saucier, J.Trilhe eds.), Proc. IFIP WG 10.5 Workshop, Grenoble, Mar. 17-19 1986, North-Holland.

[MAJ88]   A.Majumdar, C.S.Raghavendra, M.A.Breuer: *Fault-tolerance in linear systolic arrays using time redundancy*, Proc. 21th Hawaii Int'l Conf. on System Sciences, Kailua-Kona Jan. 5-8, 1988, 311-320, IEEE.

[ROS83]   A.L.Rosenberg: *The Diogenes approach to testable fault-tolerant array of processors*, IEEE-TC, Vol. C-32, N. 10, Oct. 1983, 902-910.

[ROS85]   A.L.Rosenberg: *Graph-theoretic approaches to fault-tolerant WSI processor arrays*, in: *Wafer Scale Integration*, (W.Moore, C.Jesshope eds.), Proc. Int'l Workshop, Southampton, July 10-12 1985, Adam Hilger.

The Diogenes approach considered a linear physical array onto which various target architectures (including the linear one) could be mapped. In other solutions both target and physical arrays are linear. An early example was the ALAP design [FIN77], maybe the first implemented instance of wafer-scale integration, that made use of a simple bypass scheme for avoiding faulty PEs (bypasses could be chained together). If this approach is accepted, reconfiguration criteria are necessarily simple and the obvious drawback is given by the possibility of long interconnection paths between logically adjacent PEs. Alternatively, it is possible to envision physical architectures with connectivity more complex than the linear one, from which a linear array must be extracted.

This chapter examines some solutions proposed to achieve fault-tolerance when the target architecture is a linear array and the physical architecture is a rectangular array whose interconnection network allows links to be routed along two orthogonal axes. All approaches considered here are of the incremental type: whenever a cell has been reached during construction of the linear array, its *nearest neighbors* are explored in order to extend the linear array itself by adding to it other fault-free cells. The following two considerations therefore hold:

(1) If the distribution of cells onto the chip (wafer) is not strictly rectangular but rather irregular — approximating a rectangular pattern inside a circular area, such as that appearing on a wafer at production time [AUB78] — a larger number of working cells can actually be taken into account than by limiting edge contour;

(2) if intercell connections are not simply organized along two orthogonal axes but following more complex connectivity rules (e.g., hexagonal connections), the probability of achieving reconfiguration increases. For example, consider a hexagonal interconnection pattern (figure 8.1): each cell in the array has six nearest neighbors. Thus, looking for a fault-free cell among the nearest neighbors for a given cell has a higher probability of success than in the case of rectangular interconnections, and the probability of achieving reconfiguration is consequently also higher.

Thus, the definition of the physical array given above can be considered as restrictive with respect to the more general instance. In fact, many of the proposed algorithms (and the evaluation criteria) that will be discussed in this chapter can also be extended to nonrectangular interconnection patterns, particularly, to hexagonal patterns, without any significant modifications; the limited definition of physical architectures envisioned is therefore acceptable.

The adoption of the two restrictions defined above will lead to a pessimistic evaluation of the algorithms: a safe position with regard to reliability and fault-tolerance.

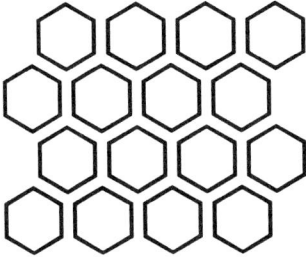

Figure 8.1
Hexagonal interconnection pattern.

An additional point concerns functionality of the array cells. Contrary to the case of more complex target array structures, when linear arrays are considered some authors (e.g., [AUB78, MAN77]) assume arrays dedicated not only to processing functions but also — or even, rather — made up of memory cells, constituting long shift registers. This assumption in turn influences the level of complexity that is considered acceptable for the algorithms envisioned. Whenever such an issue may be of relevance, we will explicitly point out whether cells are meant to have processing capabilities or whether they are generic, uncommited devices (possibly memory ones).

We may perceive the various approaches proposed for reconfiguration of linear arrays as corresponding to three classes, deriving from the conflicting requirements of higher harvesting (or spares utilization) and of better algorithm simplicity. The three classes of restructuring/reconfiguration techniques correspond to the following procedures:

(1) The array is first subdivided into *patches* of manageable dimensions, inside which reconfiguration is then performed. This is a very simple version of a *divide and conquer* philosophy: a large system is partitioned into smaller subsystems inside which 100 percent harvesting can be reached by local reconfiguration techniques, and such local subsolutions are then patched together to form a global one. Provided the *patches* are small enough, local solutions are quite simple, based on a *catalog* of predetermined cases. On the other hand, management of such an approach inevitably requires use of an external host that recognizes instances corresponding to the various fault patterns, and it is typically suited to production-time restructuring.

(2) A tree of working cells is first created, and the linear array is then mapped upon it. Harvesting is not as satisfactory as in the previous case, since each cell in the tree is required to be adjacent to at least one other cell also belonging to the tree (thus, fault-free cells enclosed by barriers of faulty ones cannot be reached), but, on the other hand, the procedure can be translated by an algorithm (rather than by a heuristic methodology).

(3) A continuous *spiral* of working cells is extracted from the total set of cells. Restrictions introduced in (2) are present here also (actually, all but the tip

cells must be adjacent to *two* other cells also present in the spiral). Harvesting
again decreases, but algorithms are simple enough to be implemented by on-
chip circuits and therefore they can be used for run-time self-reconfiguration.

It is immediately evident that these techniques are quite different from the Diogenes one. Algorithms that make up the fault-free linear array may be rather
complex, and utilization of spares may be lower — even much lower — than 100
percent, for unfavorable fault distributions. On the other hand, the interconnection network provided is usually very simple and, moreover, delays introduced in
interconnections between any pair of cells may be kept strictly bound. Both these
factors make the techniques interesting in view of run-time reconfiguration and
high-speed requirements.

(*Note*: Similar to our treatment of the Diogenes approach, we will sometimes
substitute the original figures proposed by the authors of the various proposals
with schemata that allow a better emphasis of some of the characteristics used to
perform a comparative evaluation of the different solutions.)

8.1. The patching method

The *patching method* (see [LEI86]) may be seen as a very simple version of the
divide and conquer approach, whereby the complete physical array is partitioned
into subarrays of manageable dimensions and reconfiguration is performed locally
inside such subarrays. Small target subarrays are thus created, and global reconfiguration is then achieved by connecting the target subarrays in suitable order. The
technique (not a real algorithm, since reconfiguration possibilities are exhaustively
analyzed) can be summarized as follows:

(a) The physical, rectangular array of N cells is partitioned into square *patches*
    each containing $2 \log N$ cells. The dimensions of the patches are chosen start-
    ing from the consideration that, in an array of N cells each of which has a
    50 percent probability of being faulty, the probability that $2 \log N$ cells are
    simultaneously faulty is less than $1/N$. For N sufficiently high, the proba-
    bility that each cell will contain at least one fault-free cell is therefore also
    reasonably high.

(b) The working cells in each subarray are connected to each other by a *local
    linear array* that collects all fault-free cells and that (unlike the spiral ap-
    proaches to be seen later) allows bypassing columns of faulty cells (see figure
    8.2). Given the small dimensions of the patch, the choice of the local linear
    array corresponding to the specific fault distribution can be made simply by
    enumeration in a catalog of possible paths. Thus, target linear subarrays are
    built inside each patch.

(c) Finally, all target subarrays thus built are connected to each other (see thin
    lines in figure 8.2). Connections are again chosen in a set of prespecified
    patterns.

Harvesting is 100 percent. Since from this point of view optimality is achieved, the
second factor of merit to be considered is that of the length of connection links and,

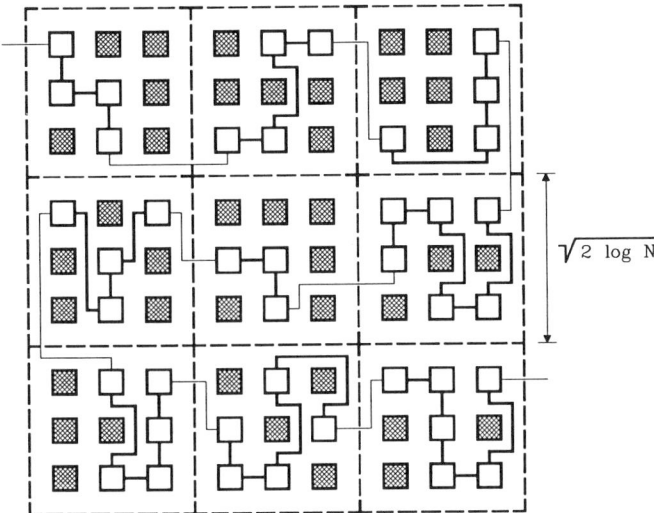

Figure 8.2

A scheme for constructing linear arrays from all live cells on a wafer with wires of length $O(\sqrt{\log N})$ and constant channel widths (from [LEI86]).

therefore, of added delays. In [LEI86] a probabilistic evaluation is presented, just as for the Diogenes approach. Since any pair of cells within a patch can be linked by a wire that is at most $2\sqrt{2\log N}$ long, the wires in a patch have length $O(\log N)$. Considering the connections between subarrays, on the reasonable assumption that each patch will contain at least one fault-free cell, these connections will require wires at most $3\sqrt{2\log N}$ long. Finally, with high probability no wire of the target array will be longer than $O(\log N)$.

The problem with the above method is that there are some fault distributions for which the bound will be surpassed. In particular, whenever single cells are present inside fault clusters, long wires will necessarily be created to insert them into a target array (the subarray inside the patch reduces then to one cell); again, when whole patches are found to be made only of faulty cells, the corresponding patch-bypassing links will have to be introduced. It is self-evident that such long links cannot be avoided, whatever the algorithm, if 100 percent harvesting is required: the alternative, as will be seen in the following sections, is to accept lower harvesting, and to balance this with bounded delays.

A further point concerns area overhead introduced by the reconfiguration technique. The a priori requirement is that it must be possible to reach each cell in a patch from any other cell in the same patch, as well as from the patch boundaries. It is obvious that a direct-link connection philosophy cannot be accepted (it would involve creation of a complete graph of interconnections inside each patch). This philosophy is discarded in favor of buses and switches in the intercell channels (a solution made evident by the example). To create a linear array one bus in each channel, together with a switch on each cell output and at each bus crossing, will support this technique: a fact that keeps area requirements acceptably low.

8.2. The tree approach

The typical solution based upon the tree approach is discussed in [FUS82]: assumptions on the physical architecture may be summarized as follows:

- Cells are organized along a rectangular array (a *nearest neighbor* mesh). Although the target architecture is a linear array, each cell has ports for interconnections in all four directions, connections being possible following both orientations of each axis;

- cells are processing elements of some inherent complexity and capacity; in particular, they are capable of executing message exchanges synchronized on a *send-receive* basis;

- an augmented interconnection network is present, enabling the creation of *return* interconnections along a line of working cells. It is possible that such return paths could be provided by the cells themselves, acting simultaneously as processing elements and as communication devices; cell functions would subsequently have to be augmented and inherent cell reliability would decrease.

A basic problem in this context is that of defining the I/O ports of the complete system. Should one single I/O port be envisioned initially in the physical array, (as the authors immediately note) the problem of reconfiguration ought to then be redefined as that of identifying one single I/O path of specified length consisting only of fault-free cells, and such a procedure could easily lead to low spares utilization. Moreover, procedures for creation of this path would also be quite complex and time-consuming. Rather, Fussell and Varman prefer to allow paths to grow in a more flexible pattern, and to foresee a subsequent technique capable of mapping the single linear array onto the complex tree structure thus obtained; this last step becomes possible by the augmented interconnection network.

In the proposed technique, a spanning tree is built starting from a fault-free cell (see figure 8.3, where the root cell is the one at the upper left corner of the array; the formal definition of a spanning tree will be given in chapter 9). The tree grows by subsequent steps: at each step, all tip cells reached in the previous step simultaneously attempt to expand the branch to which they are connected by adding to its tip a new fault-free cell not already inserted in any branch. To this end, all tip cells effect a search among unused neighbor cells, exploring the four nearest neighbors in the order North-West-South-East. Whenever no further extension of a branch is possible, the tip cell is declared to be a leaf of the tree. (The procedure described in [FUS82] is based upon a simple request-and-acknowledge protocol: other algorithms may be envisioned, reaching the same results with different performances.) The distance of a cell from the root of the tree, measured along the path that connects it to the root, thus also denotes the time (defined in terms of tree-building steps) at which it has been reached.

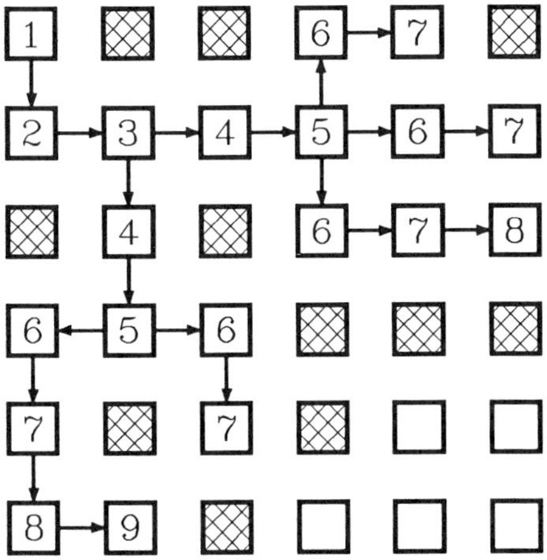

Figure 8.3
Construction of a spanning tree with fault-free cells in a rectangular physical array.

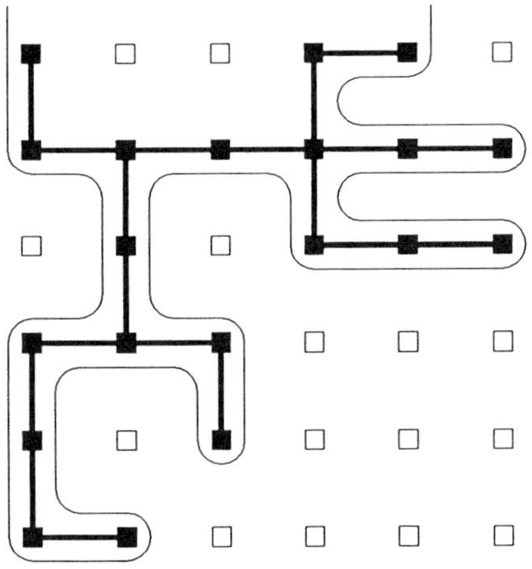

Figure 8.4
External contour of the spanning tree.

Once the spanning tree has been built, the linear array is mapped upon it by simple path-following rules, as in figure 8.4 (where the tree is drawn with bold lines). In fact, the linear path is *folded* upon the tree. That is, the external contour of the tree is first identified and this ideal line is used to complete the mapping of the target upon the physical array, as follows:

- starting from the root, the tree is explored by a simple left-turning rule, i.e., at each branching point the left branch is chosen. As long as the path gets away from the root, information transfers are effected through the cells

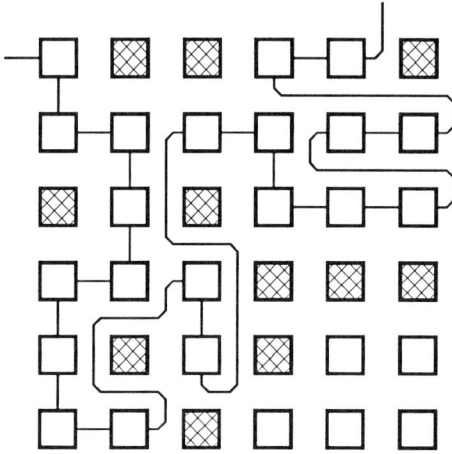

Figure 8.5
Extraction of the linear array from the spanning tree.

and data thus transferred undergo processing as requested by the application algorithm.

- Whenever a leaf is reached, the path backs up to the nearest branching point: return connections are always effected along a bus or, more generally (as far as implementation technology is concerned) along data links adjacent to the cells but not involving any processing upon the information transferred.

- If, at the branching point reached with the backup procedure, there is still one unexplored branch, this is in turn followed (and, therefore, its cells are added to the linear array that is being built). Otherwise, the path is again backed up to the next previous branching point, a process that continues until all the cells in the spanning tree have been exhaustively considered. At this point, the last leaf introduced in the path also becomes the output cell of the array (the complete linear array derived from figure 8.4 is shown in figure 8.5).

With regard to the interconnection network supporting this technique, Fussell and Varman discard the possibility of creating complex cells capable of two different data-handling functions, i.e., a *forward* one involving processing of input data and a *backward* one involving simple transmission. Besides adding to the cell complexity (therefore to inherent redundancy) this solution would involve a complex definition of the fault model: to reach reasonable reconfiguration efficiency, a cell recognized as faulty for the *forward* path (i.e., as far as its processing functions are concerned) would at the same time be required to be fault-free for the backward path. The alternative is to provide a network capable of backward information transfers. For a rectangular array of processing cells, the functions required from such a network may be supported by a structure consisting of one bus inserted in each channel between a pair of rows or columns, three switches being associated with each cell to correctly complete the information routing. It is then possible to associate, as

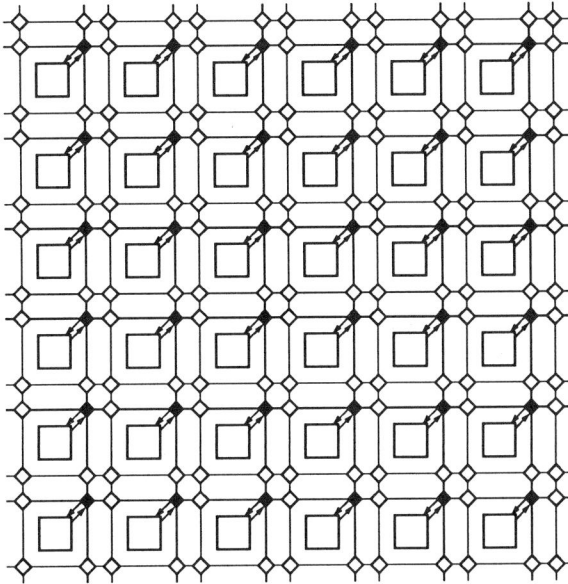

Figure 8.6
Physical switched-bus structure supporting reconfiguration of linear arrays.

in other solutions, faults with the processing cells alone, so as to guarantee fault-free return paths. (Note that the added complexity required by the augmented interconnection network is thus identical to that involved in the patching method.)

A particular implementation example can also be designed by introducing the added requirement that each PE may have only one input port and one output port. This is clearly a restriction with respect to the previous, looser assumption of four I/O ports, only two of which may be active at any time; on the other hand, it allows much more compact (and reliable) design of the PE itself. A second switched-bus system then becomes necessary to provide the four bidirectional links attached to the *virtual* PE. Finally, the structure in figure 8.6 is obtained. Square boxes represent PEs and diamonds represent switches: black diamonds are used to implement the four bidirectional links of the virtual PE and white diamonds are necessary to create the return paths.

Actual implementation of the reconfigured array in figure 8.5 then requires the switch setting (and corresponding link activation) shown in figure 8.7. The structure of figure 8.5 is not, therefore, the actual physical structure but rather a schematization useful for algorithm description. Nevertheless, it allows such straightforward reasoning that henceforth it will be referred to rather than to a physical implementation such as that in figure 8.6.

The procedure above outlined appears to be designed mostly for production-time configuration, so that the main goal is that of creating the longest possible array, or a linear array of given length from a large population of cells, rather

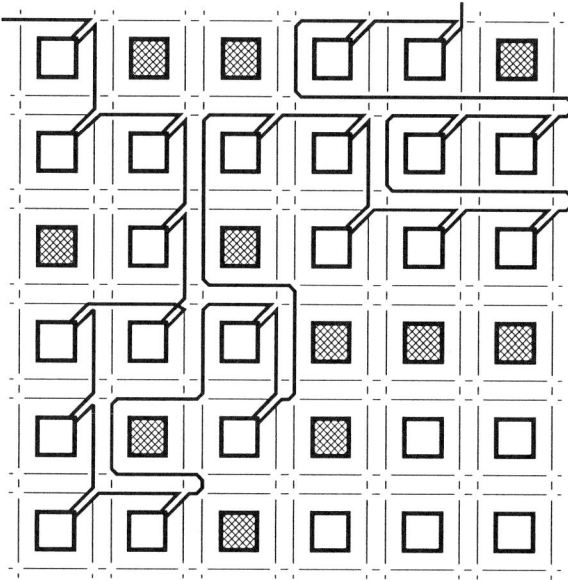

Figure 8.7
Switch and path activation implementing the structure of figure 8.5.

than overcoming individual faults in a preexisting array. Efficiency is certainly
not as high as for the Diogenes approach. As is also apparent in figure 8.5, not all
available cells can be inserted in the final working array, so that *harvesting* (i.e.,
the percentage of fault-free cells introduced into the operating array) is in general
lower than 100 percent and depends upon the number of faults as well as on the
particular fault pattern. The basic requirement for inserting a fault-free cell into
the target array is that it belongs to a *connected set*, i.e., to a set of fault-free cells
in which each cell is physically adjacent to at least one other cell of the same set
(in figure 8.5, in fact, there are two connected sets, only one of which is finally
used). If — as it is assumed in the definition of this algorithm — most fault-free
cells are located near each other, the possibility of large connected sets is realistic
and the solution has viability. On the other hand, this assumption is particularly
realistic when fault patterns correspond to *clusters of faults*, such as are likely to
appear on a wafer at production time.

In order to evaluate the performances of this procedure, direct simulation has
been applied for random distributions of faults. Simulations have been repeated
for different fault patterns, assuming a fault number variable from 1 to $N \cdot (N-1)$
in a $N \times N$ physical array. Each simulation run corresponds to one point plotted in
figure 8.8, where cardinality of the largest connected set *vs.* increasing cardinality
of the complete random fault distributions is represented: these dimensions have
been normalized against the number of fault-free cells in the array. Normalization,
in practice, corresponds to the evaluation of yield, i.e., of the percentage of fault-
free cells in a large physical array. It can be noted that for high values of yield

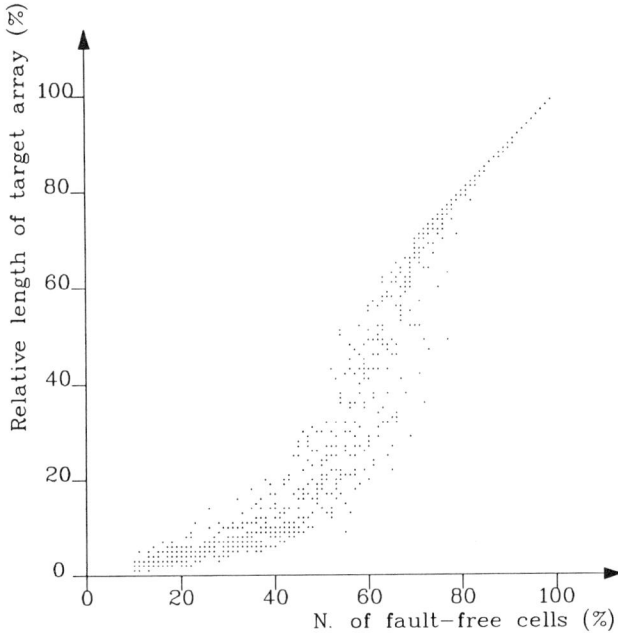

Figure 8.8

Relative cardinality of the largest connected set (i.e., length of the array) vs. number of fault-free cells available: values are normalized against total number of cells for a 10×10 physical array.

(from 100 percent down to 80 percent) the largest connected set will contain all fault-free cells. If the initial yield is in the range of from 40 percent to 80 percent, the percentage of fault-free PEs that cannot be used is fairly relevant and values of maximum length for the target array show a high statistical dispersion. When yield decreases below 40 percent, the procedure hardly becomes applicable. In this case, in fact, the probability of being able to insert in the target array only a small percentage of fault-free cells becomes quite high.

Results plotted in figure 8.8 have been derived starting from a physical array of 10×10 cells; the above considerations acquire even greater relevance when dimensions of the physical array increase (figure 8.9 was derived for a 6×6 array, figure 8.10 for a 16×16 one).

To identify the largest connected set, an initial analysis of the whole array must be performed so as to check all possible connected sets present for a given fault distribution (the root of the final spanning tree will necessarily be a cell of one such set). Such analysis must be performed by a host computer and it can be rather time-consuming: an alternative possibility is to select a predetermined cell as the starting point and then to build the target array around it.

In figures 8.11 and 8.12, for an array of 10×10 cells, the value of the connected set containing one given cell is plotted against increasing fault distributions (again,

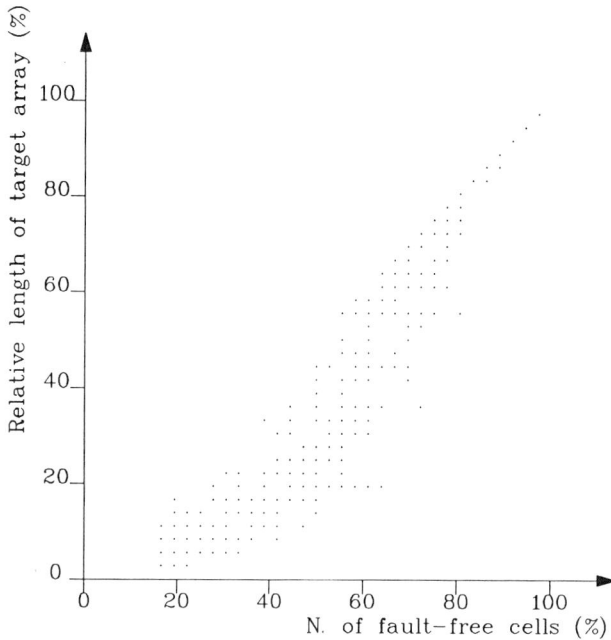

Figure 8.9

As in figure 8.8, for a 6 × 6 array.

such a number is normalized against the total number of fault-free cells). This second case has relevance with regard to the assumption that the input cell of the linear array should be in a pre-determined position (or as near to it as possible, if the nominal input cell is faulty). Figure 8.11, in particular, refers to the choice of a cell on one of the physical array's vertices as the root of the tree (simulations were performed by adopting a starting cell in or near the upper-left vertex). Such an initial choice is justifiable in view of the requirement that the first cell of the target array be as near as possible to an I/O pad on the chip. The same type of normalization adopted earlier has been used here as well. It can be noted that for high yield values (over 80 percent) results do not show any actual difference with respect to the ones given in figure 8.8. When yield decreases between 85 and 70 percent, the probability of using a high percentage of fault-free cells is high — but so too is the probability of using only a small percentage of them (this corresponds to the cloud of dots near the horizontal axis). On the contrary, the probability of using an intermediate quantity of fault-free cells (e.g. of creating a target array with half the fault-free cells) is very low. This is justified by the fact that with such yield values the probability of finding a fault pattern that isolates a small region is rather high (and so is, therefore, the probability of creating small target arrays, if the root cell lies in such an isolated region) while the number of faults is not large enough to isolate a region of intermediate dimensions.

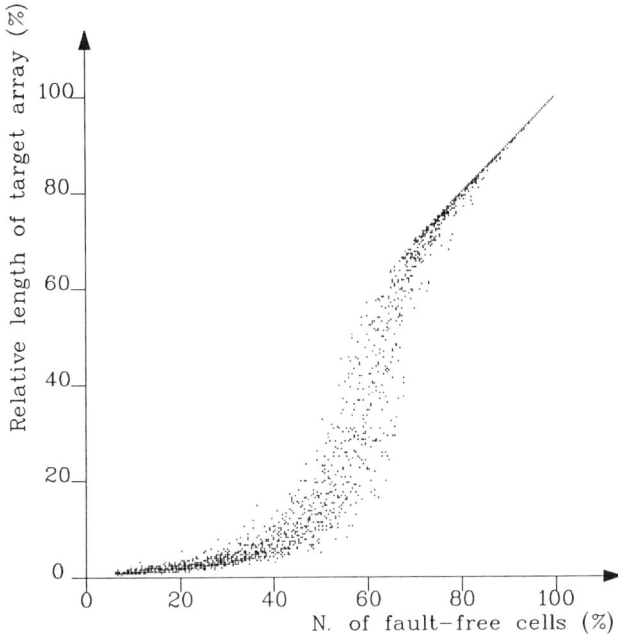

Figure 8.10

As in figure 8.8, for a 16×16 array.

Figure 8.12 plots the results obtained when the root cell is chosen at the center of the array. The previous phenomena are then less evident, since the probability of isolating very small sets of fault-free cells is higher along the border of the wafer (where the border itself, besides the pattern of faulty cells, contributes to the boundary of the set) than at the center of the wafer (where only faulty cells can act as a barrier).

The previous analysis, performed in detail for the Fussell-Varman algorithm, leads to results that are valid for the techniques described in the coming subsections [KOR81, AUB78, MAN77]. For simplicity, from now on we will always assume a fixed starting point for the target array. The area redundancy resulting from the augmented interconnection network is low, because — as shown in figure 8.6 — the network itself is quite simple; since the goal is that of harvesting rather than run-time fault-tolerance, it is impossible to speak properly of *spare cells* and to evaluate corresponding redundancy. Against these favorable performances, the authors note that delays can become quite high, since the *return* path segments can be quite long. In this context, the specific tree-building algorithm adopted can become very important. In particular, the algorithm originally described by Fussell and Varman can bring a penalty when the fault distribution corresponds to full rows or columns of fault-free cells among which faults are interspersed. Long branches of fault-free cells followed by correspondingly long return paths will then be created.

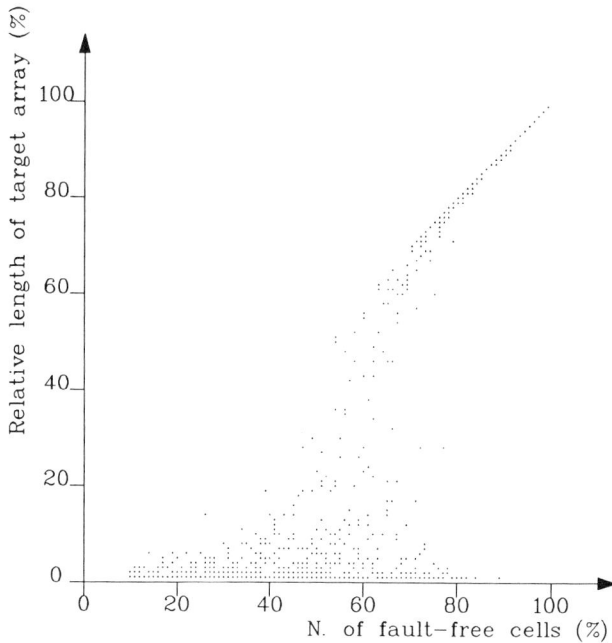

Figure 8.11

Relative length of target array vs. number of fault-free cells (normalized) for a 10×10 physical array, when the root cell is an array vertex.

Nevertheless, it has been proven [LEI86] that if path crossings are accepted when the folding procedure is applied, and if channel widths accommodating more than one bus per channel are adopted, path length can be kept within strict bounds with very high probability.

8.3. The spiral approaches

Several proposals have been published that fall within the category designated *spiral approaches*, whose main characteristics are:

- cells are distributed following a rectangular pattern and have interconnection ports in all four directions. At any time, two connections may be active;

- the spiral making up a linear array of length N consists of a set A of N fault-free cells such that each cell c_i in A is physically adjacent to one other cell in A (if c_i is a terminal cell of the array) of two other cells in A (for all other cells in the array). (An exception to this restriction is introduced by El-Gamal and Greene alone, [GRE84], whose solution will therefore be discussed at the end of the chapter.)

- When the spiral path is built onto which the array will be mapped, each cell in A is traversed exactly once and cells that become logical adjacents are also

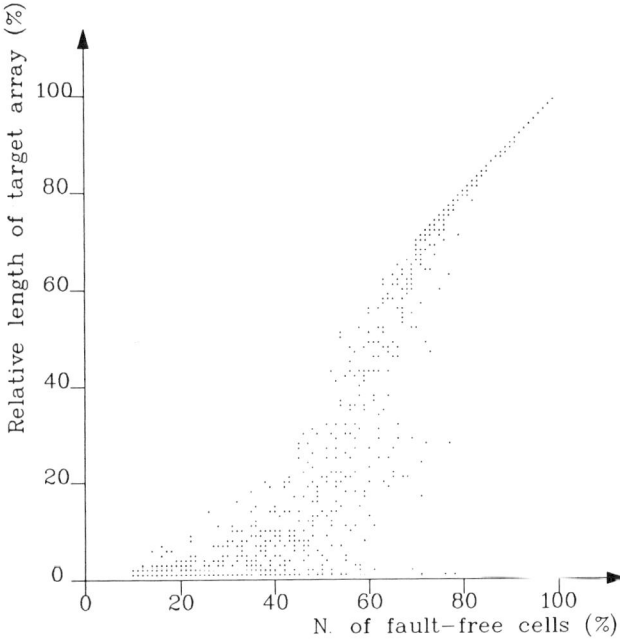

Figure 8.12
As in figure 8.11, when root cell is at the center of the array.

physical adjacents (although generally not in the same order: e.g., the left-hand logical neighbor of cell c_i is necessarily also a physical neighbor of the same cell, but it will not be required to be the left-hand neighbor). Again, only Greene and El-Gamal allow an exception on this point. No parallel paths through the same physical locations are possible, so that no point in the rectangular grid representing the physical architecture is ever traversed twice.

It is then self-evident that a spiral path, whatever the algorithm used to create it or the restrictions imposed, involves a set of cells that is a subset of one of the *connected sets* defined in the previous section.

If the basic (abstract) physical architecture envisioned in the previous section is envisioned here too, cells will be enclosed in a grid of single buses, each cell being connected by an input and an output link to one switch at the nearest bus crossing on its left-hand upper vertex (this architecture underlies the reconfiguration shown in figure 8.13). In all spiral approaches, interconnection paths follow the two rules described below:

(a) if logical cell c_i is inserted on a path, information from c_{i-1} to c_{i+1} undergoes processing by c_i (if any is performed in the array). That is, no cell acts only as a *passing* (or "connecting") element;

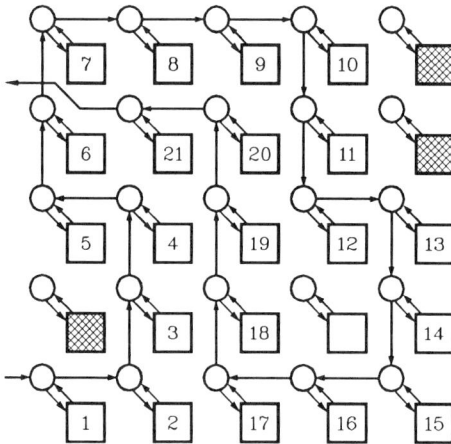

Figure 8.13

Sample spiral path: abstraction of physical interconnections. Etched cells are faulty, unconnected blank ones are unused fault-free cells.

(b) connection between two logically adjacent cells is always performed via one of the direct links attached to the cells; no additional buses are provided.

Spiral approaches have been proposed, in particular, for serial memories, and this application justifies the restriction noted here. No augmented interconnection network, therefore, need be added to that of figure 8.13.

An example of a spiral path implementing a linear array, based on fairly general principles common to most algorithms of this class, is given in figure 8.14, where a simplified conventional representation of interconnections in the basic array is used just as was done for the Fussell-Varman approach (the structure implementing the elementary connections between cells is not shown). Several general observations can be stated before examining a few specific cases in detail:

- in all solutions, a predetermined cell is chosen as *head of the spiral*. Some authors prefer a cell at the center of the array, rather than at one extreme corner. Both choices can be justified, although for quite different reasons, as noted concerning the Fussell-Varman approach. Independent of its position, the head is always assumed to be fault-free. At the beginning of reconfiguration, should it prove to be faulty, a preliminary search for the nearest fault-free cell is performed. However, since this action is not part of the reconfiguration procedure, it will not be discussed here.

- Whatever the initial choice for the head, it is necessary to *end* the spiral implementing the linear array upon an I/O pad, and the whole spiral must consist exclusively of fault-free cells: this creates an immediate problem. In fact, if the spiral head is on an external border the input pad is immediately set without problems, but the spiral tail (its farthest tip) may well be at the center of the array. This will then require introduction of an augmented inter-

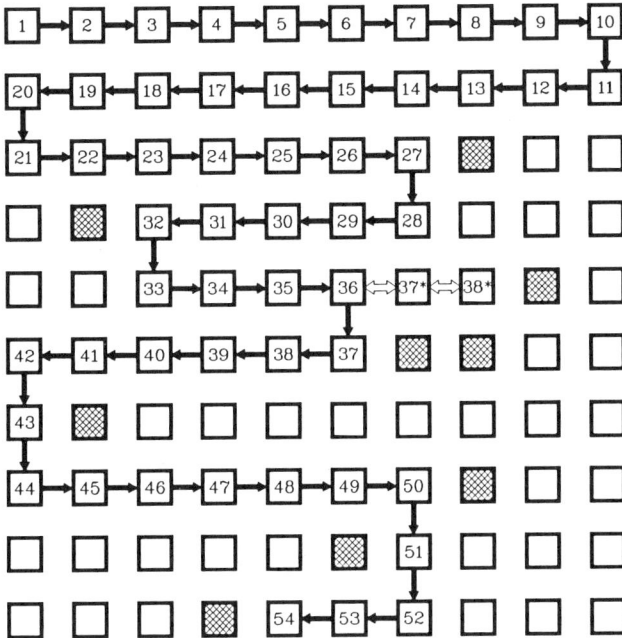

Figure 8.14

Spiral path created by Koren's algorithm.

connection network with the unique purpose of creating the final connection to the output pad. If, on the other hand, the spiral head is at the center of the array, connecting external pads is difficult for input as well as for output. While we can restrict possible positions for the head to a limited area (and therefore introduce only a few corresponding buses to implement a connection with the input pad), position of the spiral tail is as unpredictable as in the previous case and the same problems appear.

The common strategy for spiral approaches is to make the spiral grow by trying to prolong it one cell at a time. Such attempts can be made on-chip, performed by logic associated with each individual cell, or else they can be driven by a host machine (in which case, suitable commands must be propagated through the spiral when run-time reconfiguration is envisioned). Algorithms that require only a limited amount of added logic are simple but they are also not very efficient (i.e., they lead to low utilization of available cells). Highly efficient, and more complex algorithms obviously involve higher overhead.

Two separate instances can be examined:

(1) creation of a spiral of predefined dimensions inside a given array. In this case, if the length of the spiral is N and the complete array consists of $M \times M$ cells (clearly, $M \cdot M > N$), the $M \cdot M - N$ cells can be considered as *spares*. The problem is then, properly, that of reconfiguration; the final result of an

attempt at solution of any specific instance is of the go-no go type (either the array of the required dimensions is extracted, or else the fault distribution is such that *fatal failure*, i.e., impossibility of reconfiguration, is declared).

(2) Creation of a linear array of the greatest possible dimensions, given the fault distribution in the original array. This belongs rather to the end-of production restructuring class of problems, and the associated figure of merit is that of *harvesting*.

Most algorithms proposed in the spiral approach class can be used for both aims. Here, we will sometimes confine ourselves to analyzing the performances of a given algorithm for that goal for which it appears better suited.

8.3.1. The Koren algorithm

The algorithm proposed by Koren in [KOR81] is notable for its simplicity. It is suited both for restructuring and for reconfiguration. When analyzing its performance, we will consider only the first case, i.e., identification of an array of maximum length in a rectangular array of $M \times M$ cells.

Let us associate with all cells of the physical array a pair of physical indices (i, j), with i (row index) increasing from top to bottom and j (column index) increasing from left to right. Initially, the fault-free cell with lowest values of i and j, (i_m, j_m) is chosen as the spiral head. From this, the spiral is propagated until a faulty cell is found as shown in the topmost two rows of figure 8.14, i.e., based on the following rules:

- cells are connected along the first row for increasing values of j until $j = M$; then i is increased by one and connections along the new row proceed for decreasing values of j until $j = 1$ is reached.

- Repeat the above procedure alternating odd rows (like the first one) and even rows (like the second one) until either the last cell of the array is introduced in the spiral or a faulty cell is reached.

As soon as a faulty cell is found, the algorithm simply intrepets it *as if* it were a border cell, and therefore it increases the row index and inverts the sense of spiral propagation: in rows 3 and 4 this basic procedure is applied. Assume now that the target cell following such fault-avoidance procedure is in turn faulty (as happens, e.g., with row 5): the basic criterion cannot be applied, and provisions for this problem are made as follows:

- row i just completed is backtracked until a value of j is reached such that $(i + 1, j)$ is fault-free; at this point, the spiral proceeds to row $i + 1$ and cell $(i + 1, j)$.

In row 5, backing up along two positions is required.

Since in this approach the initial cell is determined a priori, without considering possible alternatives that might allow the creation of longer spirals, it is not useful to evaluate the longest arrays that can be built starting from any initial cell, as

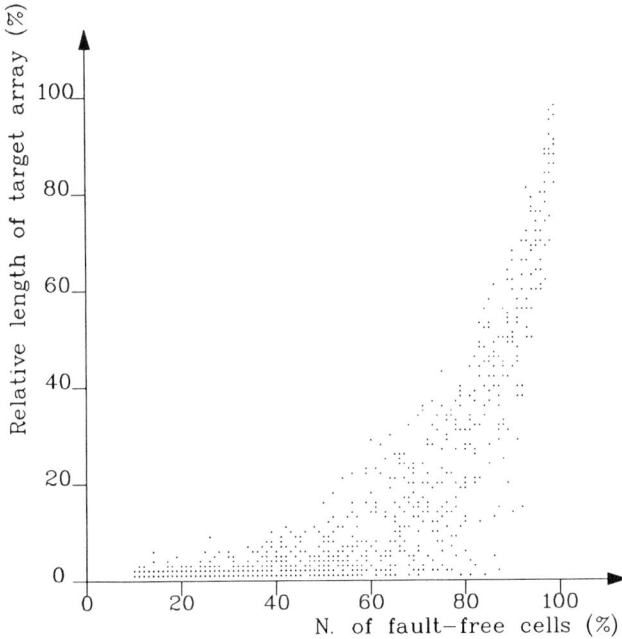

Figure 8.15

Mean length of target array vs. number of fault-free cells (normalized) for a 10×10 physical array:
Koren's algorithm.

was done for the tree approach. Rather, figure 8.15 plots the mean length of the
longest spiral, *vs.* the normalized value of faulty against faulty-free cells. Figure
8.15 was derived by direct simulation for a 10×10 physical array, and it allows us
to deduce that this method is valid only if yield is quite high (over 90 percent).
In this case, given a $M \times M$ array in which K faulty cells are present, the mean
length L of the target array can be evaluated as:

$$L = M \cdot M - M \cdot K$$

This value is obtained by noticing that one single fault at the center of any given
row i leads to the discard of $M/2$ fault-free cells both in row i and in row $i + 1$.
Harvesting is then rather low — certainly much lower than in the tree approach.
On the other hand, the interconnection network need not be augmented (except
for the problem of connecting the final tip of the spiral to the output pad) and the
path-finding algorithm is so simple that only a very limited amount of logic need
be added to the individual cells to achieve self-reconfiguration. Although Koren
foresees use of this algorithm for an array of relatively complex cells capable of
performing message exchange protocols, the algorithm can be as well implemented
by *hardwiring* the protocol with a few signal lines propagating between each pair
of rows the position of the extreme fault in the direction of spiral creation.

8.3.2. Aubusson-Catt

The algorithm proposed in [AUB78] is one of the first envisioned for reconfiguration of very large scale arrays and, contrary to the previous ones, it was explicitly foreseen not so much for processing arrays as for large linear memories. In fact, it was later applied in a prototype of a wafer-scale memory aiming at implementation of a *solid-state diskette*. Cells therefore do not a priori require any processing capacity — or, in any event, not the capacity for supporting message-exchange protocols. On the other hand, since they must be provided with added logic to support reconfiguration (at least, the added function of providing alternative connections along four directions must be allowed) they require a certain intrinsic complexity if a reasonable balance between nominal functions and added ones is to be reached. In fact, Aubusson and Catt refer to an array of *chips*, i.e., of complex subsystems on a wafer such as, for example, shift registers of fairly relevant length (1 kbits). It will be recalled that such organization on a wafer was already used for reference when discussing yield evaluation criteria.

In this algorithm, the spiral head is a cell at the center of the array, or as near to it as possible; the choice is justified by the dual purposes of increasing the efficiency of the algorithm and of increasing the probability of positioning the head of the spiral inside the largest connected set on the wafer. An attempt to create an actual spiral of working cells, reaching a given length, is then made proceeding towards the periphery of the array. In this case — typically oriented to static configuration on a wafer — the array is not necessarily of regular rectangular shape, since its external outline can be the irregular circle of a full wafer. Interconnection capacity must obviously be provided along a rigorous rectangular grid (or also — with suitable modification of the search algorithm — along an exhagonal grid).

Assume that at a given spiral-building step a spiral of length h has been already built: denote the cell at the tip of the spiral as $C(h)$. Then, the procedure can be summarized as follows:

- $C(h)$ explores the three adjacent unused cells (the fourth one being $C(h-1)$) in counterclockwise order, starting from the direction $C(h-1) \rightarrow C(h)$, looking for a fault-free cell. If no fault-free cell is found, $C(h)$ is declared to be a *pseudo-fault*, since it does not allow propagation of the spiral, and the algorithm backtracks to $C(h-1)$;

- If a fault-free cell is found, it is added to the spiral — becoming its new tip — and the previous step is repeated upon it.

An example of reconfiguration obtained by means of this algorithm is given in figure 8.16. Note that once cell $C(6)$ has been reached, the first choice for the spiral tip would be $C(7*)$: this can be added without any problem (dashed connection link), but after adding $C(8*)$ no further propagation of the spiral could be obtained. $C(8*)$ is declared to be a pseudofault and backtracking to $C(7*)$ is performed, but at this point $C(7*)$ finds around itself three faulty cells (two actual faults, one pseudofault) so in turn it is declared to be a pseudofault and further backtracking is performed: then, from $C(6)$ a different neighbor is adopted as the

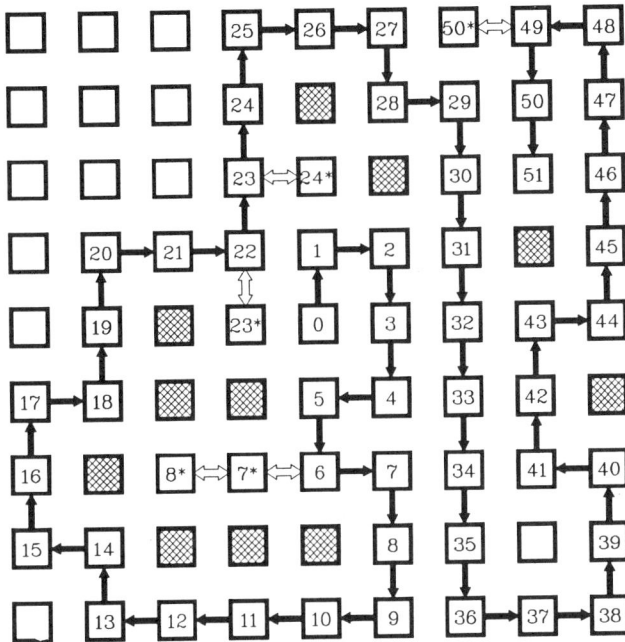

Figure 8.16
Spiral path created by Aubusson-Catt algorithm: predetermined path length.

spiral tip and the algorithm proceeds.

Let us consider the two alternative goals of restructuring, i.e., to create a spiral of predetermined length L or to extract the longest possible spiral from the physical array.

(1) In this case, the spiral length h reached at each propagation step is compared with the required value L. If L is reached, the algorithm terminates successfully. Otherwise, if a cell $c(h)$ is reached, with $h \leq L$, such that no spiral propagation is possible from it, backtraking is performed and different propagation choices are attempted. If backtracking to the head cell has finally to be performed during this attempt, fatal failure is declared.

(2) Again, the algorithm attempts all alternative propagation paths, finally backtracking to the head cell; before initiating each backtracking step, the length of the path reached is recorded by the host machine that drives the restructuring process. At the end, the longest spiral is selected.

The example of figure 8.16 refers to the first case, with a predetermined spiral length of 52 cells.

Although more complex than Koren's, this algorithm is still very simple. It can be implemented by local on-chip logic as well as by an external host controlling configuration, although the authors refer explicitly to external control (that, in

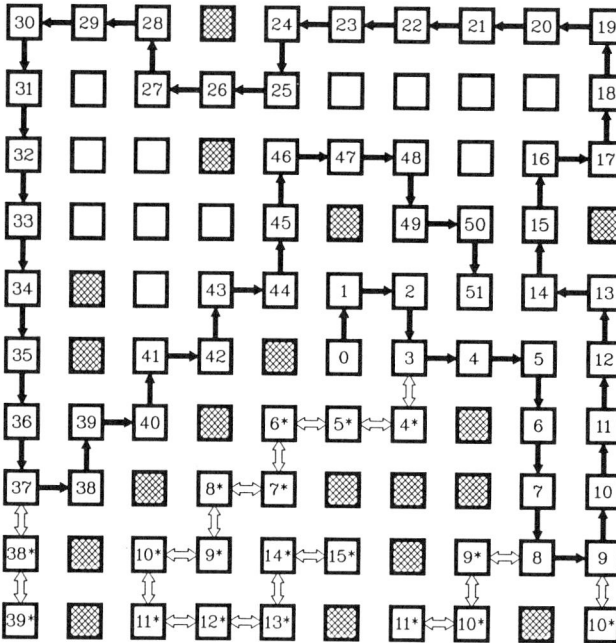

Figure 8.17

Spiral path created by Aubusson-Catt algorithm, looking for longest possible path.

the case of end-of production configuration, is probably more acceptable). No added delays are introduced, since no links longer than one unit are ever used nor are return path segments required. Thus, from this point of view, the solution performs better than the tree approach. In fact, synchronization is such that the algorithm could well be adopted also for processing arrays such as systolic ones, without creating any problem. Moreover, since no augmented interconnection network is foreseen, a fault model restricted to cell faults is quite reasonable: failure of a link is in fact totally equivalent — for our purposes — to failure of the cell from which the link originates. On the other hand, exploitation of available cells may not be very efficient: even considering the example in figure 8.17, we note that the maximum length of the final spiral is limited (and much lower than what may be obtained by the tree method, due to the large number of unused fault-free cells).

Figure 8.18 plots the results of direct simulations on a 10×10 array (the same rules adopted for previous, similar plots have been used here as well). As in the case of the tree approach, the results allow to highlight the fact that, already for relatively high yield (between 70 and 85 percent) there is good probability of achieving the creation of very short spirals. Again, this is due to the fact that choosing a fixed position for the spiral head leads to non-null probability of finding such position inside a small connected set. On the other hand, for high yield (greater than 85 percent), the mean length L of a spiral in a $M \times M$ array

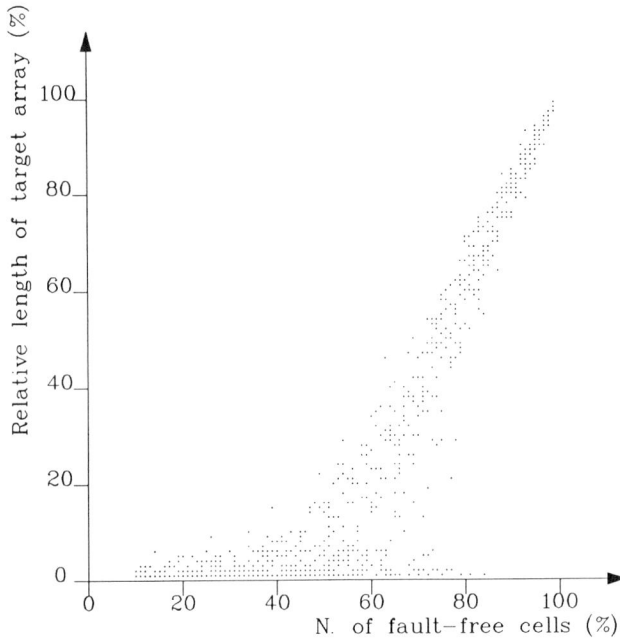

Figure 8.18

As in figure 8.15, Aubusson-Catt algorithm.

with K faulty cells can be evaluated as

$$L = M \cdot M - 2 \cdot K$$

that is, the probability that the number of unused fault-free cells equals the number of faulty cells is quite high.

For low yield values (below 50 percent) this method too, like the previous one, fails to prove efficient.

8.3.3. The Manning algorithm

The algorithm proposed by Manning in 1977 [MAN77] is again among the first reconfiguration techniques for arrays, but it is also significantly more complex than the ones discussed above.

If we return to the previous two solutions, we may note that the main drawback, leading to low harvesting, is a result of their complete *symmetry*, meaning that operation is totally independent of the present context characterizing the *tip* cell of the spiral. Both Koren's and Aubusson-Catt's techniques can be said to be *memoryless*, in that no notice is taken of the path by which the present cell has been reached nor of the particular structure of the cell's neighborhood (excepting for avoidance of faulty cells). The most relevant difference in the solution proposed by Manning is that it is adaptive, i.e., the procedure is modified depending on the position of the tip cell in the array.

To produce simulation results that could be compared to the previous one, a particular interpretation has been given here of Manning's algorithm: this also follows the necessity of translating the procedure so as to adapt it to the basic array structure of figure 8.13.

- A *border* cell is chosen as spiral head: let us denote it as $C(1)$;

- assume that a spiral segment of length h has been built and, as before, denote by $C(h)$ the cell at its tip (this cell is, therefore, at distance h from the head). Its three *free* neighbors are examined (the fourth neighbor is, as for Aubusson-Catt's method, $C(h-1)$) and their distances from the array's borders are evaluated. Denote by $C1(h)$, $C2(h)$, $C3(h)$ these three neighbors: the order of exploration is now immaterial, since subsequent ordering is based upon a measurement;

- if any of $C_i(h)$ is faulty, or if it has already been inserted in the spiral, such a cell is obviously excluded from further analysis. Otherwise, the unused, fault-free cell with minimum distance from a border is inserted as the new spiral tip, spiral length is increased by one and the algorithm resumes starting with the new spiral tip;

- if two (or more) $C_i(h)$ are characterized by the same distance from one of the borders of the physical array (i.e., if a degree of ambiguity is present), some ordering criterion must be adopted (apart from the random choice, which could always be considered). Examples presented in [MAN77] are based upon the following (arbitrary) criterion:

  - in case of ambiguity choose the cell nearest to the left border. If further ambiguity is present (clearly a possibility) choose among the two remaining cells the one nearest to the lowest border.

This represents just one of the possible rules that solve ambiguity by giving different weights to distances from the various borders. In the choice made above, weights are in order:

- the nearest border;
- the left-hand border
- the lowest border.

The example in figure 8.19 shows how the above rules are applied (figure 8.19 is based on the same fault pattern used in figure 8.17). Good harvesting can be reached: the algorithm attempts always to first of all exploit the cells on the *farthest periphery* of the array, and then it proceeds towards the center. On the other hand, it is quite clear that the overhead required for algorithm computation is much higher than in the previous cases; in fact, Manning himself does not give an evaluation of such overhead and of the area increase required if on-chip reconfiguration is envisioned (it should be emphasized that at the time of the

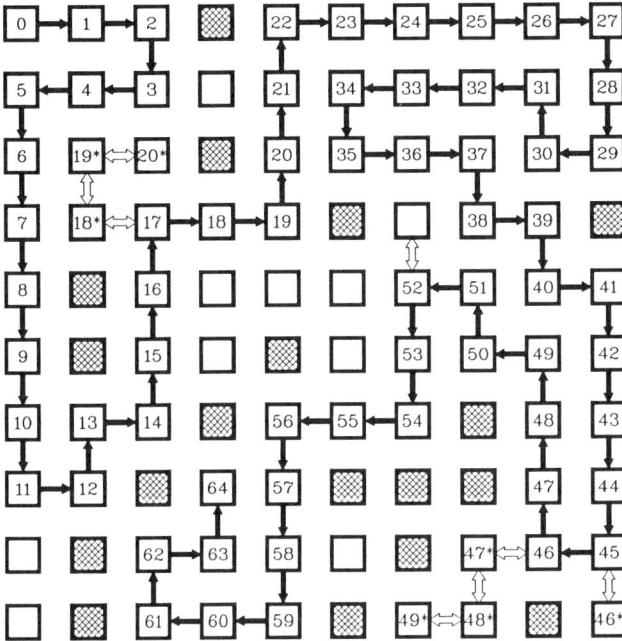

Figure 8.19

Spiral path created by Manning's algorithm.

paper's publication, technological details were far different from today) and he rather foresees two separate *functional layers* in the array, one consisting of cells and the other one controlling interconnections. It is reasonable to conclude that the procedure is acceptable for externally-driven implementation, typically for end-of production configuration: in this case, complexity of the interconnection network is not different from that required by the other approaches.

Statistical results deriving from direct simulation are given in figure 8.20, again for a 10 × 10 array.

8.4. Comparative evaluation

To compare the efficiency, in terms of probability of survival, of the several techniques discussed above, the various algorithms have been simulated using, in all instances, a rectangular 20 × 20 physical array. Previous simulations have proven that performances are reasonably independent of array size (even though, as might be expected, slightly better results are obtained for smaller arrays) so that considering only one sample array is quite acceptable.

There are some difficulties with regard to such a comparison, i.e., first the adaptability to an irregular contour of the physical array and second, to nonrectangular but still regular connectivity structures (in particular, hexagonal ones).

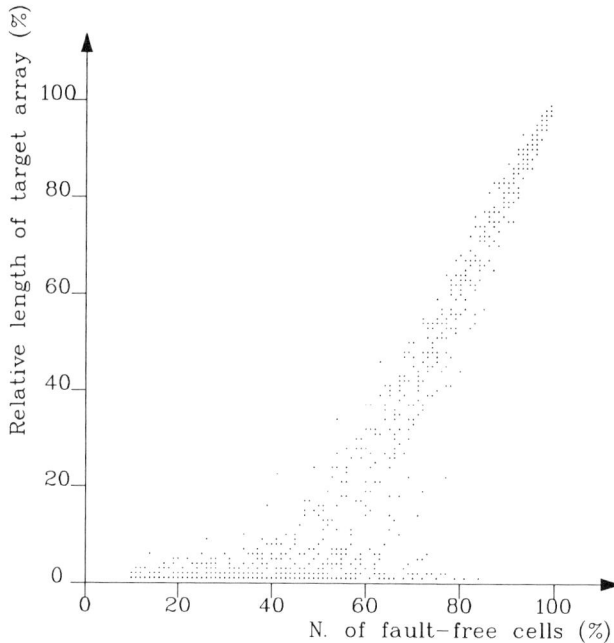

Figure 8.20

As in figure 8.15, Manning's algorithm.

Concerning the first point, while some approaches, particularly, Aubusson's and Catt's, allow immediate extension to irregular array contours, Manning's approach, which basically depends upon evaluation of distances from the borders of the array, is harder to adapt to such an instance. As for the second point, all algorithms can be adapted to hexagonal physical interconnection structures, and would in fact lead to far better performances. We may recall that — in all spiral approaches — hexagonal connections would allow exploration of a set of five, rather than three, adjacent cells as possible new spiral tips, and that the probability therefore of increasing the spiral would be correspondigly higher. Moreover, a richer interconnection system would decrease the probability that a set of faulty cells actually *isolate* fault-free cells making it impossible to reach them during reconfiguration. Results presented here are also indicative (although in a pessimistic way) for such more flexible instances.

All techniques presented refer to random faults (although they may take into account clustered defects, they are not specifically designed to meet such requirements). Therefore, simulation has been performed with random distributions of faults, assuming further that the probability of fault is identical in all points of a chip or of a wafer. Again, this is a general assumption in current literature although it is not totally realistic: when a full wafer is considered, for example, probability of defects is higher on the periphery. It has been assumed moreover that faults are mutually independent: again, this corresponds to the common fault

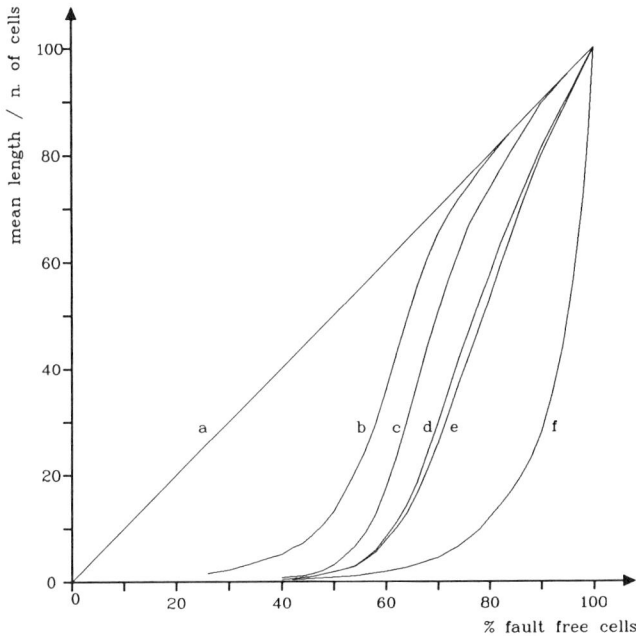

Figure 8.21

Mean length of longest linear array vs. normalized number of faults, related to: a) Diogenes approach; b) tree approach with root in the largest connected set; c) tree approach with root on the array border; d) Manning's approach; e) Aubusson-Catt approach; f) Koren's approach.

model, which excludes dependency from stress in the faults taken into account.

Figure 8.21 permits evaluation of the comparative merits of the different methodologies examined here with respect to one figure of merit, namely the mean value of the longest linear array that can be extracted from an array of 20×20 cells, for increasing numbers of faults, having normalized the number of faults so as to also take yield into account.

To obtain an *ideal* solution against which all other ones can be evaluated, the harvesting provided by the Diogenes approach has also been plotted: obviously, the curve (curve a) is simply a line slant with a 45° angle (in fact, it was seen that if harvesting is the only factor of merit considered, the Diogenes approach gives optimum results, and therefore the mean length of the target linear array always coincides with the total number of fault-free cells).

Curves **b** and **c** have been derived for the tree approach and they correspond to the two different alternatives of choice for the root cell. Curve **b** gives the best performances, in the case of a tree search, since the root of the tree is chosen inside the greatest connected set identified in the array, whatever the position of such set may be in the array; for obtaining curve **c**, it was simply assumed that the root of the tree had to be chosen among the fault-free cells nearest to the array border.

Obviously, such curves are related to the ones given in figures 8.8 and 8.11. For high yield, the tree approach achieves results quite near to the optimum, as it was to be expected since there is a correspondingly high probability of finding only one connected set in the whole array and the tree search exploits all cells inside a connected set. Clearly, if the largest connected set is chosen, performances are particularly good.

The choice of a fixed (or almost fixed) positions for the head of the linear array has been adopted for the *spiral approaches*, giving curves **d** (Manning), **e** (Aubusson-Catt) and **f** (Koren): thus, they should best be evaluated against curve **c**. Curves **d** and **e** are the best-performing ones: anyway, both are inferior to the tree approach: this can be easily justified. Consider, for example, Aubusson-Catt's method: whenever a dead end for the spiral is found, the spiral itself backtracks searching for a different region to develop itself — thus effectively discarding all cells in the dead-end region: on the contrary, in the tree approach the same cells would be inserted in a branch, so that better harvesting would be achieved. Further restrictions of Catt's algorithm justify the fact that its performances are slightly inferior also to those of Manning's algorithm: still, differences between these two approaches are not as relevant as it could be initially inferred from figures 8.17 and 8.19 (a fact that was apparent already in figures 8.18 and 8.20). Manning's algorithm, through its *adaptivity* to the environment, allows better results in particular for yields between 65 percent and 90 percent. In fact, both techniques appear to give interesting results for yields higher than 75-80 percent: the lower harvesting granted in comparison with the tree approach and with the Diogenes one should be offset by the absence of added delays, as compared with the possibly large ones introduced by the other two.

Last, curve **f** plots the performances of Koren's algorithm. As it could be foreseen given the extreme simplicity of the algorithm, performances drop very sharply for even small numbers of faults, since it is extremely probable — even for a few faults — that the *row switching* technique adopted by Koren will lead to discard a large set of fault-free cells.

8.5. References

[AUB78]   R.C.Aubusson, I.Catt: *Wafer-scale integration: a fault-tolerant procedure*, IEEE JSSC, Vol. SC-13, June 1978, 339-344.

[FIN77]   C.A.Finnila, H.H.Love: *The Associative Linear Array Processor*, IEEE TC, Vol. C-26, N. 2, Feb. 1977, 112-125.

[FUS82]   D.Fussel, P.Varman: *Fault-tolerant wafer scale architecture for VLSI*, Proc. 9th Comp. Architecture, 1982, 190-198, IEEE.

[GRE84]   J.W.Greene, A.El Gamal: *Configuration of VLSI arrays in the presence of defects*, JACM, Vol. 31, N. 4, 694-717, Oct. 1984.

[KOR81]   I.Koren: *A reconfigurable and fault-tolerant VLSI multiprocessor array*, Proc. 8th Comp. Architecture, 426-441, May 1981, IEEE.

[LEI86]   F.T.Leighton, C.E.Leiserson: *A survey of algorithms for integrating wafer-scale systolic arrays*, in: *Wafer Scale Integration*, (G.Saucier, J.Trilhe eds.), Proc. IFIP WG 10.5 Workshop, Grenoble, Mar. 17-19 1986, 177-195, North-Holland.

[MAN77]   F.Manning: *An approach to highly-integrated, computer-maintained cellular arrays*, IEEE-TC, Vol. C-26, N. 6, June 1977, 536-552.

8.6. Further readings

S.J.Hedge, R.M.Lea: *Intra-module fault-tolerance strategies for the WASP device* in: *Wafer Scale Integration*, (G.Saucier, J.Trilhe eds.), Proc. IFIP WG 10.5 Workshop, Grenoble, Mar. 17-19 1986, North-Holland. The Brunel WASP consists of a set of linear modules connected to an inter-module communication network (see Chapter 1). Fault tolerance strategies are presented both at module and at module interconnection level.

A.Rucinski, J.L.Pohoski: *A fault-tolerant distributed multiprocessor system for systolic algorithms*, Proc. ICCD-85, New York, Oct. 1985, 754-759, IEEE. The twin-layered architecture envisioned by Manning is considered here also, one layer being in charge of testing and reconfiguration while the other one is in charge of processing proper. Linear array reconfiguration is of th incremental, spiral-growing type, with neighbors of the tip within a "window" being explored for spiral propagation. Direction of exploration is random.

N.Tsuda, T.Satoh: *Hierarchical Redundancy for a linear array sorting chip*, Proc. 2nd IFIP Workshop on Wafer Scale Integration, Brunel, Oct. 1987. A multi-level approach is presented, by which at any given level "modules" are defined inside which spare cells are present and (local) reconfiguration is performed. This technique aims at confining defects by reconfiguration best suited to their dimensions. Shift switching or bypass switching is used, depending on the level.

9 GRAPH-THEORETICAL APPROACHES TO RECONFIGURATION

Most of the approaches analyzed in this book aim at the definition of *algorithms* capable of achieving restructuring or reconfiguration within given constraints and for given fault distributions. There are, however, a number of contributions whose main goal is to identify and evaluate the mathematical *complexity* of both algorithms and interconnection networks in an abstract fashion, that is, without necessarily referring to a specific technological or structural implementation. Reconfiguration algorithms examined in the other chapters are also, as a general rule, established on theoretical bases; we wish to emphasize that there is also a set of mathematical developments that allow the *complexity bounds* of wide classes of reconfiguration algorithms to be defined, abstracting from individual instances.

This chapter will identify the abstract parameters upon which evaluations are based, then summarize the mathematical tools adopted, and discuss several contributions published in this area.

9.1. Abstract evaluation parameters

The majority of reconfiguration criteria that can be seen as belonging to the *graph-theoretical* class involve the use of algorithms of relevant complexity, that could not be implemented by means of dedicated on-chip circuitry. They are thus more suitable for end-of-production restructuring or, possibly, for host-driven reconfiguration. Harvesting is the main factor of merit taken into account and is examined under two alternative viewpoints:

(1) Given a physical array containing N PEs (of which F are faulty) and with predefined connectivity rules, check whether a target array of $X \leq N - F$ with (possibly different) connectivity rules cells can be mapped upon it (i.e., evaluate the *probability of success* associated with reconfiguration);

(2) given a physical array of N PEs (of which F are faulty) evaluate the maximum dimensions of a target array with given connectivity rules that can be mapped upon it (or, in other words, evaluate as *harvesting* the percentage of fault-free cells that are actually inserted into the final target array).

In addition to this main factor of merit (evaluated quite often as a probability figure), other relevant factors taken into account are:

(a) maximum *length of connections* between logically adjacent PEs, after reconfiguration. This factor actually coincides (at least from an abstract point of view) with inter-PE communication delays, since the contributions discussed here usually abstract from reference to a specific physical implementation of the interconnection network and therefore to the nature and number of switches and multiplexers interposed upon it. Length is also related to another factor of great importance in VLSI design, namely, *locality*;

(b) *channel width*, usually defined as the number of *paths* or *tracks* that must be accommodated inside each connection channel between rows (columns) of PEs in the physical array. Once again, since no reference is made here to specific *implementations*, this number can be considered as fairly pessimistic with respect to actual layout solutions (e.g., multiple metal layers) and as an indicator of an *upper bound* when added silicon requirements introduced by reconfiguration are evaluated;

(c) number of *crosspoints*, a factor analyzed (if ever) in graph-theoretical terms but that has an obvious technological implication.

Apart from the above factors that relate to the final VLSI orientation of the problem, another figure of immediate importance is that of algorithm complexity and, consequently, of the processing time required to achieve complete connection of the target array.

It is readily apparent that comparative relevance of the above factors varies with the instance of reconfiguration taken into account. For example, the time required to perform reconfiguration has only moderate relevance when end-of-production restructuring is envisioned, since in this case the operation is performed only once during the device lifetime, and in fact follows a number of other, much lengthier, process steps. On the contrary, it is critical for run-time reconfiguration, for it actually reflects in a corresponding downtime of the system. If algorithm complexity is envisioned, its evaluation and its relevance are obvious when end-of-production and host-driven reconfiguration are envisioned since then the algorithm runs on a "conventional" computer. On the contrary, if run-time reconfiguration is autonomously driven by on-chip circuits, algorithm complexity becomes relevant (and apparent) as it is translated by *circuit complexity*; otherwise, since the algorithm is implemented in a distributed way by a highly parallel structure, evaluation of conventional algorithmic complexity becomes far from direct.

9.2. Mathematical tools

Discussion here will be limited to some basic definitions and tools; the chapter references list useful readings about the specific subject area.

- By stating that a given function $f(x)$ is $O(K(x))$ — where $K(x)$ is a predefined single-valued function — we denote the fact that $f(x)$ is characterized by an order of growth that is no larger than that of $K(x)$ for growing values of x: thus, $O(K(x))$ is an *upper bound* for $f(x)$. Formally:

  it is possible to identify a constant $c > 0$ and a constant $x_0 > 0$ such that for each $x \geq x_0$ it is $f(x) \leq c \cdot K(x)$.

- By stating that a given function $f(x)$ is $\Omega(K(x))$ for $K(x)$, c and x_0 defined as above, we denote the fact that:

  for an infinite number of values of x we have $f(x) \geq c \cdot K(x)$.

Finally, some authors also make use of the concept of *exact bound*, $\Theta(x)$ defined as follows:

- By stating that a given function $f(x)$ is $\Theta(K(x))$, with $K(x)$ defined as above, we denote the fact that:

  there are two constants $c_1, c_2 > 0$ such that, for all values of $x > x_0$, $f(x)$ is bounded above by $c_1 \cdot K(x)$ and below by $c_2 \cdot K(x)$, i.e.,:

$$c_2 \cdot K(x) \leq f(x) \leq c_1 \cdot K(x)$$

It should be emphasized that introduction of the various constants c_i in the above definitions justifies the possible (and even relevant) differences — from the actual implementation point of view — among algorithms that are proven to be characterized by the same abstract parameters: e.g., two algorithms that are both characterized by interconnection length evaluated as $O(N)$ (where N is the number of PEs in the given array) can have quite different distributions of real interconnection lengths depending, respectively, on the values of the constants c and N_0 adopted in definition of the upper bound.

As for the mathematical tools adopted by most authors, the three areas of *probability theory*, *graph theory*, and *algorithmic complexity theory* can be distinguished.

9.2.1. Probability theory

With regard to probability theory, contrary to what happens with statistics used in the case of defect distribution and yield evaluation, all authors here consistently make use of very simple probabilistic assumptions and of classical probability theory. Thus, typically, it is assumed that:

- all PEs have identical probability of being faulty (often, such probability p is taken to be equal to 0.5, a value that simplifies subsequent computations and that models a complete randomization of PEs failures, independently from technological constraints);
- all connections are fault-free;
- no mutual dependence exists as far as the probability of failure in adjacent PEs.

(No assumptions are made by any author relating to distribution of failures in time, a point justified by the fact that most authors in this area refer to end-of-production restructuring.)

9.2.2. Graph theory

With regard to graph theory, nonoriented graphs are consistently taken into account and the basic scope of reconfiguration can be defined as the problem of mapping a target graph onto a predefined topology (corresponding to the physical array), minimizing (or even, in many instances, avoiding) crossings. This type of problem relates first of all to *functional* reconfiguration, and several examples are given in discussing the *Diogenes approach* (chapter 7) and the *CHiP* structure (chapter 10).

A useful tool of graph theory, adopted for rectangular physical arrays, consists in so-called *folding* procedures applied to trees of meshes utilized to represent the arrays. Folding will be described in this chapter with specific reference to reconfiguration instances.

A further tool derived from graph theory refers to *spanning trees*, (see also chapter 8) where a spanning tree of a graph G is defined as a subgraph containing all the nodes of G and a subset of its edges such that there is exactly one path between each pair of nodes. Weights can be associated with edges and thence with paths in a spanning tree, so that optimization criteria can be easily defined.

9.2.3. Algorithmic complexity theory

With regard to *algorithmic complexity theory*, most of the problems involved can in principle be seen as either coverage or optimum association between elements of two distinct sets that would typically be NP-complete in their most general form. The *specific* algorithms, in fact, introduce constraints that allow quite substantial reduction of such complexity.

9.3. Trees of meshes and folding as tools for reconfiguration

To use trees of meshes as tools for mapping a target array onto a physical one (possibly containing faulty cells), two trees can be adopted.

First of all, the positions of the working PEs in the target array are identified, as exemplified in figure 9.1. To this end, working PEs are initially associated with the leaves of the tree of meshes related to the target array ("target tree"); this operation can also be interpreted as a *coding* one. The target array is introduced as the root of the tree: by folding the tree branches, each leaf is made to correspond to one element of the root. (In figure 9.1, the position of PEs is also given.)

Then, fault-free PEs in the physical array have to be chosen, and a correspondence between them and the cells of the target array (i.e., the leaves of the target tree) must be identified. To this end, various different methods can be adopted.

Leighton and Leiserson, for example, in [LEI85] suggest using a divide-and-conquer technique to assign to each working PE a number from 1 to M (for simplicity, assume at first that M is a power of 2). The physical array is vertically partitioned in half into two subarrays, each with $M/2$ working PEs. Recursively, code numbers from 1 to $M/2$ are assigned to PEs of the first subarray, and again from $M/2 + 1$ to M to PEs of the second subarray, and the cut is iterated; the orientation of the cut alternates between horizontal and vertical at each recursive step. The assignment is now simple: the i-th PE of the array is mapped onto the i-th leaf. This method can easily be extended to the case of arrays in which M

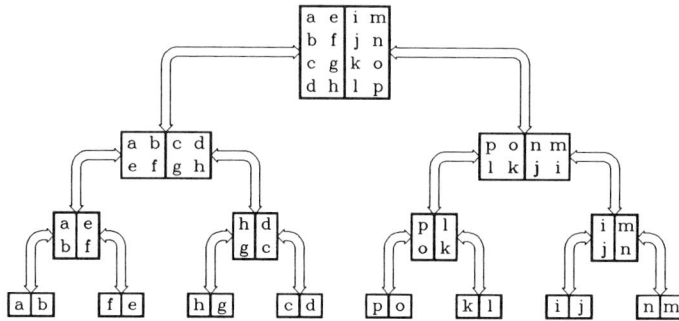

Figure 9.1

A tree of meshes: the root corresponds to the target square mesh, and the leaves to the used PEs. By folding the branches, leaves are mapped upon the places shown at the root.

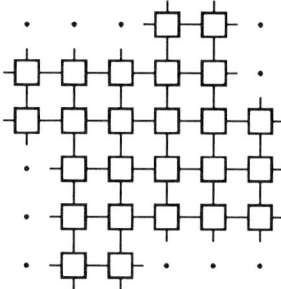

Figure 9.2

An incomplete target mesh.

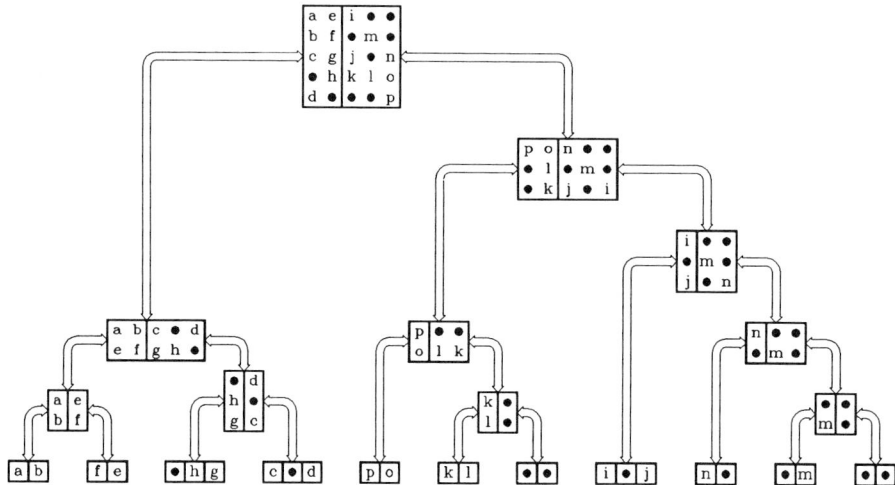

Figure 9.3

A tree of meshes where the root corresponds to a physical rectangular array, with faulty elements.

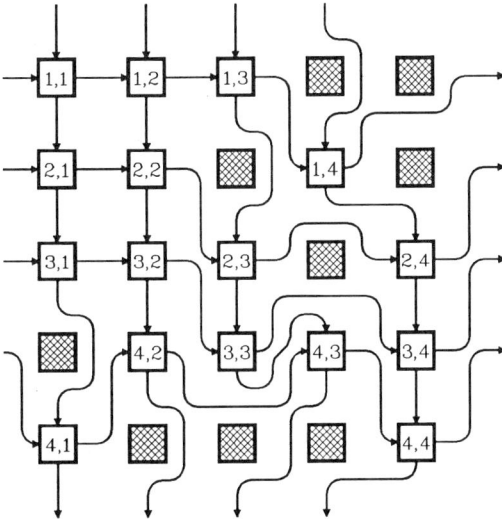

Figure 9.4

Physical array (root of figure 9.3) reconfigured by joining at the leaves the trees of meshes of figure 9.1 and 9.3.

is arbitrary, and also to incomplete target arrays like that of figure 9.2. Stated simply, the incomplete target array at the root is unfolded, by means of the tree, and working PEs are mapped upon necessary (i.e., *living*) leaves only.

This procedure suggests that the tree-of-meshes can be used to describe not only the target arrays but also the physical ones, as a means (instead of divide-and-conquer) for mapping fault-free PEs onto leaves of the target tree. The physical array, including faulty PEs, is adopted as the root of a new tree (the "physical tree"). By unfolding it, as in figure 9.3, the leaves corresponding to working PEs are found; the same figure shows how cases where the number of PEs is not a power of 2 can be dealt with.

The presence of faults in the physical array will lead to marking the leaves in the physical tree. At this point, reconfiguration becomes a problem of mapping leaves of the target tree onto fault-free leaves of the physical one.

The following distinct phases are thus involved:

(1) two trees — one corresponding to the physical array, with possible faulty leaves marked, the other one corresponding to the target array — are created;

(2) all leaves of the target tree are associated (one-to-one) with fault-free leaves of the physical tree;

(3) the physical tree is folded so as to identify the position of the target tree leaves in the physical array.

The method outlined allows a degree of freedom, which appears in the association of the target with physical leaves (step (2)); this is particularly evident when the number of fault-free cells is larger than the dimensions of the target array. Considering again the example of figure 9.3, and assuming that the target array in figure 9.1 has to be mapped upon it, we adopt the trivial solution of the orderly association of target leaves with physical fault-free leaves by quite simply going from left to right. The mapping rules in figure 9.4 are then obtained.

It is apparent from the above that techniques based on use of trees of meshes grant 100 percent harvesting. The complete definition of a specific algorithm, specifying a rational association criterion between leaves of the two trees, would allow other factors to be optimized, particularly the length of connection links.

9.4. Proposals involving abstract evaluation of performance

In principle, the first approach dealing with mathematical aspects in VLSI array reconfiguration is the Diogenes approach (chapter 7). The main problem considered there, however, was related to the possibility of mapping general graphs onto a linear array; the only complexity figure examined is the *depth* of the stack of wires superimposed onto the linear array — a figure similar, but not quite identical, to the channel width evaluated by all other authors for the more widely studied case of rectangular arrays. As previously noted, the approach does not vary with introduction of fault tolerance; its mathematical basis, therefore, will not be examined in this chapter.

9.4.1. Leighton and Leiserson's approach

Leighton and Leiserson have presented, in a series of papers, a set of evaluations of basic importance, beginning with results relating to the general problem of complexity in VLSI (see [LEI84]). Taking into account the specific problem of *restructuring* — for some different target arrays, but always starting from a rectangular physical array — they have identified corresponding complexity bounds for the interconnection network. With regard to linear arrays, the considerations outlined in chapter 8 will be summarized, while the case of rectangular arrays will be here discussed in depth.

The main cost functions adopted by Leighton and Leiserson are the length of interconnections and the channel width. In some instances, harvesting is also examined but only as a secondary factor mainly considered through its relationship with the first two ones (if a harvesting figure lower than 100 percent is accepted, optimization of the other two factors can be pushed farther). Faults are located in PEs and probability of failure is, with the exception of a few cases, taken as $p = 0.5$. Other relevant considerations relating to fault distribution (derived from this first assumption) are the following:

(a) With "high probability" (i.e., probability at least $1 - O(1/N)$ for an array of N physical cells) a given rectangular pattern of live and dead cells of size $2 \log N$ never appears in the array. This fact is instrumental for assessing the

upper bound of path length after reconfiguration since it definitely limits —
with high probability — dimensions of regions of contiguous faults.

(b) With high probability, a given rectangular pattern of live and dead cells of
size $\log N - 2\log(\log N)$ appears somewhere in the array. This is also of great
importance in light of reconfiguration algorithms and path length evaluation,
since it automatically discounts any optimistic assumption such as the presence of one single fault in a reasonably large region (as soon as dimensions of
the array increase over a few units).

In a technology-independent treatment the definition of length and width might
be somewhat tricky. The authors solve it by considering the array as made up of
square cells of unit edge (an assumption adopted throughout the present book),
and by referring to wiring as consisting of parametric *tracks*. While the number
of tracks is used to evaluate thickness of the wiring channel, when distances are
computed this number is not taken into account, and it does not affect (for computation purposes) the dimension of the cell that comprises both the individual
PE and the associated wiring.

Given such facts, in the case of target linear arrays the analysis of two algorithms, i.e., the *patching method* and the *tree method* leads to the following
conclusions:

(1) For the patching method — whose aim is to achieve 100 percent harvesting
by connecting every fault-free cell to the target array — it is found that with
high probability length of reconfigured inter-PE connections will be

$$\Omega(\sqrt{\log N})$$

Obviously, for worst-case fault distributions there *will* be inter-PE paths
whose length exceeds this bound (i.e., when very large regions are faulty)
but the previous fact allows us to discount this as having very low probability
of happening.

(2) If the requirement for 100 percent harvesting is relaxed, by accepting that
fault-free PEs isolated in large faulty regions be excluded from reconfiguration, the *tree method* can be modified by requiring that only cells allowing
connections by wires of a predetermined length d actually be introduced in
the reconfiguration tree. Relying on probabilistic considerations on distribution of live and dead cells, it can be proven that for any given d there exists
a positive constant c such that, with high probability, a fraction of at least
$1 - O(2^{-cd})$ of the fault-free cells can be connected in a tree using wires no
longer than d. The relevance of this new solution is obvious, since it allows
independence from the dimensions of the physical array. On the other hand,
the algorithm that allows creation of the tree is far from immediate.

The case of two dimensional arrays is understandably more complex. Here, in
fact, the requirement of a predefined harvesting probability (even lower than 100
percent) is not sufficient to create reconfigured arrays with interconnection links
of fixed constant length. While clusters of faulty cells can be circumvented with

high probability by a linear array with links of constant length, construction of a rectangular array around this same cluster will require either the discarding of a much greater number of fault-free cells (and therefore acceptance of a much lower harvesting than previously accepted, as will be seen in chapter 13) or else the acceptance of links whose length grows with some law related to N (as dimensions of fault clusters have also been proven to grow in relation with N).

As a consequence of the above considerations, the following theorem can be considered as basic for subsequent developments:

With probability $1 - O(1/N)$, every realization of any m-cell two-dimensional target array on an N-cell physical array has a wire of length $\Omega(\sqrt{\log m})$ for all $m = \Omega(\log^2 N)$.

Without entering into the lengthy and complex details of the proof, its outline can be summarized as follows:

(a) with high probability, the physical array contains a large number of regularly spaced square regions of $(\log m)/4$ cells, each of which is dead;

(b) any realization of a target two-dimensional array of m cells must contain a cycle of four fault-free cells that surrounds the center of one of such regions, thus requiring interconnection links of which at least one has length no lower than $(\log \sqrt{m})/2$. (Note that this second point is totally independent of the thickness of the individual link or even of the final intercell channel.)

The above is immediately useful when restructuring target arrays of fixed size; there is an interesting extension to it that can be applied in the case in which the target array can be seen as consisting of a predetermined constant fraction of the N cells in the physical array. It is proven then that:

With probability $1 - O(1/N)$ every realization of any two-dimensional array utilizing any constant fraction of the fault-free cells on an N-cell physical array has an interconnection link of length $\Omega(\sqrt{\log N})$.

This does not exclude the possibility of creating an algorithm capable of achieving reconfiguration with links of constant length. It is then necessary to renounce the requirement of utilizing a constant fraction of the live cells and to accept instead a lower utilization of spare elements.

While the interconnection length has been examined in abstract terms, i.e., without reference to a specific reconfiguration algorithm, the second factor envisioned by Leighton and Leiserson, that of channel width, is discussed with reference to a particular *divide and conquer* method. This technique operates in two stages:

(1) Recursive bisection of the physical array and count of fault-free cells in each subarray thus created, and corresponding bisection of the target array, generating target blocks containing as many PEs as there are fault-free cells inside the physical blocks. This stage ends when physical blocks contain $\Theta(\log N)$ PEs.

(2) Independent reconfiguration of each block. This can be accomplished, for example, by means of the tree-of-meshes technique, or, if blocks are small, by

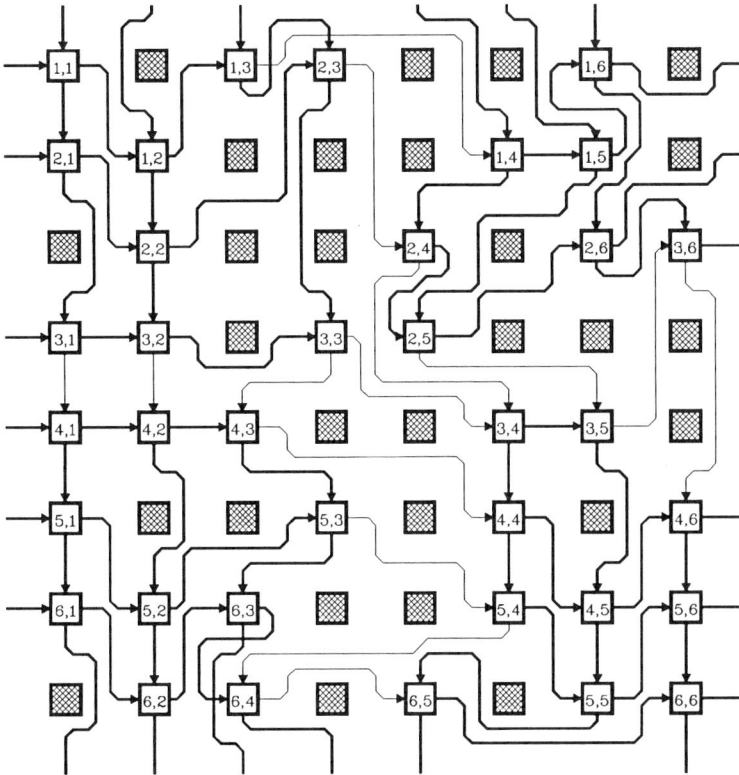

Figure 9.5
Mesh reconfigured by "divide and conquer". (See figure 9.6 for details on the method used).

table look-up (i.e., using precomputed routing). It is then possible to connect the blocks by reproducing the links identified when target array has been partitioned.

The divide and conquer algorithm allows — with great probability — the creation of a target array of the greatest possible dimensions (utilizing 100 percent of fault-free cells) with channel width $O(\log(\log N))$ and interconnection links of length

$$O(\log N \cdot (\log(\log N)))$$

The details of the divide and conquer technique can be seen more clearly through an example. Assume an initial physical array with a distribution of faults as in figure 9.5 (the physical array has 64 cells of which 28 are faulty; the reconfiguration that we will obtain is also shown). Given the requirement of a 100 percent harvesting, the target array consists then of 36 cells, and we further require that it be a 6×6 array. The configuration procedure develops by alternatively applying the bisection operation vertically and horizontally, in order to obtain at each step two partitions containing the same number of fault-free cells. Thus, the steps performed are as follows:

(1) A vertical even bisection is performed on the physical array; 19 fault-free cells are found in the left-hand half and 17 in the right-hand one. A similar bisection is performed on the target array, dividing it into two parts with 19 and 17 target cells, respectively (see figure 9.6);

(2) Each of the two halves of the physical array obtained in (1) is again evenly split into two halves, this time by a horizontal bisecting line: thus, the left-hand half previously obtained is split into one section of 9 cells and one of 10 cells. The corresponding left subarray in the target array is horizontally split in two parts of 9 and 10 cells (see again figure 9.6).

(3) A vertical partitioning is again performed, and so on recursively until each section obtained contains $\Theta(\log N)$ cells because, below this point, the distribution of cells can be arbitrarily bad. The decomposition assumed in figure 5 is in four subarrays; thin lines identify interconnections between subarrays. In figure 9.6, the final decomposition is in 16 subarrays.

(4) Each section is independently reconfigured, e.g., by means of the tree-of-meshes technique, or if small sections have been obtained, interconnections between fault-free cells can be created simply by inspection and by reference to a predefined table look-up (much as it is done in the *local reconfiguration* technique suggested by Hedlund and Snyder, see chapter 10).

(5) The previous phase has determined what fault-free cells are used, and in what position in the target array, because physical sections correspond in a precise way to sections of the target array. Thus, connections among sections of the target array can be reproduced in the physical array.

Properties of the array thus built — and in particular the bounds on channel width and interconnection length — actually derive from the characteristics of the interconnection criteria inside the sections (e.g., tree-of-meshes or table look-up) and of the interconnections among sections. It is obvious that the bounds hold with probability 1 for connections internal to the sections; for connections among sections the proof is far from immediate and the bounds are respected only with high probability (i.e., $1 - O(1/N)$). In fact, as explicitly stated by the authors, in the worst case the divide and conquer algorithm *might* require longer (even much longer) wires. To overcome these drawbacks, it is necessary to accept that only a fraction of fault-free cells be inserted into the target array. Table 9.1 (from [LEI85]) gives a good insight into the relationship between such various factors.

In [LEI85], Leighton and Leiserson suggest, for example, that the *patching method* already seen for linear arrays also be extended to the case of rectangular arrays. The same patching scheme is thus used to construct a two-dimensional array from any constant fraction (less than 1) of the fault-free PEs in a given physical array using links of length $O(\sqrt{\log N}) \cdot \log(\log N)$ and channels of width $O(\log(\log N))$. The patching method can be seen as a refinement of the *divide and conquer* technique, in which a fraction of fault-free PEs are excluded from

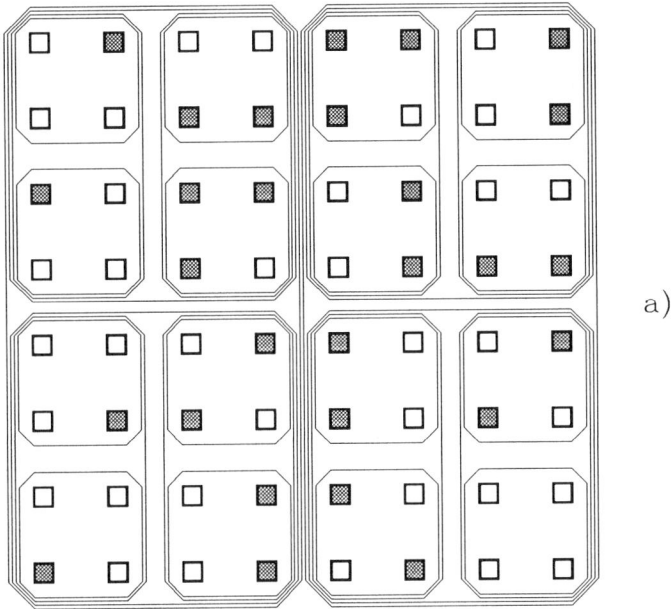

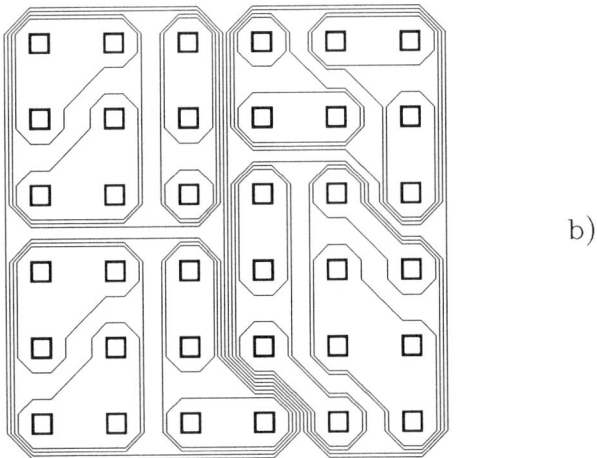

Figure 9.6

With reference to figure 9.5, partitioning of physical (figure 9.6.a) and target (figure 9.6.b) arrays, following "divide and conquer".

Table 9.1

Bounds on wire length and channel width for two-dimensional arrays (from [LEI85]).

Portion of live cells used	Wire length	Channel width
All	$O(\log_{1/p} N(s + \log_2 \log_{1/p} N))$	$O(\log_2 \log_{1/p} N)$
Constant fraction (< 1)	$O(\sqrt{(\log_{1/p} N(s + \log_2 \log_{1/p} N))})$	$O(\log_2 \log_{1/p} N)$
$\Omega(1/(\log_2 \log_{1/p} N)^2)$	$O\sqrt{\log_{1/p} N}$	1

reconfiguration at each level of recursion during application of the tree-of meshes method. Initially, the physical array is now subdivided not in just two regions, but in a number $N/c \log N$ of square regions such that each will contain $m = c \cdot \log N$ PEs. According to the initial evaluations on fault distribution, it is certainly possible to choose c sufficiently large so as to guarantee that in each region there will be with high probability at least $m' = (1/2) \cdot c \cdot \log N - (\sqrt{c}) \log N$ fault-free PEs. Inside each region we can then with high probability build a target sub-array of m' PEs whose interconnections are determined by using the tree-of meshes technique with wires of length $O(\sqrt{m} \log m)$ and channels of width $O(\log m)$. The subarrays are then interconnected into a complete array of $1/2 \cdot N(1 - 2/\sqrt{c})$ fault-free PEs with added wires whose length is again bounded by $O((\sqrt{\log N}) \cdot \log(\log N))$ and that can fit into the channels of width $\Theta(\log(\log N))$.

For such recursive technique again using the tree-of meshes basic methodology, an important point is the choice of the level of recursion at which some fault-free PEs are possibly discarded, since this may obviously affect the resulting percentage of fault-free PEs actually interconnected into the array. The authors suggest postponing such decisions until the lowest levels of recursion, provided the basic requirement of subdividing the physical array into regions containing exactly the same number of fault-free elements is always satisfied.

9.4.2. Greene and El Gamal's approach

An accurate analysis of reconfiguration problems in mathematical terms is carried out by Greene and El Gamal in [GRE84], where relevance is given to the *negative* factor of decreasing yield with increasing array dimensions. Conditions sought for such factors as length of interconnection links and channel width are related to the requirement that yield will *not* go to 0 with increasing value of array dimension N whenever the above conditions are satisfied. Again, these authors do not consider only the problem of rectangular arrays, but they also examine that of linear arrays (the corresponding results have been summarized in chapter 8), as well as the problem of configuring the so-called *selectors* across interconnection channels, i.e., the problem of connecting a set of fixed pins on one side of the channel to an arbitrary distribution of fault-free pins (i.e., PEs) in a (larger) set on the other side of the channel. This is a classical problem in routing studies: Greene and El

Gamal prove that in order to avoid — with high probability — decreasing yield with increasing array dimensions, interconnection length bounded by $\Omega(\log N)$ and channel width (in terms of tracks per channel) bound by $\Omega(\log N)$ are necessary.

The selector problem has obvious relevance for a goal that most authors do not take into account, i.e., that of designing the correct I/O structure for a complete array (linear or rectangular). Moreover, its results can be used for two subsequent problems, which we examine in order as follows.

9.4.2.1. Problem A: connection of two rows

K pairs of active elements from two parallel N-element linear arrays must be connected ($K = R \cdot N$, $R < 1$). Again, we have here two lines of *terminals* separated by a channel of variable width, just as in the selector case, but now both lines are constituted of elements of nonfixed position: as a consequence, the interconnection problem is easier. For simplicity, the authors maintain that only the distance *along* the channel (not *across* the channel) is relevant to evaluation purposes. Thus, only dimensions of the PEs (that are distributed along the channel) will be considered, while *thickness* of the single track is considered immaterial at least as far as distance computation is concerned (but not, of course, as far as channel thickness is concerned).

The relevant conclusion reached by Greene and El Gamal on this subject is that for arbitrarily large dimensions N of the physical arrays, for any $R < (1 - p)$ (p being the probability of failure of a single element of the array) for $\delta > 0$, $c < 1 - p$ and for any integer $t > 0$ such that

$$R \leq (1 - p) - \frac{p(1 - p)}{2t + 1} - \frac{(1 - p)^3 p^2 c^2}{(1 - p - c)^4 (t - 1)^2} - \delta$$

then, $K = RN$ pairs of PEs can be connected from arrays of N PEs, with a number of tracks t per channel that is $t = O(\text{constant})$, with maximum horizontal connection length d also bounded as $O(\text{constant})$, and yield $1 - O(1/N)$.

Actual pairwise connection of active elements in the two linear arrays is effected by a queueing model that is best suited for host-driven restructuring: these results, moreover, can be used (as it can be easily inferred) for interconnection of a rectangular array, as will be proven in the following.

9.4.2.2. Problem B: connection of rectangular arrays

The treatment of this problem in [GRE84] is lengthy and requires a complex mathematical background; we prefer here to give a more intuitive summary of its results, as done in [LEI86].

Interconnecting a rectangular array can be split into two subsequent phases:

(a) The physical array is first partitioned into blocks of size $1 \times c\sqrt{\log N}$, so that there are $\sqrt{N}$ rows of blocks and $\sqrt{N}/c\sqrt{\log N}$ columns of blocks (c being a constant depending on the fraction of the total array dimension, N, that we want to use for the complete array). Denote by t another constant with the same dependency (exact rules of dependency are given in [GRE83]), and mark a block as *bad*, i.e., discarded from reconfiguration, if it contains fewer than t fault-free PEs.

(b) Determine tentative rows for the array: assuming a number w of vertical tracks between each pair of columns (i.e., vertical blocks), divide them into two bundles of $w/2$ tracks each and scan the blocks, in order, from left to right and from top to bottom for each column, trying to complete the connections corresponding to individual rows. (The techniques defined for pairwise connection of two parallel linear arrays are obviously exploited here.) Chains created during this step must satisfy the constraint that no connection is longer than $k\sqrt{\log N}$, for a constant k depending on the fraction of N that we want to use in the final array. Columns of blocks are scanned from left to right and, for each column, from top to bottom: as soon as a good block is found in the leftmost columns, an attempt is made to connect it with bounded tracks to the rightmost column of the pair. If the attempt fails, the block in the leftmost column is declared to be *bad* and the attempt is repeated with the following one. In the same way, a block in the rightmost column that cannot be connected by tracks of acceptable length to any block of the leftmost column is discarded and declared to be *bad*.

An example of this procedure is given in figure 9.7 where blocks containing fewer than t fault-free PEs are crossed out and blocks that cannot be connected by the present algorithm are marked with dashed crosses. With high probability, by this procedure we obtain $(1-e)\sqrt{N}$ rows, each with $\sqrt{N}/c\sqrt{\log N}$ blocks, and finally build a $(1-e)\sqrt{N} \times 1/2 \cdot (1-e)\sqrt{N}$ array, where e is an arbitrary positive constant.

Once the rows have been created — as shown in figure 9.7, where the *logical rows* are actually marked — the second step is begun, aimed at construction of columns, as follows:

(c) conceptually, we exploit the fact that inside each block actively inserted in step (b), there are $c\sqrt{\log N}$ PEs, at least t of them being fault-free: thus, creation of columns aims at interconnecting such PEs in suitable order. Columns of PEs are established by essentially the same procedure seen before for rows, except that we scan top-to bottom and route through horizontal channels of width $w/2$. With high probability, the algorithm constructs $1/2 \cdot (1-e) \cdot \sqrt{N}$ columns: rows are now modified to include only those PEs that have been used in the columns, thus completing a rectangular array of $(1-e) \cdot \sqrt{N} \times 1/2 \cdot (1-e)\sqrt{N}$.

Further modifications to Greene and El Gamal's method are possible for achieving not a rectangular array with width-to height ratio 2 (as in the case above), but rather a square array; this last step involves a case of *functional reconfiguration* rather than of fault-tolerance.

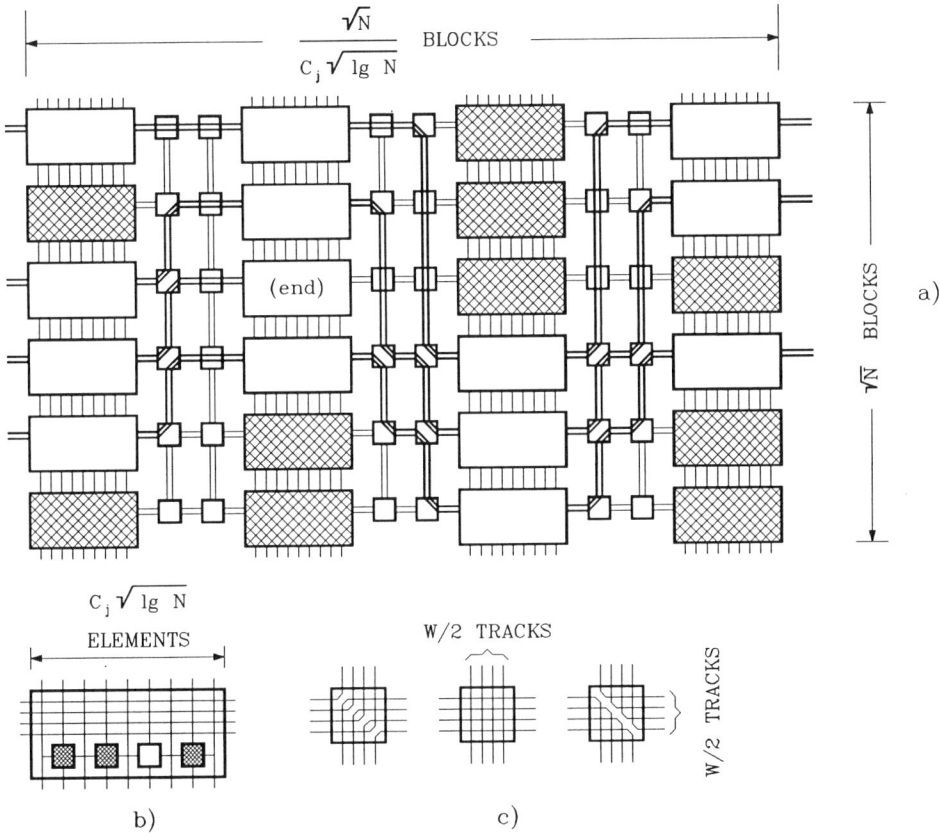

Figure 9.7
Square physical array reconfigured by Greene and El Gamal's method. Every block contains a segment of row; crossed blocks are not used. Horizontal and vertical connection lines shown represent bundles of $w/2$ tracks.

The extension is very simple (see [LEI86]) and is executed by mapping the obtained rectangular array onto the desired square array of dimensions $(1-e)\sqrt{N} \times (1-e)\sqrt{N}$. The first row of the square is embedded in the first two rows of the rectangle: the first row is completely adopted whereas only an evenly spaced portion of the second row is used. PEs of the two rows are connected in an orderly way from left to right. The second row of the square is embedded in the second and third rows of the rectangle using all the remaining cells in the second row and an evenly spaced portion of the third row and so on.

The interesting aspect of this last point is that it can be generalized to the construction of rectangular arrays with arbitrary width-to height ratio.

An interesting summary of performances, evaluated for various algorithms, is given by table 9.2 (derived from [LEI86]), where the different algorithms recalled

Table 9.2 (from [LEI86])

Bounds for One-Dimensional Arrays			
Method	Portion of cells used	Maximum wire length	Maximum channel length
patching	all	$\Theta(\sqrt{\log N})$	$\Theta(t)$
optimal	all	$\Theta(\sqrt{\log N})$	$\Theta(t)$
tree	99%	$\Theta(t)$	$\Theta(t)$
Bounds for Two-Dimensional Arrays (worst-case wafer, using all live cells)			
Method	Maximum wire length for widthless wires	Maximum channel width	Maximum wire length for unit width wires
tree of meshes	$\Theta(\sqrt{N})$	$O(\log N)$	$O(\sqrt{N}\log N)$
optimal	$\Theta(\sqrt{N})$	$\Omega(t)$	$\Omega(\sqrt{N})$
Bounds for Two-Dimensional Arrays (average-case wafer, using all live cells)			
Method	Maximum wire length for widthless wires	Maximum channel width	Maximum wire length for unit width wires
divide & conquer	$\Theta(\log N)$	$\Theta(\log\log N)$	$O(\log N \log\log N)$
matching	$O(\log^{3/4} N)$	$O(\log^{3/4} N)$	$O(\log^{3/2} N)$
optimal	$\Omega(\sqrt{\log N})$	$\Omega(t)$	$\Omega(\sqrt{\log N})$
Bounds for Two-Dimensional Arrays (average-case wafer, using 99% of the live cells)			
Method	Maximum wire length for widthless wires	Maximum channel width	Maximum wire length for unit width wires
patching	$\Theta(\sqrt{\log N})$	$O(\log\log N)$	$O(\sqrt{\log N}\log\log N)$
Greene	$\Theta(\sqrt{\log N})$	$\Theta(t)$	$\Theta(\sqrt{\log N})$
matching	$\Theta(\sqrt{\log N})$	$O(\sqrt{\log N})$	$O(\sqrt{\log N})$
optimal	$\Theta(\sqrt{\log N})$	$\Theta(t)$	$\Theta(\sqrt{\log N})$

here are compared for various fault conditions.

In the above proposals, techniques derived from graph theory were already used to achieve actual reconfiguration. A further class of reconfiguration methods that may be seen as belonging to the *mathematical approach* class are the ones in which graph theory is even more instrumental, in that the basic aim is not so much evaluation of performances and cost parameters as actual construction of a graph representing the target array. Typical of this class are the techniques based on spanning trees: one well-known algorithm, oriented to reconfiguration of linear arrays and attributed to Fussell and Varman, has been discussed in chapter 8.

9.5. References

[GRE83]   J.W.Greene: *Configuration of VLSI arrays in the Presence of Defects*, Ph.D dissertation, Stanford Univ., Stanford, CA, 1983.

[GRE84]   J.W.Greene, A.El Gamal: *Configuration of VLSI Arrays in the Presence of Defects*, JACM Vol. 31, N. 4, 694-717, Oct. 1984

[LEI84]   F.T.Leighton: *New Lower Bound Techniques for VLSI*, Math. System Theory, Vol. 17, 47-70, 1984.

[LEI85]   F.T.Leighton, C.E.Leiserson: *Wafer-Scale Integration of Systolic Arrays*, IEEE-TC Vol. C-34, May 1985.

[LEI86]   F.T.Leighton, C.E.Leiserson: *A Survey of Algorithms for Integrating Wafer-Scale Systolic Arrays*, in: *Wafer Scale Integration*, (G.Saucier, J.Trilhe eds.), Proc. IFIP WG 10.5 Workshop, Grenoble, Mar. 17-19 1986, 177-195, North-Holland.

9.6. Further readings

Algorithmics for VLSI, (C.Trullemans ed.), 1986, Academic Press.

J.D.Ullman: *Computational Aspects of VLSI*, 1984, Computer Science Press.

Two very useful introductory books devoted to the fields of algorithms for VLSI and analysis of their complexity.

Most reconfiguration techniques stress a particular figure of merit in preference to others; so-called *local reconfiguration* techniques assume structure regularity and interconnection locality as basic goals, giving them prevalence over other items such as spares utilization and area requirements. The number of spare PEs introduced for local reconfiguration is very high (increasing with N^2 for a $N \times N$ target array). While spares utilization is not very satisfactory, and the probability of fatal failure is already non-null for a number of faults much smaller than N, it is still obvious that on the other hand probability of survival is also non-null for numbers of faults increasing with N^2, i.e., with the dimensions of the target array. This last factor may be of interest for large arrays (Wafer-scale) consisting of a great number of small PEs; it will be seen, however, that the efficiency of local reconfiguration approaches actually *decreases* with increasing array dimensions.

A well-known example of such a technique will be analyzed in this chapter, as well as an extension that can be seen as a "bridging solution" between local and global reconfiguration approaches.

One of the main local reconfiguration approaches relates to an architecture that — as was seen in the Diogenes approach — was designed primarily to obtain *functional reconfiguration*, that is the *CHiP* architecture [SNY82]. The basic physical structure is a rectangular array, and reconfiguration is made possible by switched buses inserted in channels between each pair of rows and columns (see figure 10.1.a, where circles drawn at bus crossings represent switches that can be programmed to be set in one of the three states shown in figure 10.1.b). As in the Diogenes approach, the complexity of the architectures that must be mapped onto this basic structure reflects in the complexity of the interconnection network, expressed in terms of buses and switches. Thus, while the single bus network shown in figure 10.1 is sufficient to configure rectangular arrays (all switches are set to the first of the three states in figure 10.1.b) and binary trees (see figure 10.2), more complex architectures require denser interconnection networks. For example, the augmented interconnection network in figure 10.3 allows mapping of a hexagonal array (figure 10.4) and of a thorus-like array with four-neighbor connections (figure 10.5). This particular array has been designed so as to limit the length of connections between pairs of logically adjacent PEs, independently of the dimensions of the target array.

In the basic configuration now described no provisions are made for fault-tolerance. Contrary to the Diogenes approach, where the same structure provided for functional reconfiguration also immediately supports fault-tolerance, without modifications, in the CHiP case it becomes necessary to augment the interconnection network in addition to introducing spare PEs.

The primary requirement with regard to reconfiguration is that of interconnection locality: this is obtained by partitioning the $N \times N$ array into smaller

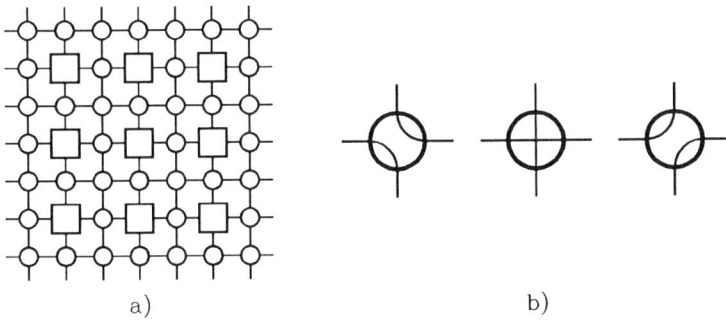

Figure 10.1

The CHiP architecture. a) Basic array. b) Possible switch settings for rectangular array mapping.

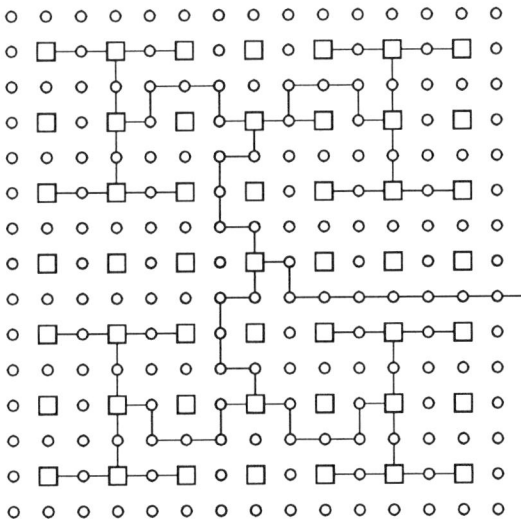

Figure 10.2

Mapping a binary tree onto a CHiP array.

$M \times M$ subarrays, introducing redundancy locally to each such subarray, and confining reconfiguration inside each individual subarray. As for the augmented interconnection network, its complexity is again related to the set of architectures to be mapped onto the (now fault-tolerant) array. Use of a twin-bus network advocated in [HED86] allows fault-tolerant linear and rectangular arrays to be mapped onto the physical array. More complex basic interconnection structures would also require quite complex augmented networks to support fault-tolerance, and therefore these instances have not been discussed in current literature. Added complexity might in fact become such as to make it debatable whether this type of approach would actually be cost effective.

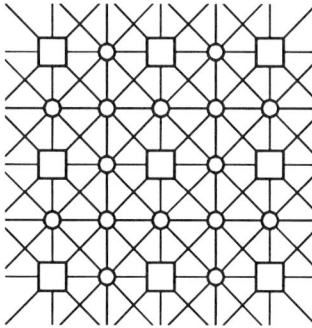

Figure 10.3
Augmented CHiP network supporting exhagonal and thorus array.

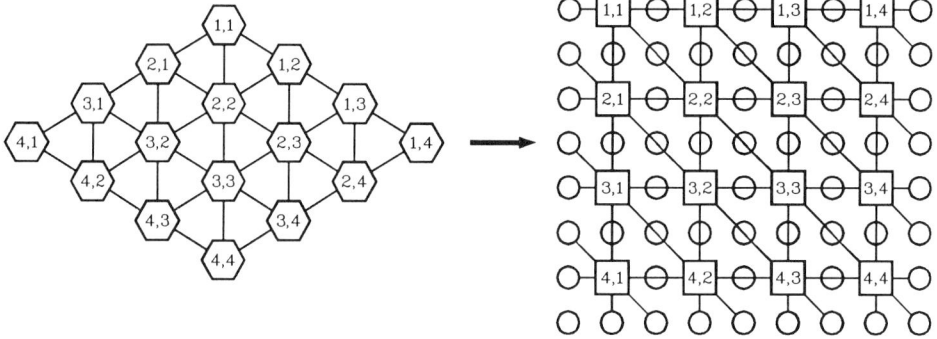

Figure 10.4
Hexagonal array mapped onto the structure in figure 10.3.

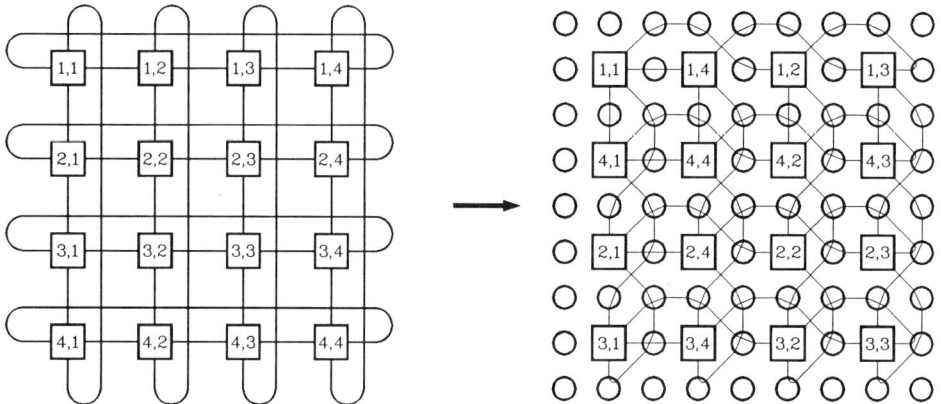

Figure 10.5
Thorus array mapped onto the structure in figure 10.3.

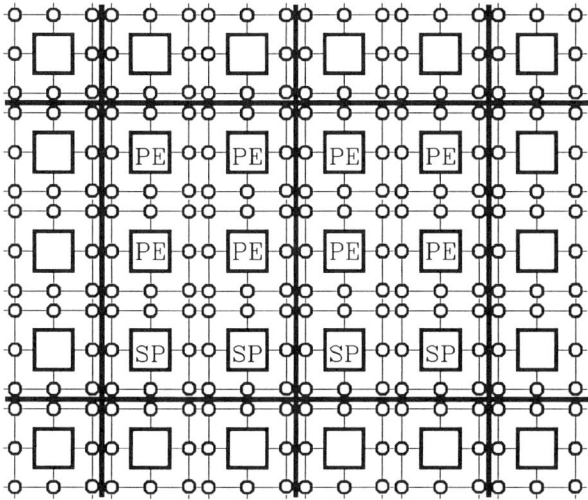

Figure 10.6

Block organization for local reconfiguration.

Dimensions of the subarrays directly relate to the maximum length that inter-connection links can reach as a consequence of reconfiguration. The proposal made in [HED84] therefore partitions the nominal array into 2×2 subarrays and adds to each subarray a spare *sub-row* of 2 PEs (see figure 10.6. where two subarrays, or *blocks* with 4 nominal cells marked PE and two spares marked SP are shown); this choice guarantees that the longest links will not exceed length three times the basic one (see the example in figure 10.7.). As for reconfiguration inside the individual subarray, given the limited dimensions now considered, it is possible to take into account all distributions of F faults $(1 \leq F \leq M)$ inside the $M \times (M+1)$ subarray, and to list in a dictionary the reconfiguration patterns corresponding to each such distribution and involving a suitable set of spares. Thus we are not in the presence of an actual *algorithm*, formally defined, but rather of a technique. (The method proposed is, from this point of view, quite similar to the *patching approach* seen for linear arrays in chapter 8.) Since the technique is foreseen for off-line, static configuration, it can be simply assumed that all possible configuration patterns are stored in a dictionary in the memory of the host machine that controls configuration.

The approach is characterised by obvious simplicity, and the same is true for the interconnection network supporting it; against these favorable points, a low spares utilization must be emphasized. It is quite evident that clustering of just $M+1$ faults inside one given $M \times (M+1)$ subarray will lead to the impossibility of reconfiguration, not only for the individual subarray but also for the whole array, i.e., to *fatal failure*. Evaluation of probability of survival against distribution of faults is plotted in figure 10.8 for target arrays of different dimensions (6×6 to 30×30). On the horizontal axis, the reference variable has been chosen to be the

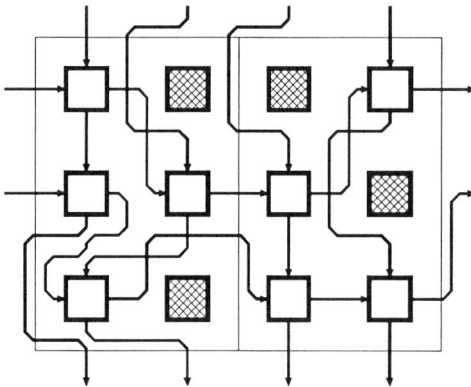

Figure 10.7
Example of local reconfiguration.

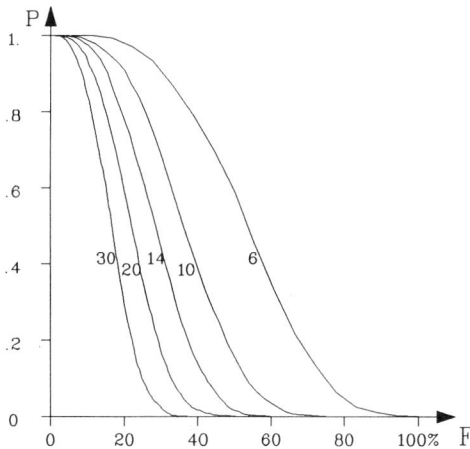

Figure 10.8
Probability of survival P vs. number F of faults (normalized with respect to the number of available spares) for target arrays of varying dimensions.

number of faults normalized with respect to the number of available spares.

From figure 10.8 it can be deduced that spares utilization decreases with increasing array dimensions; since spares utilization is one of the relevant factors of merit (recall that both yield and reliability decrease when area increase is excessive in comparison to its actual utilization) this can be considered as a negative factor for the local reconfiguration approach. Simulations have led to the conclusion that probability of survival depends not (as might be expected) on the total number of spares, therefore on $N \cdot N$, but rather on $N^{1.2}$, that is on a factor nearer the efficiency of other approaches that require a number of spares proportional to N only.

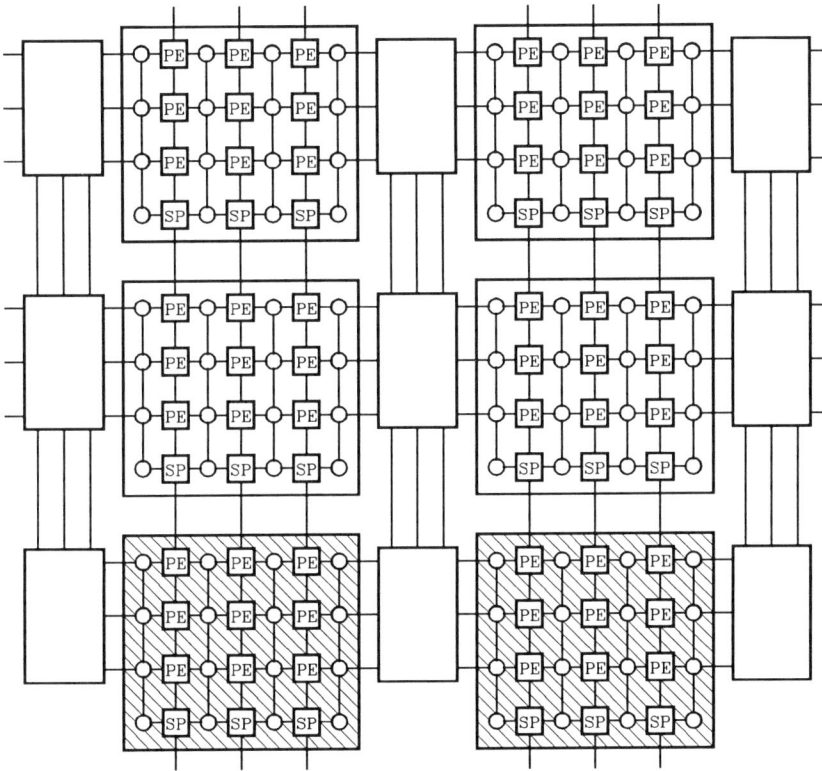

Figure 10.9
Structure for two-level reconfiguration.

The relevance of area increase here involved allows us to conclude that the *local reconfiguration* approach may be favored when the dimensions of the array as well as of the individual PEs are sufficiently small to offset weight of the spare PEs by the simplicity of the interconnection network.

An approach that can be seen as bridging the local-reconfiguration proposal and the simplest among global reconfiguration techniques is suggested in [WAN87]. The main consideration leading to modify the purely local reconfiguration technique is that the impossibility of reconfiguration inside one single subarray leads to fatal failure. The solution proposed is actually a hierarchical one, involving both reconfiguration *inside* the blocks and, at array level, *among* the blocks. To this end, again, the $N \times N$ nominal array is partitioned into blocks of $M \times M$ PEs and each block is provided with local redundancy (a spare sub-row). Moreover, a full row of blocks (also provided with internal redundancy) is added to the whole array. Referring to the example in figure 10.9, the whole array is partitioned into blocks consisting each of nine nominal cells (marked PEs) and of three spare cells (marked SP); of the six blocks in the array, four are nominal ones and two are spare ones. Locally, within each block, reconfiguration is enacted by a column-bypass technique; i.e., whenever a PE is found to be faulty inside a subarray, it is bypassed by a bypass link and the spare in the same column is inserted into operation. Fatal failure can be induced by even simpler fault distributions than in the previous case: whenever more than one fault is present in a column of any given block, the whole block itself must be declared faulty. To overcome this drawback, the authors introduce the row of spare blocks (the last row in figure 10.9), and again adopt the same reconfiguration criterion — bypassing, this time, a whole faulty block rather than a single faulty PE.

With regard to the interconnection network required to support this reconfiguration policy, we should distinguish reconfiguration internally to the individual blocks and externally, at array level.

(1) Referring to reconfiguration inside the individual block, the proposal published by Wang, Cutler and Su in [WAN87] introduces one horizontal bus and one vertical bus per channel because bypass of a faulty PE is effected by using a *pass-through* link. Such a solution is not homogeneous with all other reconfiguration techniques, and would also involve a very specific fault model since a faulty PE should be still properly working as a communication channel. To make a comparison with all other criteria, it is necessary to introduce *two* vertical buses in each channel (one of them used for bypasses) while the single horizontal bus is sufficient.

(2) Referring to array-level reconfiguration, a pass-through links traversing faulty blocks *is* a completely acceptable solution, since such a link is implemented by means of the vertical buses internal to the faulty block itself (the faulty block, being discarded, undergoes no internal reconfiguration). On the other hand, reconfiguration along columns of blocks requires $(M + 1)$ vertical buses in each vertical channel between two columns of blocks.

As a consequence, the overall structure is less regular than that involved in the previous approach. In figure 10.10, an example is given for a target array of 6×6 cells and a physical array of 4×3 blocks, each of them consisting of 3×2 cells.

Locality is once again kept high: the longest link is due to full-block bypass, and its length is therefore related to the dimensions M of the block itself.

The probability of survival for this *two-level approach* is given in figure 10.12. As in figure 10.8 for Hedlund's approach, the probability of survival has been plotted *vs.* the number of faults normalized against the number of spares: again, simulations have been run for target arrays of, respectively, 6×6, 10×10, 14×14, 20×20 and 30×30 PEs (see figure 10.11).

To obtain a meaningful comparison between the two local reconfiguration approaches, this second technique has been evaluated starting from physical blocks of 3×2 PEs just as for the first technique: larger subarrays (3×4) actually decrease performances.

As in the previous case, probability of survival has been found to depend not on N^2 but on a power of N nearer to 1: worse results appear here than

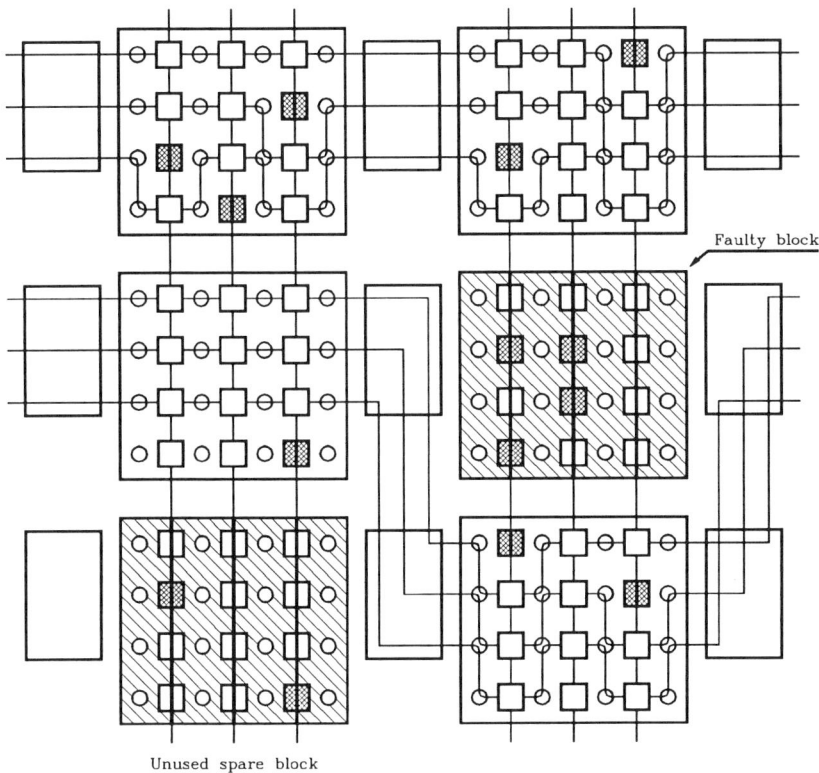

Figure 10.10

Example of *two-level* reconfiguration.

in figure 10.8 because the larger number of spares reflects on the normalization adopted on the x axis. Area requirements, while low as far as interconnection network is concerned, are rather high as regards spare PEs, whose number is now $N \cdot (M + 1 + N/M)$.

While Hedlund's approach fails only when three faults are found inside one block, the approach of [WAN87] already fails when two faults are found in the same column inside one block. This drawback is overcome by introducing the second reconfiguration level, at the added cost of a larger set of spares. Curves in figure 10.12 give the final results. The variable on the x axis here is the number of faults normalized against the *total* number of PEs in the physical array. From this point of view, the algorithm of [WAN87] appears to be slightly more efficient than Hedlund's. The inverted evaluation with respect to the results given in figures 10.8 and 10.11 is a result of the fact that in figures 10.8-10.11 spares utilization also was actually taken into account, while here the reference rather to total area gives some insight into yield (at production time) or reliability (at run time) performances.

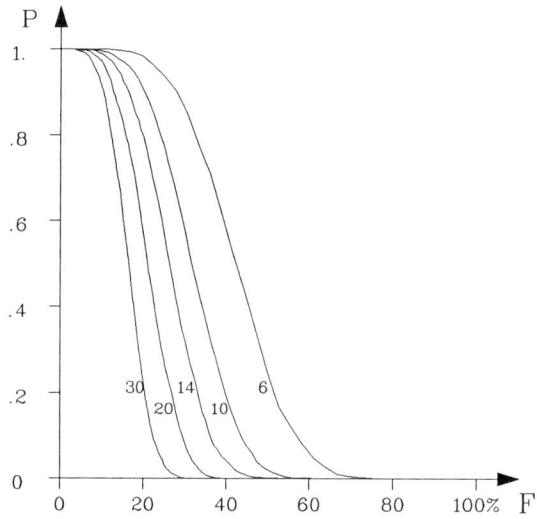

Figure 10.11
As in figure 10.8, for two-level reconfiguration.

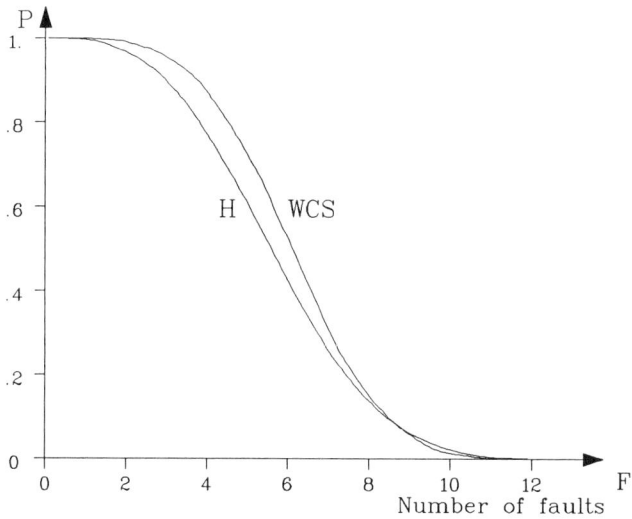

Figure 10.12
Comparison between Hedlund's (H) and Wang-Cutler-Su (WCS) reconfiguration methods.

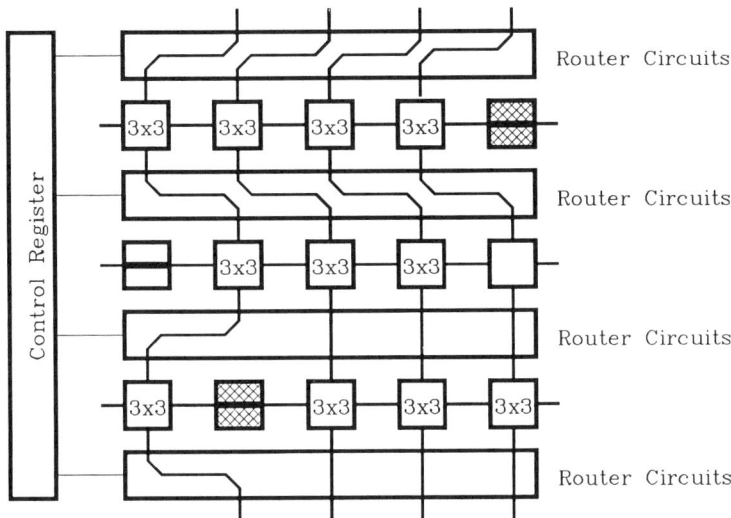

Figure 10.13

Example of reconfiguration following [HWA86].

Some further considerations are in order concerning the two-level approach: since it involves an actual simple algorithm, it is also suited for dynamic reconfiguration — and, given the very simplicity of the algorithms, it can be driven by elementary on-chip circuits controlling reconfiguration. In fact, area requirements due to such circuits are low enough so as not to influence in a relevant way the evaluations previously performed. The *mixed-mode* solution, moreover, could also be extended to other reconfiguration algorithms — the proposal discussed here is not actually dependent on the particular criterion envisioned inside the blocks or at array level — but in that case, of course, the increased complexity of interconnection networks should also be taken into account and the interconnection locality would depend on the algorithm chosen.

Another technique that bridges local and global reconfiguration approaches partitions the rectangular arrays not into subarrays of pre-defined dimensions, but rather into simple substructures whose dimensions are (at least partially) variable with those of the complete array. The most typical instance of this philosophy subdivides the array into *rows*, provides configuration locally inside each row by means of an immediate algorithm, and then supports correct interconnection between rows by means of suitable router circuits inserted between each pair of rows. (It may be noted that, in a more general and abstract way, Greene and El Gamal's approach was oriented along the same lines [GRE84].)

An example of this approach is suggested in [HWA86]. Referring to the array in figure 10.13, spare PEs are added as the leftmost column. Inside each row, a simple bypass criterion is used to avoid faulty PEs; between each pair of rows, the router circuits grant correct connections between fault-free PEs. Contrary to the previous local techniques, redundancy here increases with N for an $N \times N$ array: on the other hand, two faulty PEs in one row already lead to fatal failure. (The authors imply that the technique will be used for relatively small chips implementing *subarrays* rather than arrays, so that finally an array-level solution not far from Hedlund's one may be envisioned, with total redundancy still increasing with N^2.)

10.1. References

[GRE84]   J.W.Greene, A.El Gamal: *Configuration of VLSI Arrays in the Presence of Defects*, JACM Vol.31, No.4, 694-717, Oct. 1984.

[HED84]   K.S.Hedlund, L.Snyder: *Systolic architecture — A wafer-scale approach*, Proc. ICCD 84, New York, Oct. 1984, 604-610, IEEE.

[HED85]   K.S.Hedlund: *WASP — A wafer-scale systolic processor*, Proc. ICCD 85, New York, Oct. 1985, 665-671, IEEE.

[HED86]   K.H.Hedlund: *The design of a prototype WASP machine*, in: *Wafer Scale Integration*, (G.Saucier, J.Trilhe eds.), Proc. IFIP WG 10.5 Workshop, Grenoble, Mar. 17-19 1986, North-Holland.

[HWA86]   J.H.Hwang, C.S:Raghavendra: *VLSI implementation of fault-tolerant systolic arrays*, Proc. ICCD 86, New York, Oct. 1986, 110-113, IEEE.

[SNY82]   L.Snyder: *Introduction to the Configurable Highly Parallel Computer*, IEEE Computer, Vol.15, No.1, Jan. 1982, 47-56.

[WAN87]   M.Wang, M.Cutler, S.Y.H.Su: *On-line error detection and reconfiguration with two-level redundancy*, Proc. COMPEURO 87, Hamburg 1987, 703-706, IEEE.

10.2. Further readings

D.Gordon: *Efficient embedding of binary trees in VLSI arrays*, IEEE-TC, Vol. C-36, N. 9, Sept. 1987, 1009-1018. The paper deals with *functional* reconfiguration of switched-bus interconnected arrays. In particular, embedding of complete binary trees in square- or hexagonally-connected arrays is considered.

P.A.Ivey, M.Hutch, T.Midwinter, P.Hurat, M.Glesner: *Design of a large SIMD array in Wafer Scale technology*, Proc. 2nd IFIP Workshop on WSI, Brunel, Sept. 1987. A hierarchical, two-level approach is adopted. Subarrays ("chips") consist of a *linear* organization of PEs, and reconfiguration is of the bypass type. At a higher (wafer) level, switched buses are used to support reconfiguration.

N.Tsuda, T.Satoh: *Hierarchical redundancy for a linear-array switching chip*, Proc. 2nd IFIP Workshop on WSI, Brunel, Sept. 1987. A general introduction of a multiple-level hierarchical approach, followed by an application-specific example. Relationships between hierarchical levels, amount of spares and yield are evaluated.

11 GLOBAL RECONFIGURATION TECHNIQUES ROW/COLUMN ELIMINATION

The set of reconfiguration algorithms commonly identified as *row/column elimination techniques* are characterized by several useful characteristics that can be briefly summarized as follows:

- predefined bounds on interconnection links;
- predefined bounds on inter-PE communication delays (obviously a consequence of the previous point);
- simplicity of reconfiguration-supporting network.

On the other hand, the complexity of reconfiguration algorithms varies from extreme simplicity (and low efficiency in terms of spares utilization) to relevant complexity (balanced by good spares utilization). The common constraint to all methods of this class can be stated as follows:

- All rows (columns) of the reconfigured array will have identical length and, therefore, identical timing. Since in the target array all rows (columns) involve identical numbers of cells, the above constraint refers to the total sum of interconnection links and of switches present on the global path corresponding to the (reconfigured) row (column).

Undoubtedly, the strongest point concerns communication delays between processor pairs; it is therefore of interest to evaluate, among the various alternative communication modes, the ones for which such constraint is most important. As outlined in chapter 1, we classify arrays, with regard to communication modes, respectively as *asynchronous*, *synchronous*, and *systolic*. It is then necessary to analyze for which class the possibility of predetermining delay bounds can be a main figure of merit, dominant with respect to other ones.

Asynchronous arrays are typically exemplified by wavefront-computation arrays with monodirectional information flow, using combinational PEs. The whole array is itself again a combinational network. Information is fed to the external input pins as signal levels, and a known amount of time must elapse before *stable* signals are available at the output pins. Assuming that the individual PE requires a processing time t_0 to produce valid outputs, and that both outputs are generated at the same time, a delay t_0 must be accounted for with respect to the time at which *all* its inputs are valid. In a nonreconfigured (fault-free) array, signal propagation is such that uniform delays will be created on all propagation paths and therefore PE (i, j) will make its outputs available (valid) at time $t_0 \cdot (i + j)$. On the other hand, if reconfiguration has been performed delays will become different on different propagation paths, and some means of synchronization could then be required to guarantee reliable array operation (a simple solution is presented in [DIS86]).

Asynchronous wavefront-computation arrays of this type are best suited to relatively small architectures, in which the amount of spare processing cells must of necessity be kept rather low (e.g., one spare row and one spare column) and for which row/column elimination techniques (better oriented to *harvesting* problems in very large structures with possibly high numbers of spare rows and columns) are not preferred.

Synchronous arrays are characterized by the distribution of one clock signal throughout the whole array. It is therefore necessary for the inputs to the individual cell to be valid during a well-defined *window* corresponding to the clock signal. This in turn leads to requiring a predetermined maximum delay not so much at *array* level (i.e., with respect to global input-output relations) but rather with reference to the pairwise communications between logically adjacent cells. To guarantee correct communications between cells, latches must be inserted on cell inputs and outputs.

In principle, then, the problem of modified delays due to reconfiguration could be solved quite simply by evaluating the clock frequency with respect to the maximum delay that can be introduced in intercell communications — and this in turn would allow design of even very large (wafer-scale) arrays without introducing particular difficulties. Actually, clock distribution in such large arrays is a relevant problem (see for example [FRI86], since *clock skew* will easily become evident (in addition to the obvious reliability bottleneck resulting from a single clock distribution structure). If an alternative is foreseen in which the clock is distributed along the same directions as information, identical propagation delays will appear for both information and clock and the delay bounds introduced by the row/column elimination techniques will become useful.

Systolic arrays are characterized by extremely strict synchronization requirements — requirements that lead in some reconfiguration techniques to the introduction of registers along the modified interconnection paths so as to guarantee that the intercell delays are kept to a value multiple of the basic clock cycle (this was already made evident in the case of linear arrays through the criterion proposed by Kung and Lam, in which bypass registers rather than simple bypass links were introduced). In this specific case, delay bounds become of extreme importance and the row/column elimination techniques acquire corresponding relevance.

Finally, it should be stressed that, whatever the particular type of array taken into account, row/column elimination techniques, while not granting simplicity of *algorithm* (quite the contrary, as shall be seen, for high-performing ones) do grant relevant simplicity for *control* of switches in the interconnection network.

In this chapter, we shall examine the most elementary technique (simple row elimination), and then discuss a number of row/column elimination algorithms characterized by different performances and complexity.

11.1. Case A: Row elimination

The row elimination technique is trivial: whenever a row contains at least one faulty cell, the whole row is bypassed and a spare one is inserted in operation.

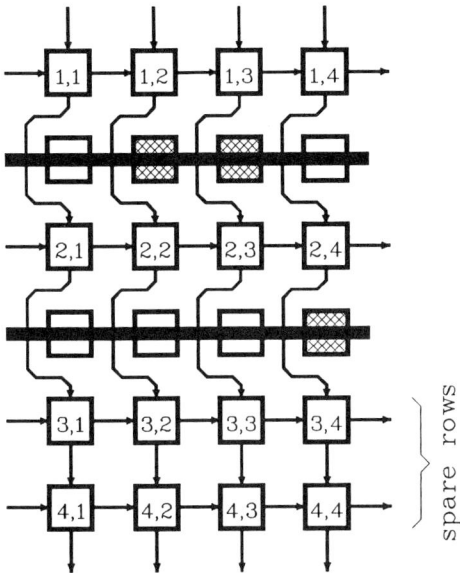

Figure 11.1
Example of row elimination.

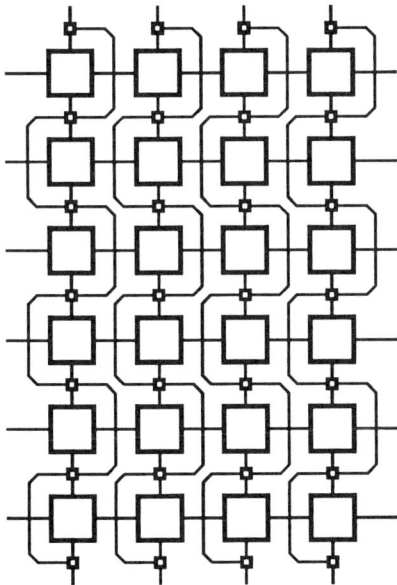

Figure 11.2
Row elimination supported by switched-bus interconnection network (small squares represent switches).

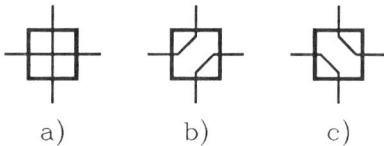

a)   b)   c)

Figure 11.3
Switch settings for the structure in figure 11.2.

Thus, if by (i', j') we denote logical indices associated with any cell and by (i, j) its physical indices, and if we conventionally assign null logical indices to nonoperating cells, the results will be that:

- if row i contains no faulty cell, for each cell in row i:

  $j' = j$

  $i' = i - F(i)$, where $F(i)$ is the number of rows preceding i in the direction of reconfiguration and containing each at least one faulty cell;

- if row i contains at least one faulty cell, for each cell in row i, let $i' = j' = 0$. Cells in unused spare rows will also have null logical indices; a simple example is given in figure 11.1, where spare rows are added at the bottom and reconfiguration proceeds downwards.

The corresponding interconnection network is also very simple (see figure 11.2): a single bus is inserted between each pair of rows and columns and a single switch is associated with each cell. Switches, in turn, have two states only — denoted as 0 and 1 — with different settings for even and odd rows. Settings are shown in figure 11.3, where state 0 corresponds to through connection (figure 11.3.a), state 1 for even-row bypassing (leftward turn) is given in figure 11.3.b, and state 1 for odd-row bypassing (rightward turn) is given in figure 11.3.c. Each switch is set to 0 if it is between two rows characterized by identical state (i.e., either both operating or both excluded from operation), and to 1 if it is between two rows characterized by different states (one operating, the other one excluded). No problem arises for connection of the array rows with the I/O pads: simple switch settings allow the completion of such connections.

Given such switch structure and the reconfiguration algorithm, it is possible to envision on-chip dynamic self-reconfiguration driven by simple controlling circuits. Assuming a single error signal for each row, set this signal to 1 whenever at least one cell in the row is found to be faulty; the control circuit is then just a XOR gate on row-error signals.

Alternatively, multiplexer-based schemes may be envisioned with different distributions of multiplexers to support arbitrary row bypasses ([MCC83, MOO84]). Again, the structure is very simple and control signals are purely local.

Such overall simplicity — involving not only architecture and control signals but mainly the reconfiguration algorithm as well — is the only factor favoring this approach, utilized column-wise instead than row-wise in the MPP architecture (see [BAT80]). To achieve reasonable probability of survival a large number of

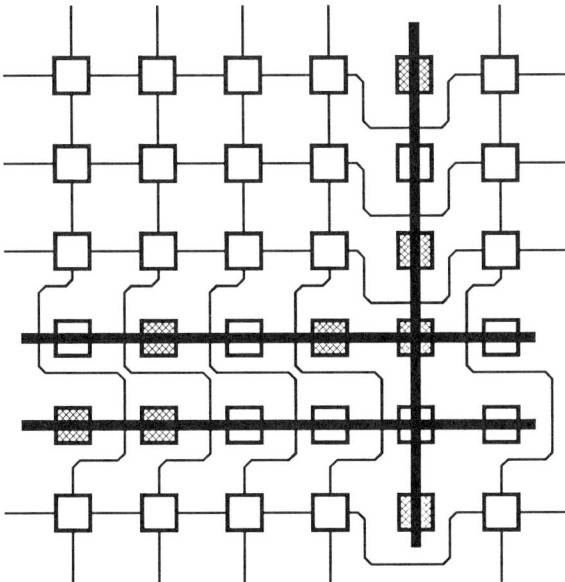

Figure 11.4

Example of row-column elimination minimizing total number of eliminated rows+columns.

spare rows must be foreseen and, in any event, spares utilization is the lowest of any reconfiguration technique for two-dimensional arrays.

11.2. Case B: Row and column elimination

In this second class of approaches, the original (nominal) rectangular array is provided with a number of spare rows as well as of spare columns. If the basic approach described in the previous section is adopted, given any faulty cell (i,j) (provided, of course, there are still spare rows and columns available) the fault can be overcome by substituting either row i with a spare row or column j with a spare column. (See for example figure 11.4.) It now becomes possible to look for optimum substitution criteria, so that spares utilization and probability of survival are maximized (within other predetermined constraints). As a consequence, reconfiguration algorithms become very complex — it has been proven in [KUO87] that the problem is NP-complete — and they are therefore suited for static restructuring or at most for host-driven reconfiguration (both computation time and computation complexity do not allow run-time self-reconfiguration).

A general outline of the problem and of the related figures of merit is as follows:

All faulty PEs must be eliminated by eliminating a suitable set of rows and columns containing them. The cost of this operation, defined as the number of fault-free PEs eliminated by the reconfiguration, must be minimized.

The above is valid both for run-time reconfiguration (where the number of spare rows and columns available is predetermined) and for end-of production restructuring.

A formal statement of the optimization problem can be derived. Assume that a physical array of R_p rows and C_p columns is initially available, and that F faults are present in it. The elimination of a row results in the elimination of C_p PEs (some of which will be faulty); the subsequent elimination of a column leads to discarding further $R_p - 1$ PEs, of which some will be faulty. If by f we denote the number of faulty PEs present in the row and column eliminated, the total number of fault-free PEs thus eliminated is

$$C_p + R_p - 1 - f$$

If an algorithm is found that by elimination of r rows and of c columns leads to discarding all the F faulty PEs, the total number of fault-free PEs discarded is

$$r \cdot C_p + c \cdot R_p - r \cdot c - F \qquad [11.1]$$

Several different instances can be considered. The first one is that of reconfiguration proper, with R_s spare rows and C_s spare columns available; a target array with $R_p - R_s$ rows and $C_p - C_s$ columns must be extracted from the physical array. Minimization of equation [11.1] is then subject to the constraints:

$$\begin{cases} r \leq R_s \\ c \leq C_s \end{cases}$$

If the number of spare rows and columns is very small compared to R_p and C_p, and so is the number of faults F, equation [11.1] can be simplified so that the expression to be minimized becomes:

$$r \cdot C_p + c \cdot R_p \qquad [11.2]$$

If the physical array is a *nearly square* mesh (meaning that $C_p - 1 \leq R_p \leq C_p + 1$) a reasonable requirement is that of extracting again a nearly square mesh. It must therefore be $c - 1 \leq r \leq c + 1$.

In this case, minimization of equation [11.1] can be approximated to minimization of:
$$2 \cdot r \cdot C_p - r^2 - F \qquad [11.3]$$
i.e., quite simply, minimization of r with the constraint $r \leq R_s$.

If, on the contrary, starting from a nearly square mesh with a very low number of faults, a rectangular array can be obtained, equation [11.1] can be approximated as

$$(r + c) \cdot R_p \qquad [11.4]$$

Table **11.1**

	r_1	r_3	r_4	r_5	r_6	c_1	c_2	c_4	c_5
1,5	x	.	.	.	.	.	.	.	x
3,5	.	x	.	.	.	.	.	.	x
4,2	.	.	x	.	.	.	x	.	.
4,4	.	.	x	.	.	.	.	x	.
4,5	.	.	x	.	.	.	.	.	x
5,1	.	.	.	x	.	x	.	.	.
5,2	.	.	.	x	.	.	x	.	.
6,5	.	.	.	.	x	.	.	.	x

and minimization must be applied to $r + c$.

In the case of end-of production restructuring, if the goal is to extract from the physical array a working logical array as large as possible, the same equations considered before are still valid, with the only provision that the individual bounds on r and c do not exist anymore.

The example in figure 11.4 presents a choice of rows and columns that have been eliminated so as to minimize equation [11.4].

The choice of rows and columns that must be effectively deleted (discarded from the working array) is a problem for which different solving algorithms, developed in different application areas, can be used. Some of the most relevant instances will be briefly examined.

An *association matrix* M_a is built in which any given row i corresponds (one-to-one) to a faulty cell, while columns correspond, in order, to the rows $r_1, ..., r_N$ and to the columns $c_1, ..., c_N$ of the physical array. Thus, row i represents the i-th faulty cell in the physical array, with coordinates (j, k). In matrix M_a, there are two marks in row i: one in correspondence of array row r_j, the other one in correspondence of array column c_k. Matrix M_a for the example in figure 11.4 is given in table 11.1.

At this point, a set of columns of M_a must be chosen so as to cover all its rows, a row of M_a being said to be covered iff at least one of its marks is comprised in one of the chosen columns. This covering is effected by considering that:

(a) whenever a mark in position $M_a(l, m)$ is chosen for coverage, M_a is immediately simplified by deleting its column m and all its rows (including l) that have a mark in this same column. For example, choice of fault $(4, 2)$ in table 11.1 involves deletion of matrix column marked r_4 and of matrix rows $(4, 2)$, $(4, 4)$, $(4, 5)$. The rationale for this is that elimination of the array's row asso-

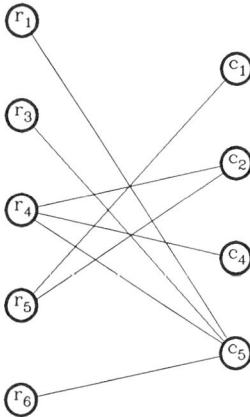

Figure 11.5
Bipartite graph derived for example in figure 11.4.

ciated with column m in M_a eliminates not only the chosen faulty cell l but also all other faulty cells in the same row of the array (a very similar property holds for the elimination of an array's column).

(b) When all marks in M_a have been deleted by the above procedure, matrix M_a is said to be covered and the procedure terminates. Optimization consists in choice of the minimum cost coverage.

A second representation makes use of a *bipartite graph*, rather than of an association matrix. For a graph G, let:

- (V) the set of nodes;
- (E) the set of edges;
- $f(V) = E$ the function establishing connectivity among nodes.

For the bipartite undirected graph defining the reconfiguration problem, V is subdivided into two subsets:

- (V_r): each node in (V_r) represents a row of the array;
- (V_c): each node in (V_c) represents a column of the array.

Function f is defined so as to connect nodes in (V_r) with nodes in (V_c) with the following rule:

- a node $v_i \in (V_r)$ is connected to a node $v_j \in (V_c)$ if and only if cell (i,j) is faulty.

The bipartite graph in figure 11.5 has been derived for the example in figure 11.4.

When this representation is used, the reconfiguration algorithm can be stated as follows:

(1) elimination of a row r_i or of a column c_j leads to deletion of node r_i or, respectively, node c_j from the corresponding vertex set, and of all edges ending

on r_i (or c_j);

(2) the procedure terminates when all *edges* of the graph have been deleted.

Algorithms based on the representation by means of a bipartite graph have been discussed, e.g., in [KUO87] and in [CLA83].

Minimization criteria defined for equations [11.1] – [11.4] can be applied to both representations.

Whatever the representation of the problem adopted, it is first necessary to examine the particular constraints introduced for the definition of a particular algorithm. These, in turn, correspond both to physical feasibility and to optimization of some specific figure of merit. The various row/column elimination techniques are characterized by

- different figures of merit adopted

- different suboptimal heuristic techniques used to minimize computation complexity. In our definition, a reconfiguration algorithm is *optimal* if it is certain to find the solution, whenever such a solution exists.

The simplest approach makes use only of the physical feasibility constraint, without introducing any figure of merit. All possible solutions then — if more than one can be found — are equivalent, as long as there are spares available. This clashes with the basic aim of row/column elimination techniques, that is to keep communication delays within predefined bounds and to guarantee that delays along all rows (columns) are identical. Therefore, this first approach must be evaluated against the delay distributions it causes.

(a) When strictly systolic arrays are considered, elimination of (possibly multiple) rows does not cause malfunctioning problems since synchronized latches are inserted along the interconnections between processing elements: thus, all signals are still synchronized.

(b) In the case of synchronous arrays without latches, a misaligning of signals through the array can occur. Since use of all available spare rows and columns is foreseen, provisions should be made in the design phase for the worst possible misalignment. Assuming that each PE is capable of realigning its own output signals with the system clock, the worst case here occurs whenever a number of *adjacent* rows (columns) are eliminated.

Coming now to the more general case and considering then the various figures of merit adopted by different authors, the main ones are:

(1) Balancing the use of spare rows and columns. This figure of merit is interesting in the case of production-time restructuring, with harvesting as the main goal rather than for run-time reconfiguration. Such balancing will then lead to a final array whose two dimensions have been reduced as equally as possible, so that the ratio between the two dimensions will (ideally) be kept unchanged.

(2) Minimizing $r + c$. Besides production-time restructuring, this can also be useful for *incremental* run-time approaches that do not completely reassign all substitutions whenever a new fault appears (as is done in global techniques),

but rather take into account previous reconfiguration steps. The number of available spares at each step then evidently affects capacity of reconfiguration when new faults appear.

(3) Minimizing some *function* $f(r, c)$. This function is related to the technology used; if the cost envisioned is simply that of PEs, instances presented in equations [11.1] — [11.4] are again found. If, on the contrary, specific technological characteristics must be taken into account, more complex functions may appear. Thus in [DAY85], the technology supporting restructuring of a memory is such that invoking a spare row requires programming 8 links, while invoking a spare column requires programming 15 links. Therefore, the cost function in a square array may be taken as $Cost = 8 \cdot r + 15 \cdot c$.

Given the NP-completeness of the problem (even in the simplest case, when physical feasibility only is required) the various authors have chosen different ways for dealing with it. Basically, three alternative approaches are apparent in the literature.

(a) An optimum solution is sought. In this case, the NP-complete problem is attacked, and the techniques differ in the use of additional figures of merit besides physical feasibility and in the number of steps required to reach the (possible) solution (or to state whether no solution exists).

(b) Particular conditions are detected that allow to determine coverage of a *subset* of faults either independently of the coverage of other faults or a priori. Coverage of the remaining faults will again involve an approach of type (a), but the *dimensions* of the problem have been reduced.

(c) An approximate, suboptimum solution is sought, balancing lack of optimality with lower (possibly much lower) computation complexity. Solutions of this type may differ with regard to performances (harvesting or probability of survival, depending on the particular instance envisioned) and computation complexity.

Some techniques belonging to each of the above three classes will now be analyzed. The final choice of one specific solution depends here also on the trade-offs between a number of factors. In addition to the usual ones, the number of spares should also be taken into account. Three different situations are possible:

(1) Number N of PEs is very high (even in the case of WSI, this leads to assuming relatively small, simple PEs such as memory cells or very simple arithmetic devices in a large parallel arithmetic structure). Algorithmic complexity of an optimum solution may then become so relevant with respect to other factors such as spares utilization as to suggest the choice of nonoptimal algorithms characterized by *polynomial* complexity (obviously, with low polynomial degree!). A problem of this type is outside the scope of the present book, since *processing arrays* have in our case been defined as consisting of relatively complex PEs.

(2) The total number of PEs, while high, is not as large as in the previous case and, most important, although the number of spare rows and columns provided

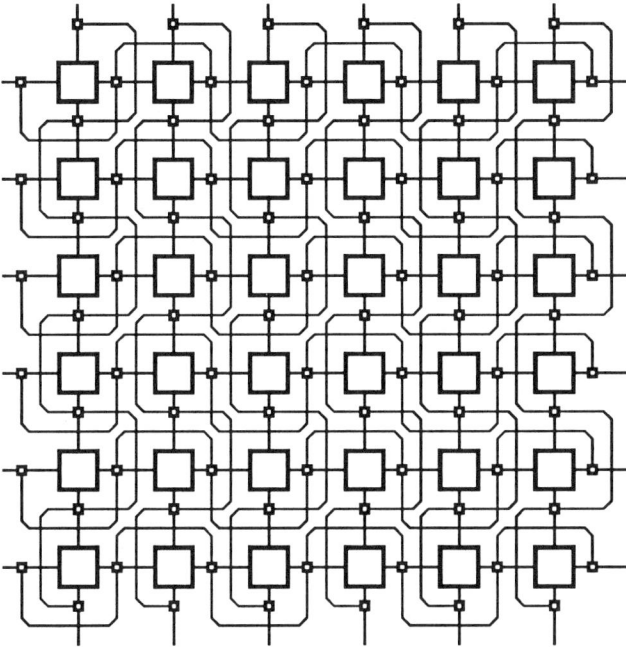

Figure 11.6

First interconnection network supporting row-column elimination algorithms.

is large (this is a basic condition for application of row/column elimination techniques) it is still significantly lower than the physical dimensions of the complete array. This definition is self-evident when reconfiguration aiming at a target array of predetermined dimensions is envisioned; yet, it may be easily extended to the case in which highest harvesting is sought. Reference is then to instances in which the target array finally extracted after reconfiguration must have dimensions comparable to those of the initial physical array. An underlying assumption in this case is that complexity of individual PEs leads to silicon area requirements much higher than those associated with switches and buses (such an assumption could hardly be made in the previous case).

The number of PEs in this case is large but such that even an NP-complete problem can be tackled; on the other hand, efficient exploitation of spares must be granted. Therefore, search for an optimum solution and consequent algorithm optimization is a reasonable goal.

(3) The total number of available PEs is small. This case certainly does not lend itself to row/column elimination techniques; these would grant probability of survival quite lower than criteria such as index-mapping ones (discussed in the next chapter) or even local reconfiguration.

Two alternative structures can be considered with reference to the augmented interconnection network required to support this class of algorithms (note that we refer only to the interconnection network, not to its control signals):

(a) Two buses are introduced in each channel between pairs of rows and columns, and two switches are associated with each PE; the corresponding structure is shown in figure 11.6. One vertical bus and one horizontal bus, together with a switch, are dedicated to routing the horizontal signals, while the other two buses together with the remaining switch are reserved for vertical signals. Physical separation of devices supporting the two types of signals can be justified when information carried by such signals has a different characteristic (e.g., different word length). As it happened for the simple row elimination case, each switch has two settings only: the different treatment of odd/even rows presented in the case of row elimination must also be adopted here for odd/even columns. (For the sample structure in figure 11.4, these rules have been adopted.)

(b) One bus is inserted in each channel, and three switches are associated with each PE. The structure associated is shown in figure 11.7, while the same example represented in figure 11.4 has been treated again, with the new rules, in figure 11.8. In this case, distinction between vertical and horizontal buses as associated with signals has been overlooked; moreover, each switch has three different settings, and setting rules are slightly more complex than in the previous case.

A comparison between the two solutions is not possible here, since the choice is strongly technology-dependent (involving dimensions of buses and, even more, structure, technological implementation and cost of the switches). As for the simple row-elimination technique, multiplexer-based structures could also be envisioned, but no simplification of the structure would be consequently achieved.

Several typical examples will now be considered, all of which can be supported by the classes of interconnection networks described above.

11.3. Optimal algorithms

These are exhaustive techniques looking for *all* possible solutions. We shall first describe one, without analyzing complex figures of merit; having determined the various solutions, they will be compared so as to choose the one with lowest cost. Several conditions that allow simplification of the method will then be described.

See the example in figure 11.9: the physical array has $R_p + R_s = C_p + C_s = 10$, while $R_s = C_s = 3$. Fault distribution involves 12 faults, coded in order as $a, b, \cdots, n$. (This same example has been solved by Kuo and Fuchs in [KUO87].) To reach the optimum solution, a binary tree is built as in figure 11.10.

Tree construction is incremental and it proceeds by the following rules:

(1) associate the *root node* of the tree with the complete set of faults;

(2) arbitrarily select one fault inside the "node-set" for coverage. Two branches depart from the node: the leftmost branch corresponds to coverage of the

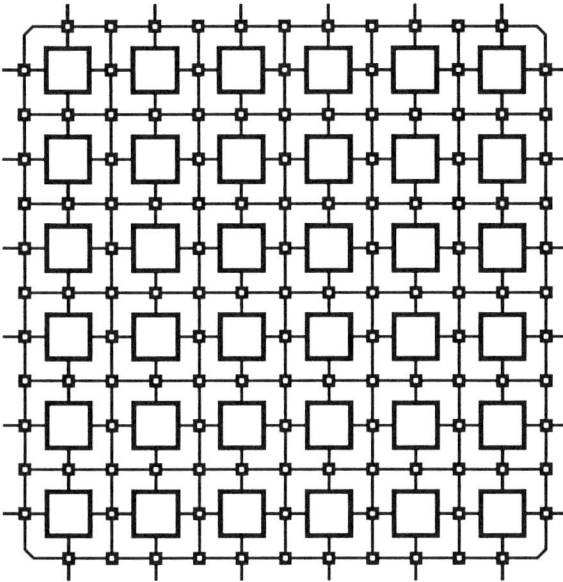

Figure 11.7
Second interconnection network supporting row-column elimination algorithms.

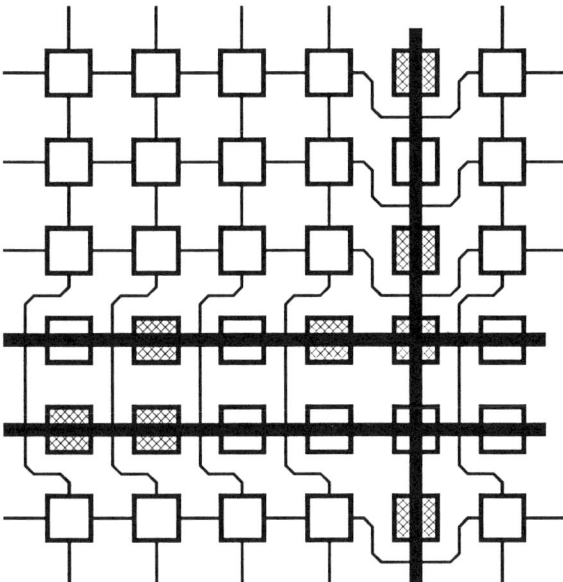

Figure 11.8
Implementation of example given in figure 11.4 by means of the interconnection network in figure 11.7.

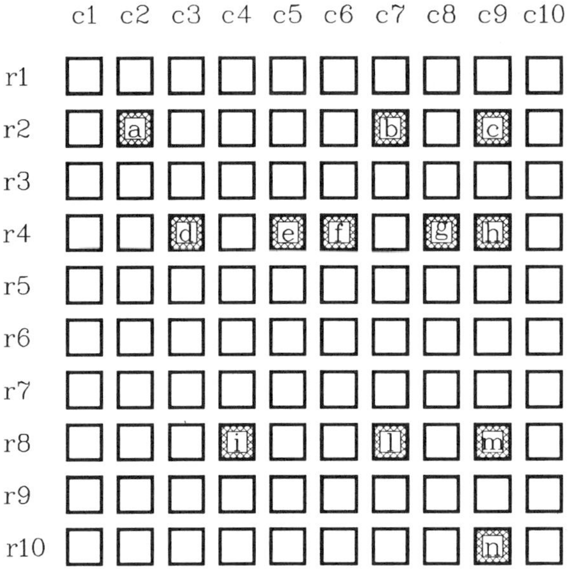

Figure 11.9
Sample physical array: etched cells represent faulty PEs.

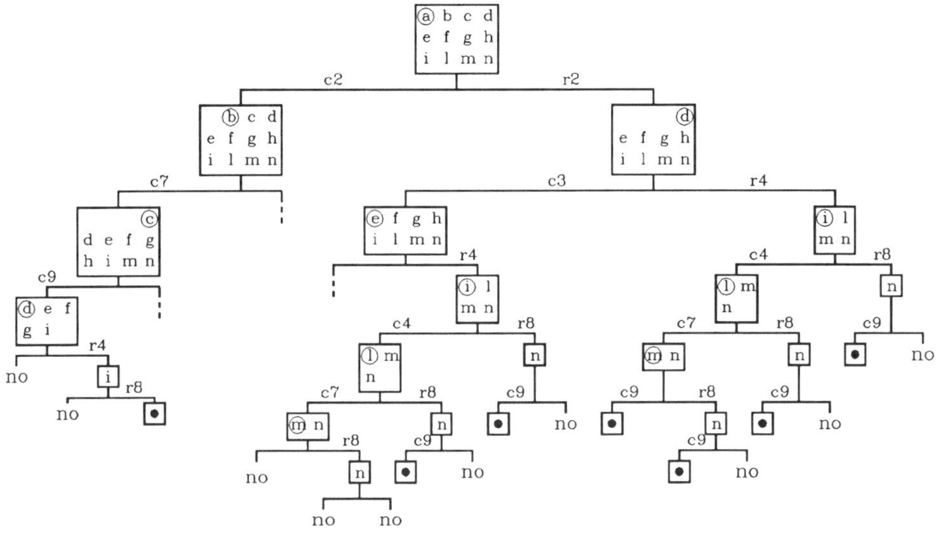

Figure 11.10
Binary tree built for solution of example 11.9, following Kuo-Fuchs technique.

fault by row elimination, the rightmost one to coverage by column elimination. Either alternative leads to a subset of the remaining non covered faults, reduced with respect to the set associated with the origin node. Introduce a new node at the end of each branch, associated with each corresponding subset of faults. Thus, in the tree of figure 11.10 the leftmost branch from the root corresponds to the choice of column 2 to cover fault a (marked with a circle in the root node) and the rightmost branch to the choice of row 2 to cover the same fault. The leftmost branch leads to a node associated with a fault set from which only a has been deleted, while the rightmost branch leads to a node associated with a fault set from which faults a, b, c have been deleted.

(3) Starting from the new nodes now added to the tree, the procedure is iterated from step (2) until one of the following conditions is met:

- the new node to be inserted is associated with the empty set (all faults have been covered):

- one of the bounds initially introduced (in particular, availability of spare rows/columns) has been reached.

Thus, referring again to figure 11.10, if at each step the rightmost branch is chosen, rows r_2, r_4, r_8 are eliminated in sequence until a node corresponding to the single fault n remains to be covered. While choice of row r_{10} would there require more than the allotted 3 spare rows, choice of column c_9 allows us to reach an empty node, thus identifying a solution.

Figure 11.10 already presents a number of possible alternative solutions: they can be considered all equivalent, if the only constraint is that of staying within the predetermined number of spare rows and columns. If, on the other hand, we wish to minimize the total number of spare rows and columns used, solution (r_2, r_4, r_8, c_9) is the optimum. Obviously, different choices will lead to the construction of different trees and (possibly) to identification of different solutions; in fact, it is even possible that some of the various trees will allow no solution.

It may be useful — particularly for increasing array dimensions — to find conditions that allow simplification of the problem (even though it still is NP-complete, we will reach an algorithm that for a percentage of possible cases is less costly). To this end, fault distributions are analyzed to verify whether some particular rows and columns can be chosen *a priori*, on the basis of specific conditions, without affecting the final optimality of the algorithm (i.e., without affecting the possibility of finding a solution whenever it exists). Such conditions are the following:

(a) there is a row r_k (column c_h) containing a number of faults larger than the number of spare columns (rows) available. Elimination of r_k (c_h) is then mandatory, since otherwise it would be necessary to eliminate all columns (rows) in which r_k (c_h) has faulty PEs, and the number of available spares would not be sufficient. This policy is known as *must repair*. In the example being considered, choice of r_4 and of c_9 is mandatory, given the restriction to

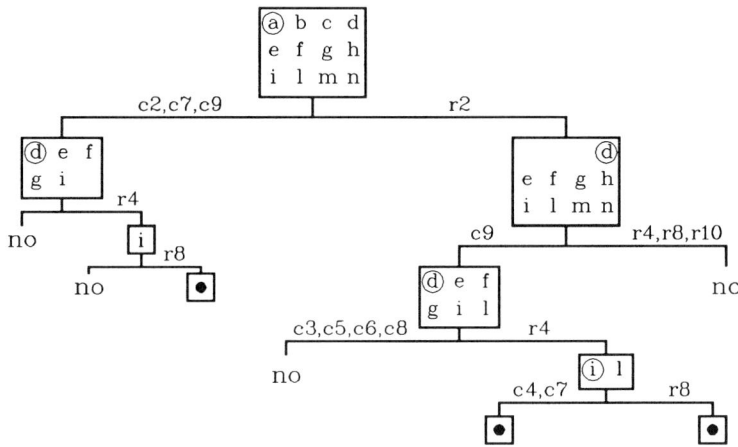

Figure 11.11
Binary tree obtained by simplification rules.

three spare rows and three spare columns. Note that, in the example of figure 11.10, this leads to a relevant simplification, since only four faults (a, b, i, l) still have to be covered;

(b) assume that there is a faulty PE f_i that does not share either row or column with any other fault: either choice (row or column) then has the same cost. Solve the covering problem for *all* faults but f_i: at the end, one of the following instances will be present:

  - no spare row or column is available: fatal failure has been reached;

  - only unused spare rows are available: the spare row covering f_i is chosen;

  - only unused spare columns are available: as above, the spare column covering f_i is chosen;

  - both spare rows and spare columns are available: any of the two choices can be made, indifferently;

Choosing rows and columns on the basis of the above conditions allows simplification of the array that is still to be reconfigured, so that the general algorithm will thereafter be applied to a smaller array.

Consider now the case in which there is a fault (i, j) such that no other faults are present in column j; again, the tree built to represent possible alternatives can be simplified:

(1) examine the two alternative coverages for (i, j), first, elimination of row i, and second, elimination of column j;

(2) first try coverage by elimination of row i: the matrix is simplified by deleting row i and the general problem is presented again, without any modification;

(3) then try elimination of column j: in general, other faulty devices $f_1, f_2 \cdots f_k$

will be present in row i and if column j has been chosen it still is necessary to cover all other faults $f_1, \cdots, f_k$ — typically, by choosing the corresponding columns, since otherwise choosing row i to cover one of the faults $f_1, f_2, \cdots, f_k$ would lead us back again to the previous alternative, but at a greater cost. Again, complexity of the algorithm can thus be reduced by selecting row i.

The complete tree of choices in figure 11.11 has been derived based on this last property alone (in order to verify the reduction of tree complexity on the simple example proposed, we did not use here the *must repair* rules previously recalled). In column c_2 there are no faulty cells besides cell a, while in row r_2 there are also faulty cells **b** and **c**; therefore, the first choice starting from the root node involves elimination either of r_2 or of all three columns c_2, c_7 and c_9 at once. If the first alternative is selected, the subsequent choice of fault **n** is adopted, that can be solved by elimination of either c_9 or r_{10}. For **n**, we find no other faults in r_{10} and other noncovered faults (corresponding to rows r_4 and r_8) in column c_9. If elimination of the three rows r_4, r_8 and r_{10} was performed here, the number of available spare rows would be exceeded, so that elimination of c_9 is the only viable alternative.

All three simplification rules listed above can be iteratively applied until no further simplification is possible: the general technique must then be adopted.

11.4. Suboptimal fast algorithms

In some cases (particularly, when the dimensions of the array are very large) it may be preferrable to look for techniques that — while not granting optimality as does the general technique presented above — are much less complex. We shall now examine a few such suboptimal criteria, and their motivations.

11.4.1 Case (a): "Repair-most" criteria

This philosophy is briefly cited in [KUN84], [KUO87]; its outline can be described as to look for rows and columns with the greatest numbers of faults, and eliminate them first.

The technique appears to be intuitively reasonable (note that it automatically implies the *must repair* condition previously recalled); it does not, however, grant optimality. Consider the example in figure 11.12: if the above rule is used, column c_1 and row r_{10} are eliminated first — and this leaves us to cover the remaining faults with the row sequence r_1, r_2, r_3 and with the column sequence c_7, c_9, c_{10} (or with equivalent solutions). There is a less costly solution — the optimum one, on the basis of spare cost — using $r_1, r_2, r_3, c_7, c_9, c_{10}$. Moreover, if only three rows and three columns are available, the *simplified* technique would not have allowed us to reach it.

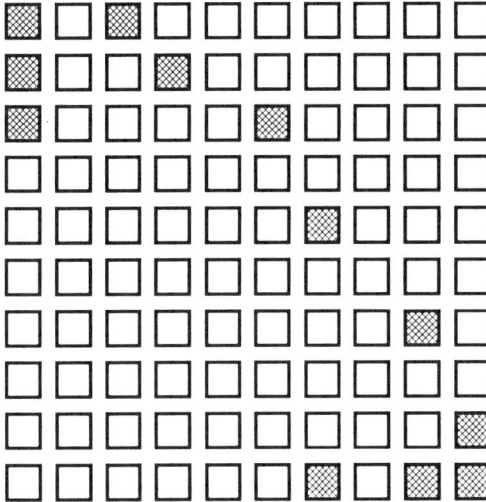

Figure 11.12
Sample array to be solved by repair most technique.

11.4.2 Case (b): "Leading Elements" coverage

A further nonoptimal algorithm has been described in [WEY87]: it has been defined for independent (arbitrary) numbers of spare rows (R_S) and spare columns (C_S); but for simplicity we will consider here $R_S = C_S = S$.

Rows are numbered in increasing order from top to bottom and columns from left to right. A faulty cell (i, j) is said to be a *leading element* (LE) if:

- i is the lowest index of a row containing faults as yet not covered;
- j is the lowest column index of a faulty cell in row i.

Thus, in the sample array of figure 11.13.a, the first LE is cell (1,1). The first LE identified is covered by elimination of *both* the row and column covering it. Subsequent LEs are found, in order, by the algorithm, and their coverage is again effected by elimination of row and column (positions and order of selection of the various LEs are shown in figure 11.13.a). If we assume only three spare rows and columns, the simple rule above for the given example would not allow reconfiguration. Yet, subsequent reductions are possible. Covering initially achieved may be redundant (see figure 11.13.b): e.g., cell (3,2) is in row 3 and column 2, both of which do not contain any other faulty cell, and elimination of row 3 only (figure 11.13.c) or of column 2 only (figure 11.13.d) would be sufficient. Note also that the first row in figure 11.13.b again covers faulty cells already covered by columns c_1, c_3, c_6: thus, elimination of r_1 is again redundant (see figure 11.13.c). Such considerations, performed after the preliminary LE sequencing, allow us to reach acceptable solutions such as those in figures 11.13.c or 11.13.d.

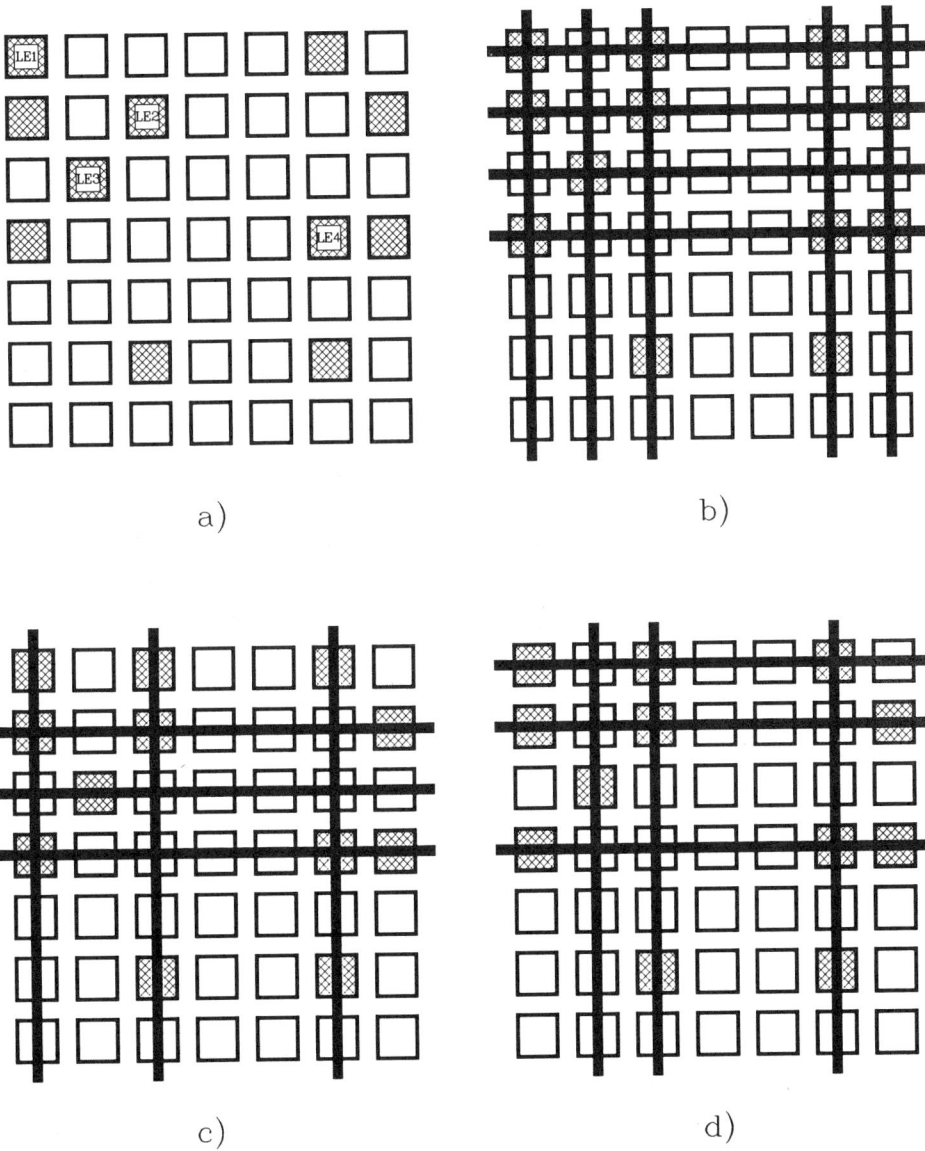

Figure 11.13

Leading Elements technique. a) Sample array and determination of Leading Elements. b) Initial elimination. c) Simplification of redundant eliminations. d) Alternative simplifications.

11.5. References

[BAT80] K.E.Batcher: *Architecture of a massively parallel processor*, Proc. 7th Symp. Computer Architecture, May 1980, 168-173, IEEE.

[CLA83] K.L.Clarkson: *A modification of the Greedy algorithm for vertex cover*, Information Processing Letters, Vol. 16, Jan. 1983, 23-25.

[DAY85] J.Day: *A fault driven comprehensive redundancy algorithm*, IEEE Design and Test, Vol. 2, N. 3, June 1985, 35-44.

[DIS86] F.Distante, M.G.Sami: *A protocol for asynchronous wavefront-computation arrays*, in: *Systolic Arrays*, (W.Moore, A.McCabe, R.Urquhart eds.), Proc. First Int'l Workshop on Systolic Arrays, Oxford, July 2-4 1986, Adam Hilger.

[KUN84] H.T.Kung, M.S.Lam.: *Fault-tolerant VLSI systolic arrays and two-level pipelining*, Journ. Parall. and Distr. Process., Aug. 1984, 32-63.

[KUO87] S.Y.Kuo, W.K.Fuchs: *Efficient Spare allocation for reconfigurable arrays*, IEEE Design and Test, Vol. 4, N. 1, Feb. 1987, 24-31.

[MCC83] J.V.McCanny, J.G.McWhirter: *Yield enhancement of bit-level systolic array chips using fault-tolerance techniques*, Electronic Letters, N. 19, 1983, 525-527.

[MOO84] W.R.Moore: *Switching circuits for yield enhancement of an array chip*, Electronics Letters, N. 16, 1984, 667-669.

[WEY87] C.L.Wey, _ F.Lombardi: __ *On the repair of redundant RAMS*, IEEE TCAD, Vol. CAD-6, N. 2, 1987, 222-231.

[FRI86] J.Fried: *Power clock and signal distribution in the WASP image processing device*, in: *Wafer Scale Integration*, (G.Saucier, J.Trilhe eds.), Proc. IFIP WG 10.5 Workshop, Grenoble, Mar. 17-19 1986, North-Holland.

11.6. Further readings

J.A.B.Fortes, C.S.Raghavendra: *Dynamically reconfigurable fault-tolerant array processors*, Proc. FTCS-14, June 1984, IEEE.

J.A.B.Fortes, C.S.Raghavendra: *Gracefully degradable processor arrays*, IEEE-TC, Vol. C-34, N. 11, 1985, 1033-1044.

In these two papers, the problem of array reconfiguration is considered simultaneously with that of reorganization of the algorithm to be mapped upon the array. Various row and column elimination schemes are considered, and their performances are compared, in particular with reference to increase of execution time for the algorithm computed by the array.

J.Hopcroft, R.Karp: *An $n^{5/2}$ algorithm for maximum matching in bipartite graphs*, SIAM J. Computing, Vol. 2, Dec. 1973, 225-231.

M.G.H.Katevenis, M.G.Blatt: *Switch design for soft-configurable WSI systems*, Proc. 1985 Chapel Hill Conf. on VLSI, 197-218. Attention is given in particular to design of switches supporting run-time reconfiguration.

12 GLOBAL MAPPING:
INDEX MAPPING RECONFIGURATION TECHNIQUES

An entire class of reconfiguration techniques can be described on the basis of one unifying philosophy, independent of the number and distribution of spares, the structure of the supporting interconnection network, and even — at least, in principle — of array connectivity (although in the present chapter we shall explicitly deal with rectangular, four-neighbor arrays). Such approaches consider reconfiguration as the novel mapping of array functions — represented by *logical indices* associated with the single PEs functions — onto the working cells present in the array (and identified by their *physical indices*). Thus, in a way, reconfiguration is achieved through a *deformation* of the initial array, circumventing faulty cells and distorting intercell connections. Algorithms are then identified by rules creating such novel mapping and by constraints imposed on the distorted links (a typical constraint being length bounds). Figures of merit used for their evaluation will be (besides link length, i.e., communication delay) the capacity of overcoming fault distributions for increasing numbers of faults and the simplicity of the algorithm itself as compared with its effectiveness.

Since, in this chapter, we will deal with rectangular arrays, each PE will be identified by a pair of indices: the *physical coordinates* in the case of physical indices and the *logical coordinates* of the function in the case of logical indices.

The fault model adopted by all techniques discussed here is the most usual one, i.e., by which faults are located only in PEs, following a random distribution, while the interconnection network and the circuits that control reconfiguration are assumed to be fault-free. No assumptions are made concerning distribution of faults in time: the global mapping philosophy outlined above makes it inherently useless to introduce restrictions such as the appearance of one single fault between two subsequent reconfigurations. Such limitations are in fact characteristic of *incremental* techniques; when global ones are considered — such as those discussed here — the *whole* distribution of faults is taken into account to perform the reconfiguration, so that appearance of new faults at run-time involves a completely new reconfiguration action. Ultimately, even transient faults might be accepted, simply by providing for periodical clearings of fault information (whether stored in local memories or in external storage) and repetitions of the testing phase preceding reconfiguration.

12.1. Global reconfiguration for purely systolic arrays

An initial approach is discussed in [KUN84] with reference to purely systolic arrays, involving as such stringent synchronization constraints. As a consequence, reconfiguration involves not only modified interconnection paths but also insertion on such paths of additional delays. Whenever stringent limits on speed reduction are present, it is therefore necessary to adopt *asynchronous*, wavefront computation

arrays, that do not require insertion of such delays. This was already emphasized when discussing Kung and Lam's approach for linear arrays (from which the
philosophy here described derives).

For linear systolic arrays (chapter 7) the distribution of delays was fairly complex when bidirectional information flow had to be accommodated, and restrictions
on allowable fault distributions had to be taken into account as well.

Given the primary scope of keeping synchronization intact, the philosophy
suggested by Kung and Lam gives little relevance to *locality* as a factor of merit.
Delay and timing aspects are accounted for even though intercell links may become
quite long; locality, on the contrary, is a primary factor in wavefront arrays, where
the length of intercell links directly affects delays.

Another main cost factor, i.e., silicon area increase, is satisfactorily dealt with,
since limited channel width is required for the augmented interconnection network
(this approach deals with *harvesting* rather than with *probability of survival*, so
that it is not meaningful to evaluate area requirements introduced by spares).

In [KUN84] a basic philosophy is introduced first, and two solutions of increasing complexity and flexibility, based upon it, are then suggested. Following
the authors' example, we shall develop one of the two in particular detail, outlining
the second.

Several definitions are needed prior to formally stating the reconfiguration philosophy and the algorithms:

(1) correct array operation requires that each cell will receive both its inputs *at
the same time*: in turn, the cell will autonomously generate its outputs, and
generation of both outputs will again take place at the same time;

(2) let (i', j') be the *logical* indices of any given PE in the array: $(i', (j-1)')$ and
$((i-1)', j')$ are the logical indices, respectively, of its *logical* horizontal and
vertical predecessors, i.e., of the fault-free PEs providing its inputs. Denote
now by (r_1, s_1) the *physical* indices of the PE upon which $(i', (j-1)')$ is
mapped, (r_2, s_2) the ones of the PE upon which $((i-1)', j')$ is mapped, and
(r_3, s_3) the ones of the PE upon which (i', j') is mapped.

Then, the *distance* between (r_1, s_1) and (r_3, s_3) (measured with reference to
the horizontal data path connecting them) is evaluated as:

$$h = [r_3 - r_1] + [s_3 - s_1]$$

In the same way, distance between (r_2, s_2) and (r_3, s_3) (measured with reference to the vertical data path connecting them) is evaluated as

$$v = [r_3 - r_2] + [s_3 - s_2]$$

(Note that such definition of distance is rather different from the ones generally adopted by other authors or for other algorithms described later in this
book.) Then:

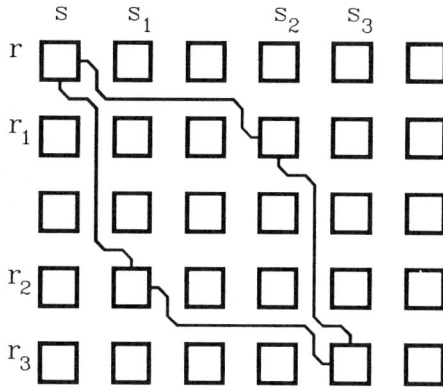

Figure 12.1
Sample array supporting the Kung-Lam min-cut theorem for rectangular arrays.

*whenever two cells are connected by a data path (either horizontal or vertical)
such that the associated distance is d, it has to be assumed that along such
data path there will be exactly d registers, each of them providing a unit delay.*

Physically adjacent cells are at distance 1: this evaluation accounts for the
fact that results produced by a cell are used (i.e., processed) by the successor
cell in correspondence of the following clock cycle (i.e., processing by the
individual PE is assumed to require one clock cycle). In correspondence, a
unit delay register is introduced on each direct link.

(3) The direct relation between distances and structure of data paths defined
in (2) is true only in the case of interconnection paths made of horizontal
segments oriented rightward and/or of vertical segments oriented downward
(see figure 12.1). The following necessary condition therefore holds:

$$r_3 \geq r_1 \qquad r_3 \geq r_2$$

$$s_3 \geq s_1 \qquad s_3 \geq s_2$$

(4) Given all the above assumptions, the following theorem, derived from the cut
theorem discussed in chapter 7, holds:

*Provided all data paths follow only a rightward-downward orientation, then
each PEs in the array receives both its inputs (horizontal and vertical) simultaneously.*

Refer to figure 12.1; let δt denote the unit time interval (corresponding to the
clock cycle). Let (r, s) be the physical indices of a fault-free cell with which
logical indices $((i-1)', (j-1)')$ have been associated: it generates its outputs
at time t_o. Logical successor cell $(i', (j-1)')$ receives its vertical input at time
$t_o + \delta t \cdot (r_1 - r + s_1 - s - 1)$, and it generates its horizontal output after a
unit time interval t. In the same way, cell $((i-1)', j')$ generates its vertical
output at time $t_o + \delta t \cdot (r_2 - r + s_2 - s)$.

Finally, the horizontal input of (i', j') is present at time

$$t_o + \delta t \cdot ((r_1 - r) + (s_1 - s) + (r_3 - r_1) + (s_3 - s_1) - 1) = t_o + \delta t \cdot ((r_3 - r) + (s_3 - s) - 1)$$

while the vertical input is present at time

$$t_o + \delta t \cdot ((r_2 - 3) + (s_2 - s) + (r_3 - r_2) + (s_3 - s_2) - 1) = t_o + \delta t \cdot ((r_3 - r) + (s_3 - s) - 1)$$

that is, at the same time as the horizontal one.

The above synchronization rules set the constraints that allow definition of a reconfiguration algorithm, mapping the target rectangular array onto a physical array in presence of any given distribution of faults:

12.1.1. First reconfiguration algorithm of [KUN84]

(1) All logical indices of the first target row are mapped onto all (and only) the fault-free cells of the first physical row. Implicitly, this sets a limit to the target array's dimensions.

(2) All logical indices of the first target column are mapped onto all (and only) the fault-free cells in the first physical column (as in (1), this gives a further bound to the dimensions of the target array).

(3) Assume that logical indices (i', j') have to be mapped onto a still undefined physical cell, while its logical predecessors have already been allocated. Let $s_3 = max(s_1, s_2)$ and $r_3 = max(r_1, r_2)$ (see the previous definitions for indices (r_i, s_i)): then, (i', j') can be mapped onto any fault-free cell (r, s) such that $r \geq r_3$, $s \geq s_3$.

(4) By iterating step (3) until it becomes impossible to complete a new row and/or column, all logical indices can be mapped onto fault-free cells and a target array is built.

Because of step (3) — which allows a choice among a number of possible alternatives — the above is still not a deterministic algorithm but rather a *philosophy*: final definition of a complete algorithm can be reached by introducing further constraints, e.g., which require that at each step the choice minimizing incremental delay (i.e., minimizing $r_3 + s_3$) be adopted. (Even such a constraint will leave a measure of ambiguity in determining reconfiguration for most fault distributions.) Requirements of this type, being of an *incremental* value, will not lead to a global optimization, be it in terms of harvesting, or of total delay: still, reaching such global optimization would involve exploring all alternatives, e.g., with a *branch and bound* technique, with an obvious increase in processing time.

The procedure — with the added restriction of minimizing incremental delays — can be exemplified by referring to figure 12.2. Assume that logical pairs of indices $(2, 3)$ and $(1, 4)$ have been already mapped, respectively, onto physical cells $(2, 4)$ and $(1, 5)$ and that the logical pair of indices $(2, 4)$ must now be allocated

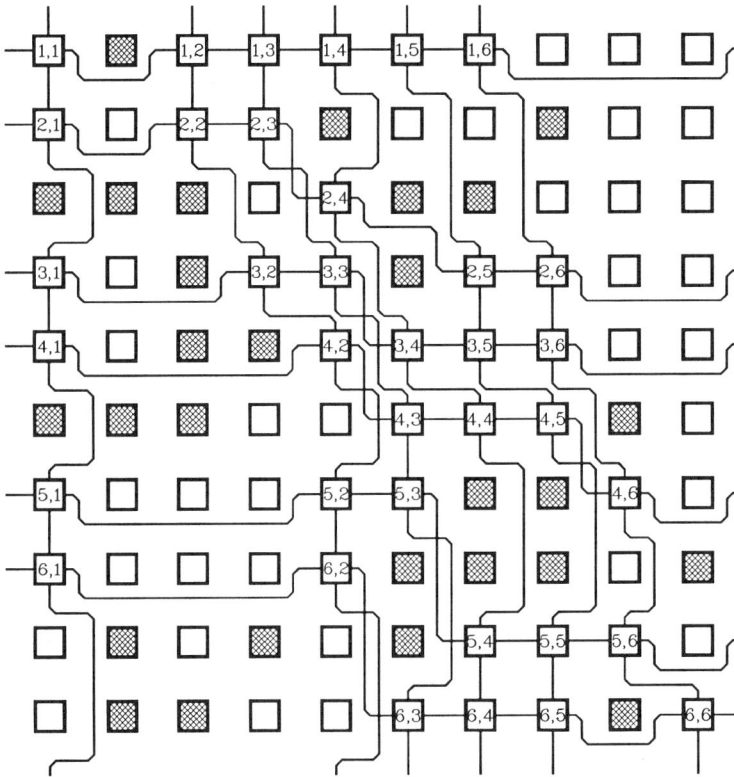

Figure 12.2
Reconfiguration of a rectangular systolic array by means of Kung-Lam first algorithm.

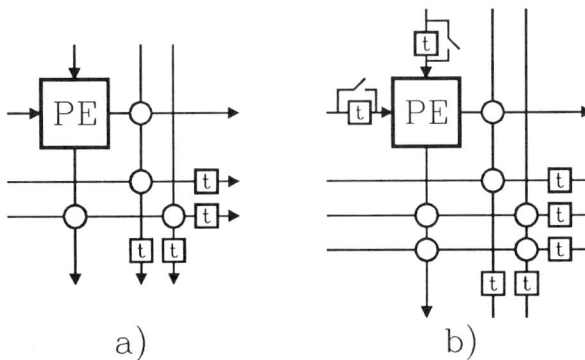

a)   b)

Figure 12.3
Interconnection network supporting reconfiguration respectively a) for algorithm 12.1.1 and b) 12.1.2.

onto a physical cell. Rule 3 above states that such mapping can be performed onto any fault-free cell with a physical row index greater than 2 and a physical column index greater than 5: among these, physical cells (3,5) and (2,6) lead to minimum-delay paths with identical delays, and as such are equivalent. The concept of *equivalence* here is very limited, referring to step (3) of the procedure only but *not* to the capacity of reaching solutions with identical harvesting. On the contrary, whenever alternatives are present, different arbitrary choices will in general lead to different harvesting figures after all index-mapping iterations have been completed for the remaining cells. Whatever the choice, it is also self-evident that harvesting can be quite low (all unmarked cells in figure 12.2 correspond to unused fault-free cells).

With regard to the interconnection network supporting this algorithm, the number of buses required is related to the allowable fault pattern: thus, in the example represented in figure 12.3.a, two buses in each connection channel are sufficient, while four switches and four unit-delay registers associated with each PE allow the creation of reconfigured paths with correct delay distributions. No on-chip circuits for control of reconfiguration need be taken into account, since this technique is obviously designed for end-of production restructuring and is therefore externally driven.

12.1.2. Outline of the second reconfiguration algorithm of [KUN84]

To at least partially overcome the drawback of an unsatisfactory harvesting, Kung and Lam suggested in the same paper a modification of the basic philosophy that relaxes the constraint on data path orientation. This second approach in fact relaxes the constraint that segments of reconfigured data paths should be oriented in the directions of wavefront propagation only; fault-free cells that would have been overlooked by the first approach can thus be used now for index mapping.

Besides leading to a more complex and even less deterministic technique (the number of choices at each step obviously increases with respect to the previous case), relaxation of the restriction due to point (3) in 12.1.1 causes the cut theorem to no longer be satisfied. Identical delays along data paths are therefore not created automatically during the reconfiguration procedure and programmable delays must be inserted along the interconnections. Assume again that logical indices (i', j') have to be mapped onto a physical cell; and consider the two input data paths:

$$((i-1)', (j-1)') \mapsto ((i-1)', j') \mapsto (i', j')$$

$$((i-1)', (j-1)') \mapsto (i', (j-1)') \mapsto (i', j')$$

Additional delays will have to be inserted along the *shortest* data path (i.e., the one that would be associated with a lower delay).

As is self-evident from figure 12.4, where this second approach is used to achieve restructuring against the same fault distribution already used in figure 12.2, harvesting now reached is higher than in the previous case. On the other

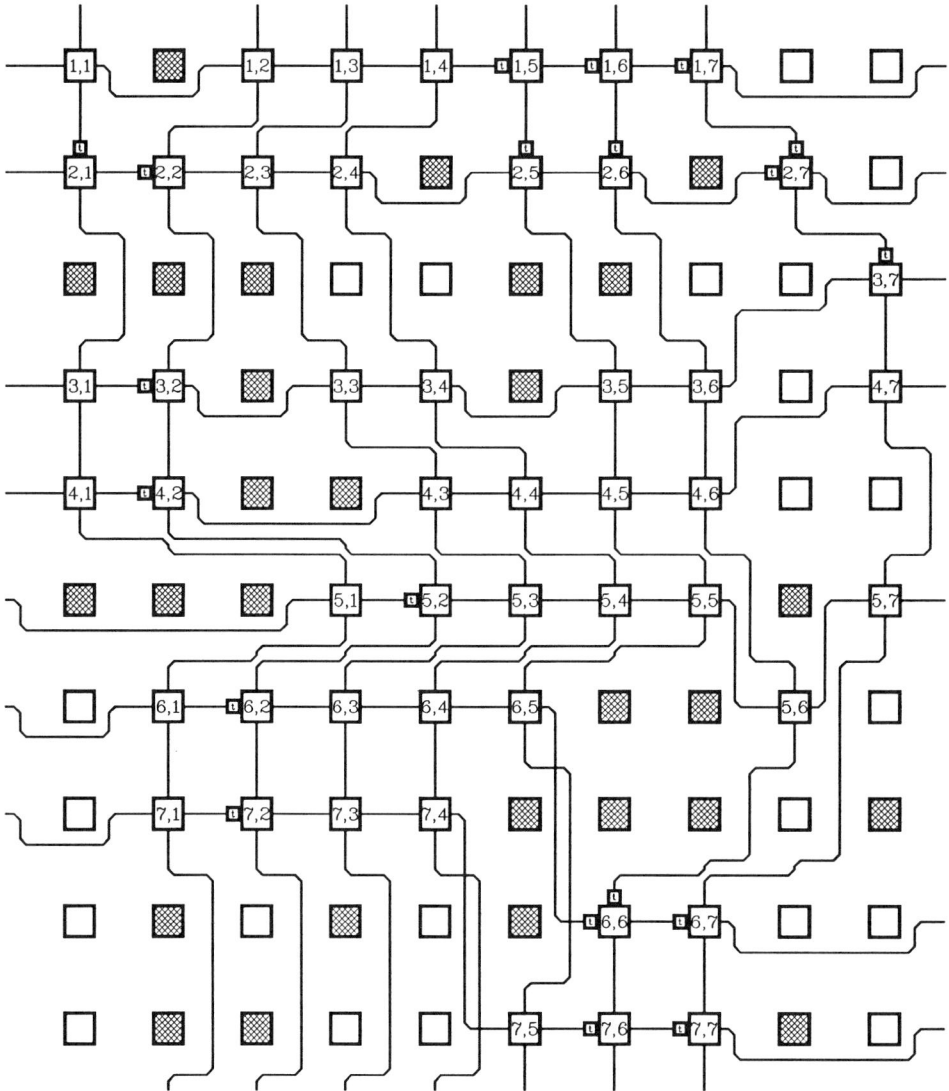

Figure 12.4

Reconfiguration of a rectangular systolic array by Kung-Lam's second algorithm (the fault distribution is identical to that in figure 12.2).

hand, in addition to complexity of the algorithm, structure complexity also increases in a relevant way as a result of the necessity of programmable delays, as may be seen from figure 12.3.b.

In this case also, complexity of the augmented interconnection network is related to allowable fault patterns: the example shown in figure 12.4 requires three horizontal buses and two vertical ones.

12.2. The general index-mapping approach

A number of reconfiguration algorithms relying on the index mapping philosophy have been defined beginning from a formal approach that then allows techniques of varying complexity and performances to develop [SAM83, NEG85, SAM85, NEG86, SAM86a, SAM86b, SAM87]. We shall first define the basic approach, and then state several algorithms within such a framework. Reference is made to *run-time* reconfiguration: dimensions of the target array are a priori determined and a number of spare cells organized into regular patterns of spare rows and columns are introduced. The physical array consists of R_p rows and C_p columns: the target array consists of $R_t \times C_t$ cells, where

$$R_p \geq R_t$$

$$C_p \geq C_t$$

(the more general requirement $R_p \cdot C_p \geq R_t \cdot C_t$ would involve also the possibility of *functional* reconfiguration by mapping a rectangular array of arbitrary dimensions onto the given physical one, and therefore it is not adopted by the greatest majority of authors). There are $R_p - R_t$ spare rows and $C_p - C_t$ spare columns: spare elements are not active as long as there are no faults in the array, and they are introduced in operation as a consequence of the *deformations* following reconfiguration. [1] Whatever the algorithm, only $Fm = R_p \cdot C_p - R_t \cdot C_t$ faults can be tolerated at most. In any event, there will usually be a non-zero probability that reconfiguration will fail even for particular distributions of $F < Fm$ faults, because of constraints introduced when defining the algorithm (particularly, constraints concerning locality of interconnections and algorithm complexity). We will evaluate then as a global figure of merit the *probability of survival* provided by the various algorithms, i.e., the probability of achieving reconfiguration against increasing numbers of faults.

---

[1]   In this chapter, as in the following ones, the following conventions will be adopted:

- row indices are ordered in increasing order from top to bottom and column indices from left to right;

- spare rows are added at the bottom of the array, spare columns on the right border, unless otherwise specified.

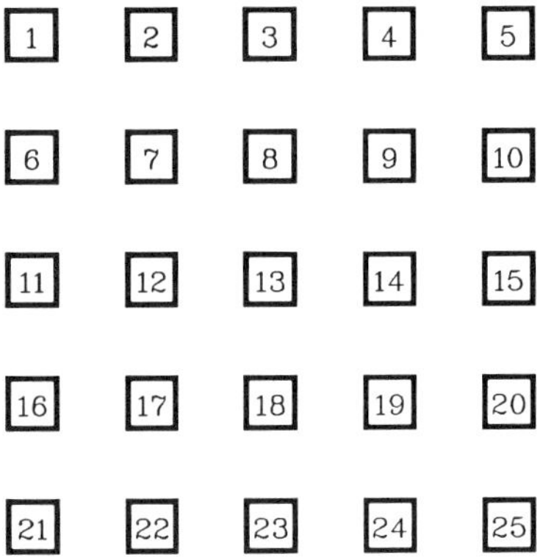

1	2	3	4	5
6	7	8	9	10
11	12	13	14	15
16	17	18	19	20
21	22	23	24	25

Figure 12.5
Ordering codes for a sample physical array.

The aim of reconfiguration can be stated as:

- Define the target array as a $R_t \times C_t$ *target matrix* T, whose entries correspond — in an ordered way — to the pairs of logical indices associated with the cells of the target array. Then create a *logical matrix* LG with R_p rows and C_p columns, where

$$LG(i,j) = T(i,j) \quad for \quad i \le R_t, \quad j \le C_t$$

$$LG(i,j) = (0,0) \quad for \quad i > R_t, \quad j > C_t$$

- Define the physical array as a $R_p \times C_p$ *physical matrix* whose entries are pairs of indices corresponding to the coordinates of the individual PEs in the array. Reconfiguration is achieved by mapping all indices of the target array cells onto fault-free PEs of the physical array so that each pair of logical indices will appear once and once only: this is obtained through a *deformation* of the logical matrix LG.

The above problem can also be defined as a coverage problem, by introducing a general mapping table Mg created as follows:

- associate an ordering number to all physical PEs; the ordering is totally arbitrary (e.g., rowwise as in figure 12.5);

- identify logical cells by their pairs of indices in the target array;

- introduce a column in table Mg for each physical PE and a row for each logical cell;

- mark entry $Mg(k,h)$ if and only if the logical cell associated with row k can be mapped onto physical PE h: our mapping problem is then reduced to that of covering table Mg by selecting suitable marks so that:

    - in each selected row there will be one only selected mark (each logical cell must appear once and only once in the final mapping)

- in each selected column there will never be more than one selected mark (otherwise, the physical PE corresponding to that column would be called upon to perform more than one different function at the same time).

In the absence of faults and of any constraint on location of and interconnection between physical PEs, any logical cell can map onto any physical PE, so that each entry $Mg(k, h)$ will be a mark. In the presence of faults, but again without any constraint, the general mapping problem previously defined leads to the introduction of a mark in any position $Mg(k, h)$ such that h corresponds to a fault-free PE (i.e., no marks are present in columns corresponding to faulty PEs). Such totally free mapping would, of course, lead to 100 percent survival against up to Fm faults.

In figure 12.7, a sample 5×5 physical array is shown, the target array being a 4×4 one. Matrix Mg is given in figure 12.6. A solution has been achieved by observing constraints (1) and (2) and performing the selection of marks in an otherwise random way: the selected marks are denoted by S in figure 12.6. The mapping of logical onto physical indices derived as a consequence of covering performed on Mg is given in figure 12.7.

Let us now introduce the basic factor of merit considered in the present section, i.e., *locality*, defined as follows:

- assume an initial fault pattern and a corresponding reconfiguration achieved by a given algorithm;

- let (i, j), (i_x, j_x), (i_y, j_y) be the physical indices of three fault-free PEs and let (i', j'), $(i', (j+1)')$, $((i+1)', j')$ be — in order — the logical indices associated with them.

- define distances between pairs of cells as:

$$Dx(i', j') = [i - i_x] + [j - j_x] - 1$$

$$Dy(i', j') = [i - i_y] + [j - j_y] - 1$$

Again, as in all other instances, it can be seen that width of interconnection channels is not taken into account for evaluation of distances; the maximum distances for all pairs i', j' are defined as:

$$Dx' = \max[Dx(i', j')]$$

$$Dy' = \max[Dy(i', j')]$$

- It is possible to define the maximum distances Dx and Dy obtained by a given reconfiguration algorithm for all possible fault patterns: they correspond to the maximum physical length of the horizontal and vertical connections between cells, using the cell side as unit length. A reconfiguration algorithm satisfies the locality condition if and only if the generated distances are not greater than the two predefined bounds. (Note that the bounds are fixed independently of array dimensions.)

	1	2	3	4	5	6	7	8	9	10	11	12	13	14	15	16	17	18	19	20	21	22	23	24	25
1,1	x	S	.	x	.	.	x	.	.	x	x	x	.	.	x	x	x	.	x	x	x	.	x	x	x
1,2	x	x	.	S	.	.	x	.	.	x	x	x	.	.	x	x	x	.	x	x	x	.	x	x	x
1,3	S	x	.	x	.	.	x	.	.	x	x	x	.	.	x	x	x	.	x	x	x	.	x	x	x
1,4	x	x	.	x	.	.	x	.	.	x	x	S	.	.	x	x	x	.	x	x	x	.	x	x	x
2,1	x	x	.	x	.	.	x	.	.	S	x	x	.	.	x	x	x	.	x	x	x	.	x	x	x
2,2	x	x	.	x	.	.	S	.	.	x	x	x	.	.	x	x	x	.	x	x	x	.	x	x	x
2,3	x	x	.	x	.	.	x	.	.	x	x	x	.	.	S	x	x	.	x	x	x	.	x	x	x
2,4	x	x	.	x	.	.	x	.	.	x	x	x	.	.	x	x	S	.	x	x	x	.	x	x	x
3,1	x	x	.	x	.	.	x	.	.	x	S	x	.	.	x	x	x	.	x	x	x	.	x	x	x
3,2	x	x	.	x	.	.	x	.	.	x	x	x	.	.	x	x	x	.	x	x	x	.	S	x	x
3,3	x	x	.	x	.	.	x	.	.	x	x	x	.	.	x	x	x	.	x	x	x	.	x	S	x
3,4	x	x	.	x	.	.	x	.	.	x	x	x	.	.	x	x	x	.	x	x	x	.	x	x	S
4,1	x	x	.	x	.	.	x	.	.	x	x	x	.	.	x	x	x	.	x	x	S	.	x	x	x
4,2	x	x	.	x	.	.	x	.	.	x	x	x	.	.	x	x	x	.	x	S	x	.	x	x	x
4,3	x	x	.	x	.	.	x	.	.	x	x	x	.	.	x	x	x	.	S	x	x	.	x	x	x
4,4	x	x	.	x	.	.	x	.	.	x	x	x	.	.	x	S	x	.	x	x	x	.	x	x	x

Figure 12.6

Reduced coverage matrix after elimination of marks in columns corresponding to the faulty PEs in figure 12.7.

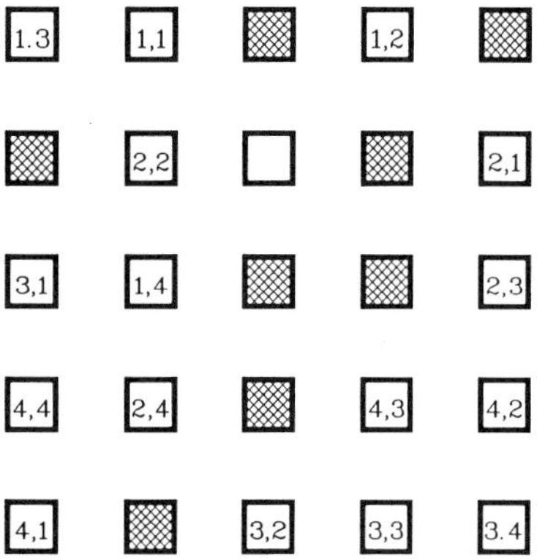

Figure 12.7

Logical indices for a target array mapped onto a physical one in presence of faults. The mapping derives from the solution shown in figure 12.6 (etched cells are faulty, the blank one is an unused spare).

Assuming one bound d for both vertical and horizontal distances, the coverage problem with locality constraint can now be formulated as follows:

- matrix Mg must be covered in such a way as to guarantee that no distance between logically adjacent cells will be greater than d.

The problem is NP-complete, unless further constraints are introduced. Let us first assume that no additional information is available to provide a guideline for the coverage procedure. Initially a *complete* matrix Mg is then created, where only columns corresponding to faulty PEs do not contain any mark. Moreover, if no further constraints are provided, each row will certainly contain more than one mark (each logical cell can be covered by more than one physical PEs) so that no predetermined initial choice can be made. A *sequential* coverage procedure must then be adopted, i.e., one in which one mark at a time is considered. Assume thus that the mark corresponding to mapping of logical cell (i', j') onto physical PE (h, k) has been chosen. Rows corresponding to logical adjacents of (i', j') are now examined and all marks in such rows that correspond to physical PEs whose distance from (h, k) is greater than d are deleted (under the locality constraint, their choice would not be acceptable). When mapping of $(i', (j+1)')$ and $((i+1)', j')$ is afforded, then, only a subset of all mappings previously possible will be considered.

The coverage procedure outlined above enables us to see at once that at each step of the procedure itself (i.e., every time an association logical cell-physical PE is made) the coverage matrix Mg is modified. It is quite obvious that each choice influences — by modifying Mg in a peculiar way — subsequent possible choice. Thus, locality is in this context a *dynamic* property, typical not simply of the fault distribution but also (and even more) of the order following which subsequent choices are made. This is actually a consequence of the fact that while the overall constraint (i.e., locality) as well as the basic concept underlying reconfiguration, is *global*, the coverage procedure as described is itself *incremental*: therefore, even the possibility of finding a solution becomes dependent upon the order in which choices are made. If an optimal algorithm is required — meaning by this that the algorithm must always find a solution whenever one exists — all possible orderings should be tried.

An example of a reconfiguration technique that can be seen as an instance of dynamic coverage of matrix Mg (even though the constraints introduced there do not depend on locality) is given by the Kung-Lam approach described previously.

A tighter bound — leading to simpler, although less effective algorithms — can be defined through a concept of *static locality*. While in the previous, general definition it was required that logically adjacent cells would map onto physical PEs characterized by bounded distance d, this second definition requires that any pair of logically adjacent cells (i', j'), (h', k') map onto two physical cells each distant not more than $d/2$ from the *original* physical positions (i.e., respectively, from (i, j) with $i = i'$, $j = j'$ and (h, k) with $h' = h$, $k' = k$). It is immediately apparent that such a new condition is sufficient to guarantee that the distance *between* the two physical cells after reconfiguration will also be no greater than d: it is not, yet, a *necessary* condition. It is thus more restrictive than the previous one, and

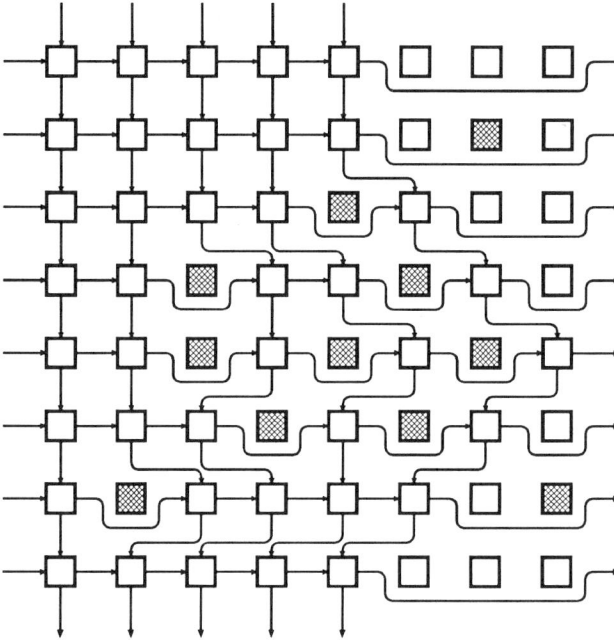

Figure 12.8
Example of reconfiguration with maximum distance 1 (except for I/O connections).

effectively limits the possibilities of reconfiguration (and, therefore, of coverage for Mg). Figure 12.8 represents a possible reconfiguration for a sample 5×5 target array starting from a 8×8 physical array, with upper bound for distance equal to 1; here, *static* locality is satisfied (except for connections to I/O pins).

This restricted definition of locality can also be described with reference to *adjacency domains* and *inverse adjacency domains* (both evaluated with reference to a specific algorithm), defined as follows:

(a) given a reconfiguration algorithm, described as a *deformation* of the original logical matrix (in the absence of faults), it can be said that as a consequence of reconfiguration each cell with logical indices (i', j') undergoes a *displacement* from its original physical position (i, j) $(i = i', j = j')$ onto a PE with physical indices (h, k);

(b) for every physical position (i, j), the associated *adjacency domain* (related to the given algorithm) consists of all physical positions (including (i, j) itself) such that the logical indices (i', j') initially associated with (i, j) can be mapped onto them as a consequence of an allowable reconfiguration. Figure 12.9.a is an example of adjacency domain for an algorithm that allows the assignment of the functions corresponding to logical indices (i', j') either to physical PE (i, j), $i = i'$, $j = j'$, or to one of its adjacents $(i, j+1)$, $(i+1, j)$;

(c) for every physical position (i,j) the *inverse adjacency domain* of (i,j) (again, related to a given reconfiguration algorithm) consists of all physical positions (h,k) (including (i,j) itself) such that logical indices (h',k') initially associated with (h,k) can be mapped onto (i,j) as a consequence of an allowable reconfiguration. Figure 12.9.b gives the inverse adjacency domain related to the adjacency domain in figure 12.9.a.

If reconfiguration is performed by granting displacements only inside adjacency domains the bound on distance between the physical cells upon which two logical adjacents may map can be determined as follows. Consider the initial, fault-free array in which physical and logical indices coincide: refer to cells (i,j), $(i,j+1)$ with $(i = i', j = j')$. Whatever the subsequent fault distribution and related reconfiguration, the physical cells onto which logical indices (i',j') and $(i',(j+1)')$ may map can be found only inside the adjacency domain, respectively, of (i,j) and $(i,j+1)$. Let AD_1 and AD_2 be such domains, and let $AD = AD_1 \cup AD_2$ be their union; the maximum distance between the two target cells is then the maximum diameter of AD. Thus, the locality condition is satisfied. A consequence of this restricted definition of locality, involving *static adjacency domains*, is that the number of spare rows and columns introduced in the physical array is in turn limited, with an easily defined relationship to the dimensions of the adjacency domain: thus, the adjacency domain of figure 12.9.a allows utilization of only 1 spare row and 1 spare column.

Reconfiguration algorithms that satisfy this restricted locality condition can be determined following two different approaches: by coverage algorithms operating upon coverage matrix Mg and by definition of matrix operators that create *deformation lines*, capable of relating spares to faulty cells in a unique way (and, therefore, of identifying in a unique way a global reconfiguration). We shall consider these two approaches, examining some relevant algorithms corresponding to each one and discussing possible implementations of the augmented interconnection networks that support them.

12.3 The coverage matrix philosophy

The problem can now be stated in a very simple way:

(1) Create the coverage matrix by taking into account the restrictions introduced by the adjacency domain: thus, every row will contain only (at most) as many marks as there are cells in the adjacency domain. For simplicity, consider a very small adjacency domain as in figure 12.9.a, consisting of only three cells. Consequently, only one spare row and one spare column can be added to the original array. Assuming the logical array in figure 12.10 and the physical array in figure 12.5, the coverage matrix in figure 12.11 is created for the fault-free original array. Thus, for example, logical cell $(2,1)$ can map either onto its original position (physical cell 6) or undergo a displacement of one position to the right (physical cell 7) or of one position downwards (physical position 11).

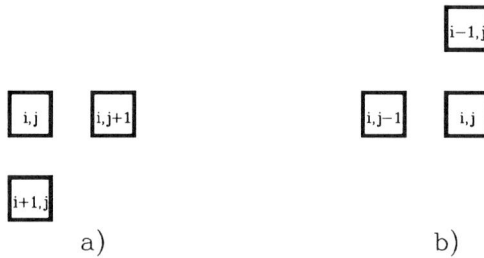

Figure 12.9
a) Sample adjacency domain. b) Corresponding inverse adjacency domain.

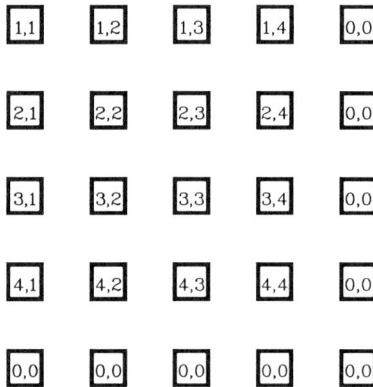

Figure 12.10

Initial mapping of a target array onto a fault-free physical array with one spare row and one spare column.

(2) Given the fault distribution, delete from matrix Mg all marks in the columns corresponding to faulty PEs: thus, assuming that PEs 4, 13, 14, and 17 are faulty, matrix Mg is modified as in figure 12.12.

(3) Identify a coverage that allows mapping of the logical array onto the fault-free cells. This can be immediately reduced to a *complete matching* problem, in which all logical cells have to be mapped onto fault-free PEs. Efficient matching algorithms have been presented in the literature; particular characteristics of the problem considered here allow an a priori simplification of Mg, so as to reduce the overall computation requirements. Such are the following:

  (a) Fatal failure can be stated whenever there is at least one row in Mg containing no marks; in this case, it is impossible to map the corresponding logical cell onto any acceptable physical PE.

  (b) Whenever a row R_j contains one single mark (*essential row*) in a column C_i, the mapping of logical cell R_j onto PE C_i is immediate and necessary. Mg can be consequently simplified by deleting the essential row R_j and the related column C_i (the proof is trivial). Thus, in our example, rows

	1	2	3	4	5	6	7	8	9	10	11	12	13	14	15	16	17	18	19	20	21	22	23	24	25
1,1	X	X	.	.	.	X	.	.	.	.	.	.	.	.	.	.	.	.	.	.	.	.	.	.	.
1,2	.	X	X	.	.	.	X	.	.	.	.	.	.	.	.	.	.	.	.	.	.	.	.	.	.
1,3	.	.	X	X	.	.	.	X	.	.	.	.	.	.	.	.	.	.	.	.	.	.	.	.	.
1,4	.	.	.	X	X	.	.	.	X	.	.	.	.	.	.	.	.	.	.	.	.	.	.	.	.
2,1	.	.	.	.	.	X	X	.	.	.	X	.	.	.	.	.	.	.	.	.	.	.	.	.	.
2,2	.	.	.	.	.	.	X	X	.	.	.	X	.	.	.	.	.	.	.	.	.	.	.	.	.
2,3	.	.	.	.	.	.	.	X	X	.	.	.	X	.	.	.	.	.	.	.	.	.	.	.	.
2,4	.	.	.	.	.	.	.	.	X	X	.	.	.	X	.	.	.	.	.	.	.	.	.	.	.
3,1	.	.	.	.	.	.	.	.	.	.	X	X	.	.	.	X	.	.	.	.	.	.	.	.	.
3,2	.	.	.	.	.	.	.	.	.	.	.	X	X	.	.	.	X	.	.	.	.	.	.	.	.
3,3	.	.	.	.	.	.	.	.	.	.	.	.	X	X	.	.	.	X	.	.	.	.	.	.	.
3,4	.	.	.	.	.	.	.	.	.	.	.	.	.	X	X	.	.	.	X	.	.	.	.	.	.
4,1	.	.	.	.	.	.	.	.	.	.	.	.	.	.	.	X	X	.	.	.	X	.	.	.	.
4,2	.	.	.	.	.	.	.	.	.	.	.	.	.	.	.	.	X	X	.	.	.	X	.	.	.
4,3	.	.	.	.	.	.	.	.	.	.	.	.	.	.	.	.	.	X	X	.	.	.	X	.	.
4,4	.	.	.	.	.	.	.	.	.	.	.	.	.	.	.	.	.	.	X	X	.	.	.	X	.

Figure 12.11
Coverage matrix for a fault-free array corresponding to the adjacency domain shown in figure 12.9.

	1	2	3	4	5	6	7	8	9	10	11	12	13	14	15	16	17	18	19	20	21	22	23	24	25
1,1	X	X	.	.	.	X	.	.	.	.	.	.	.	.	.	.	.	.	.	.	.	.	.	.	.
1,2	.	X	X	.	.	.	X	.	.	.	.	.	.	.	.	.	.	.	.	.	.	.	.	.	.
1,3	.	.	X	.	.	.	.	X	.	.	.	.	.	.	.	.	.	.	.	.	.	.	.	.	.
1,4	.	.	.	.	X	.	.	.	X	.	.	.	.	.	.	.	.	.	.	.	.	.	.	.	.
2,1	.	.	.	.	.	X	X	.	.	.	X	.	.	.	.	.	.	.	.	.	.	.	.	.	.
2,2	.	.	.	.	.	.	X	X	.	.	.	X	.	.	.	.	.	.	.	.	.	.	.	.	.
2,3	.	.	.	.	.	.	.	X	X	.	.	.	.	.	.	.	.	.	.	.	.	.	.	.	.
2,4	.	.	.	.	.	.	.	.	X	X	.	.	.	.	.	.	.	.	.	.	.	.	.	.	.
3,1	.	.	.	.	.	.	.	.	.	.	X	X	.	.	.	X	.	.	.	.	.	.	.	.	.
3,2	.	.	.	.	.	.	.	.	.	.	.	X	.	.	.	.	.	.	.	.	.	.	.	.	.
3,3	.	.	.	.	.	.	.	.	.	.	.	.	.	.	.	.	.	X	.	.	.	.	.	.	.
3,4	.	.	.	.	.	.	.	.	.	.	.	.	.	.	X	.	.	.	X	.	.	.	.	.	.
4,1	.	.	.	.	.	.	.	.	.	.	.	.	.	.	.	X	.	.	.	.	X	.	.	.	.
4,2	.	.	.	.	.	.	.	.	.	.	.	.	.	.	.	.	.	X	.	.	.	X	.	.	.
4,3	.	.	.	.	.	.	.	.	.	.	.	.	.	.	.	.	.	X	X	.	.	.	X	.	.
4,4	.	.	.	.	.	.	.	.	.	.	.	.	.	.	.	.	.	.	X	X	.	.	.	X	.

Figure 12.12
Reduced coverage matrix after elimination of marks in columns corresponding to faulty PEs.

3,3 and 3,2 are both essential rows; row 3,3 has a mark only in the column corresponding to PE 18, and row 3,2 has a mark only in column 12. This means that the two logical cells can be mapped, respectively, only onto PE 18 and PE 12; the mapping is carried out and the two columns can be subsequently deleted, since no further assignment is possible for the two PEs (see figure 12.13.a).

(c) If a column C_i contains a single mark in row R_j (*singular column*), whenever the required complete matching exists, there is a solution in which the logical cell R_j maps onto the physical PE C_i.

The proof is simple. By assumption, R_j contains at least one other mark in another column (otherwise case (b) would be incurred). Let C_k be such column, and let a solution in which R_j maps onto C_k. If this solution is modified by substituting C_k with C_i, its validity is not impaired since no other mappings are made impossible (C_i could not be associated with any other row). A solution in which C_i is associated with R_j thus certainly exists if the complete matching is possible. Identification of singular columns allows to simplify Mg by deleting the singular columns and the related rows.

(d) Simplification of Mg can create *pseudo-essential rows* (rows in which one mark *remains* after simplification steps) and *pseudo-singular* columns. Steps (b) and (c) can be iterated for such rows and columns until no further immediate association is possible. In our example, row 4,2 is a pseudo-essential one; the only possible association with PE 22 is performed, see figure 12.13.b.

The matrix in figure 12.13.b can be fully covered by applying the above rules. Figure 12.14 gives one of the possible solutions, while figure 12.15 gives the final mapping of logical indices.

Whenever Mg cannot be completely solved by steps (a) through (d), a complete matching algorithm is applied. To this end, a bipartite graph $G(S, T, E)$ is built, in which the logical cells are associated with the source nodes S, the physical PEs are associated with the target nodes T, and an edge $e \in E$ exists from a source to a target node iff there is a mark in the corresponding position of the (simplified) matrix Mg. The complete matching is then a subset of the edge set E such that no two edges share either a source or a target node and that all nodes in S are matched with nodes in T.

Complete matching algorithms have been presented in the literature (see, e.g., [MIC80]) granting a complexity that in our case is $O((N^2)^{1.5}) = O(N^3)$. Such a value is acceptable for end-of production configuration or for off-line reconfiguration.

In [DIS88] the further problem of evaluating the number of buses per channel supporting such an approach has been examined; an upper bound has been found for this number that is related to the dimensions and shapes of the adjacency domains and is independent of the dimensions of the array.

	1	2	3	4	5	6	7	8	9	10	11	12	13	14	15	16	17	18	19	20	21	22	23	24	25
1,1	X	X				X																			
1,2		X	X				X																		
1,3			X					X																	
1,4					X				X																
2,1						X	X				X														
2,2							X	X																	
2,3								X	X																
2,4									X	X															
3,1												X				X									
3,2											S														
3,3																		S							
3,4														X			X								
4,1															X				X						
4,2																					X				
4,3																		X				X			
4,4																		X	X				X		

Figure 12.13.a
Selection of marks in essential rows.

	1	2	3	4	5	6	7	8	9	10	11	12	13	14	15	16	17	18	19	20	21	22	23	24	25
1,1	X	X				X																			
1,2		X	X				X																		
1,3			X					X																	
1,4					X				X																
2,1						X	X				X														
2,2							X	X																	
2,3								X	X																
2,4									X	X															
3,1												X				X									
3,2											S														
3,3																		S							
3,4														X			X								
4,1															X				X						
4,2																					S				
4,3																		X				X			
4,4																		X	X				X		

Figure 12.13.b
Selection of marks in pseudo-essential rows.

	1	2	3	4	5	6	7	8	9	10	11	12	13	14	15	16	17	18	19	20	21	22	23	24	25
1,1	S																								
1,2		S																							
1,3			S																						
1,4				S																					
2,1					S																				
2,2						S																			
2,3							S																		
2,4								S																	
3,1											S														
3,2												S													
3,3																S									
3,4															S										
4,1														S											
4,2																						S			
4,3																							S		
4,4																								S	

Figure 12.14

A possible covering solution.

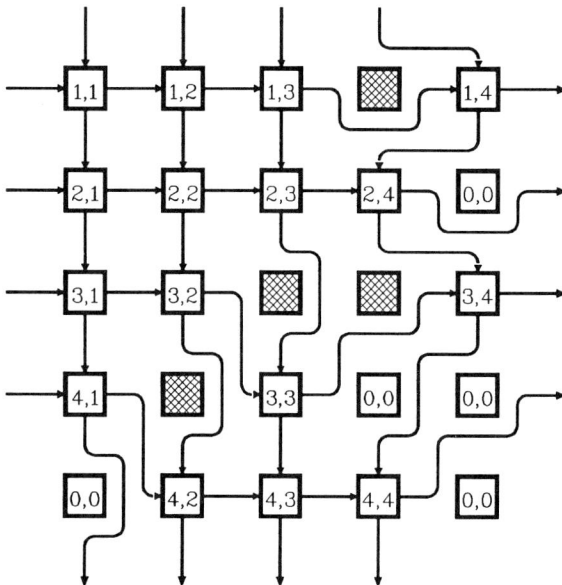

Figure 12.15

Final index mapping.

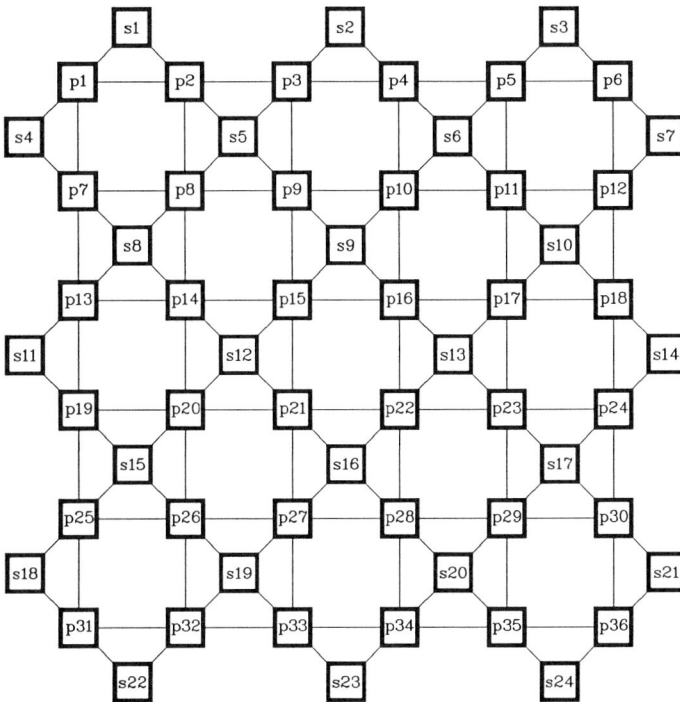

Figure 12.16

Interstitial redundancy. P and S denote respectively nominal PEs and spares.

12.4. The deformation line philosophy

Rather than considering the total mapping of logical onto physical array, an alternative approach consists in creating correspondences between faulty elements and spares so as to achieve the reconfiguration.

A simple solution of this type was presented in [SIN85]. Spares are inserted in the channels between pairs of rows and/or columns, by a criterion that the author defines *interstitial redundancy*. Different distributions lead to different redundancy percentages, ranging from 25 percent upwards; an example corresponding to 50 percent redundancy is given in figure 12.16. Redundancy increases here with N^2, i.e., with the total number of cells in the logical array, as in the local methods; the augmented interconnection network consists of direct links between each spare and all nominal PEs which it can substitute during reconfiguration. Thus, in figure 12.16, each spare can be used to substitute any of the four nominal PEs directly connected with it and each nominal PE can be substituted by one of the two spares directly connected with it.

Reconfiguration is performed by adopting here also a complete matching algorithm. A bipartite graph is derived in which the source nodes correspond to

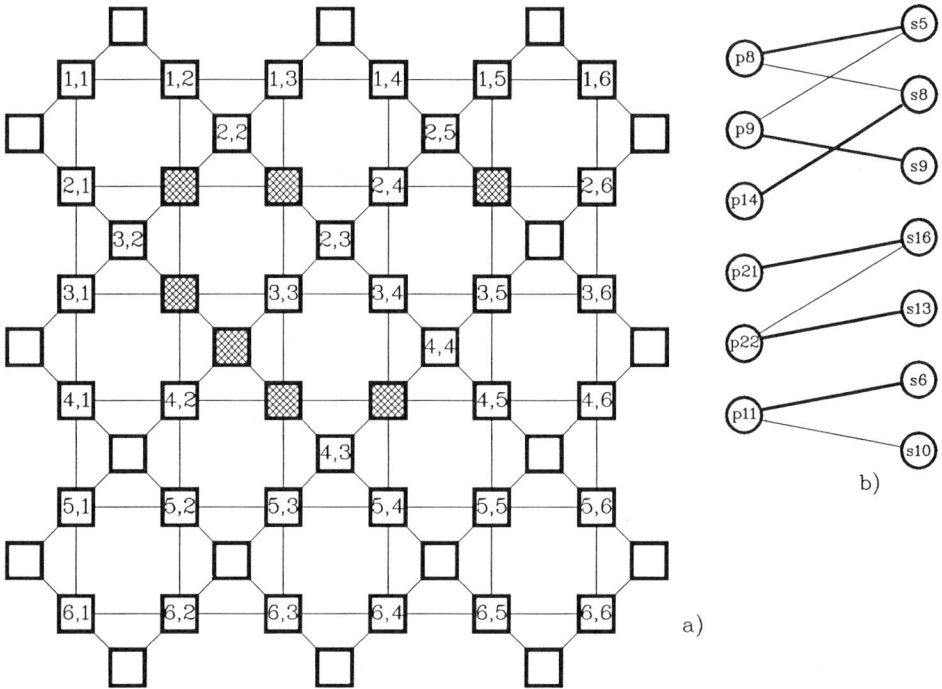

Figure 12.17

a) Fault distribution and index mapping. b) Bipartite graph: nodes on the left and right side represent respectively faulty nominal PEs and fault-free spares. Thick edges denote the solution shown in a).

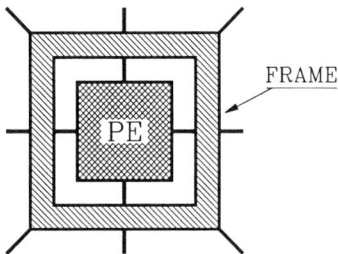

Figure 12.18

The "chip frame" interconnection structure. The PE is connected via nominal inputs and outputs to the frame, which in turn allows also diagonal links.

faulty PEs and the target nodes correspond to the subset of fault-free spares that can substitute them; an edge connects a source to a target node if there is a direct link between the two cells (see in figure 12.17.a a sample fault distribution and in figure 12.17.b the corresponding bipartite graph).

The augmented interconnection network as described in [SIN85] is oversimplified. If the elementary structure as given in figure 12.16 was adopted, the fault model should require that a failed processor keep correctly acting as a connection element, and its switching capacities should be fairly complex since a number of alternative pass-through paths should be allowed. A solution avoiding such unrealistic fault model could be to use a *frame* around each PE, following the philosophy described in [FRA87] (see figure 12.18), where *pins* extending from the frame allow implementation of all possible paths beside the frame. The obvious consequence would be a far greater increase in the complexity of the interconnection network.

A general approach to global reconfiguration is based upon the concept of creating a *correspondence* between each faulty PE and one associated spare. This does not mean that an immediate one-to-one substitution is made, as in [SIN85], but rather that a series of coordinated deformations or *displacements* is used so that finally one particular spare is introduced into operation as a consequence of each fault. Such correspondences may be graphically represented by a set of *deformation lines* creating the correspondence between fault and spare. Thus, given a fault distribution, the solution consists in identification of the corresponding set of deformation lines (if it exists) that creates a unique correspondence between faults and spares.

In order to create such deformation lines, suitable *operators* are defined. These are applied upon matrices that identify the physical array, the fault distribution, and the constraints concerning locality and priority between possible alternatives of the deformation lines, so as to create a final matrix that specifies the mapping of logical indices upon the physical array. Matrices P (physical indices) and LG have already been introduced: $LG(i,j) = i', j'$ denotes the pair of logical indices mapped upon cell (i,j).

The distribution of faults is described by means of a $R_p \times C_p$ matrix F (*fault matrix*) where $F(i,j) \neq 0$ if cell (i,j) is faulty. The constraints due to the algorithm such as locality and priority are described by a set of *deformation matrices DG_r*, all with $R_p \times C_p$ elements. Then, denoting by LG_0 the *initial state* of the logical matrix and by LG_f the *final state* of the logical matrix (corresponding to the complete mapping), reconfiguration results from the application of a sequence of matrix operators $R_1...R_f$ to deformation matrices $DG_1...DG_f$, starting from LG_0, until the final logical matrix LG_f is reached.

Each operator makes use of an associated deformation matrix and of an adjacency domain; the innermost operator of the sequence then operates upon the initial logical matrix and produces an intermediate logical matrix, upon which the second operator acts, and so on, until the final logical matrix is reached. In the most general case (r operators) we have therefore:

$$LG_f = R_{r-1}(DG_{r-1}, R_{r-2}(DG_{r-2}, (...R_0(DG_0, LG_0))))$$

Deformation matrices are built by using the following rules:

(1) for each operator R_S, adopt a coding for the cells in the adjacency domain AD_S and in the inverse adjacency domain IAD_S, starting with code 0 for the central cell (i, j) of AD_S. If $AD_S(k)$ is the code associated with the cell in AD_S having indices $i + x$, $j + y$, the coding of IAD_S must be such that $IAD_S(k)$ be the code of the cell in IAD_S having indices $i - x$, $j - y$. We will also use the simplified notation:

$$LG_S(k) = LG_S(i - x, j - y)$$

$$DG_S(k) = DG_S(i + x, j + y)$$

(2) each entry in $DG_S(i, j)$ then denotes the cell k in the IAD_S of (i, j) such that entry $LG_S(k)$ consists of the pair of logical indices to be transferred into $LG_{S+1}(i, j)$;

Correct reconfiguration results in a distribution of entries in LG_f such that:

- each pair of logical indices (corresponding to the target array) appears once and only once in the final LG;

- there are as many null entries as there are spares;

- all entries corresponding to positions either of faulty PEs or of unused spares are null.

While in principle sequences of any length are possible, in practice we consider only sequences of one or two operators (note that even single-operator algorithms may be considerably complex, since more complex deformation matrices can be defined). If multiple-operator algorithms are adopted, deformation lines will be created as the *composition* of line segments created by each operator; the *complete* line must satisfy the two following conditions:

(1) target points of the deformation lines must coincide with spare cells;

(2) origin points of the deformation lines must coincide with faulty cells;

Referring to a two-operator algorithm, the complete deformation line results from the composition of a first segment starting from the faulty cell c_1 and of a second line segment ending on spare cell c_S. The effect of the first (innermost) operator is to transfer logical indices as follows:

$$\begin{cases} c_{S-1} \mapsto c_S \\ c_{S-2} \mapsto c_{S-1} \\ \dots \\ \dots \\ c_v \mapsto c_{v+1} \\ 0, 0 \mapsto c_v \end{cases} \qquad [12.1.a]$$

so that c_v acts as a *virtual fault*. Subsequently, the second operator effects the following transfer of indices:

$$
\begin{cases}
c_{v-1} \mapsto c_v \\
c_{v-2} \mapsto c_{v-1} \\
\cdots \\
\cdots \\
c_1 \mapsto c_2 \\
0,0 \mapsto c_1
\end{cases}
\qquad [12.1.b]
$$

so that here c_v acts as a *virtual spare*.

A third condition for the composite deformation line is then:

(3) the target or/and origin points of intermediate segments will be virtual faults or spares.

A basic rule is that no other cell in the second line segment, besides c_1, can be faulty: otherwise, correct reconfiguration would be impossible. On the contrary, the first segment can contain faulty cells without impairing the correctness of the reconfiguration. Moreover, two *different* deformation lines may intersect (i.e., share one PE) or even share full segments, provided the shared cell or segment belong to different operators.

Let us consider a few selected algorithms, exemplifying some particular points.

12.4.1. Algorithm 1

This simplest case is quite similar to that in [HWA86], discussed in chapter 10; we examine it here because it can be described in terms of general operators and deformation lines. Its definition is as follows:

- one column of spares is introduced;

- the adjacency domain consists only of cells (i, j) and $(i, j+1)$; reconfiguration is performed *rowwise*, bypassing a faulty PE and introducing in operation the spare cell on the same row (see figure 12.19). The corresponding formal definition is

$$
LG_1(i,j) = R_0(DG_0(i,j), LG_0(i,j))
$$

i.e., a single operator performs the index mapping through use of a deformation matrix DG_0 defined as follows:

- if $DG_0(i,j) = 0$, and one (only one) value of k exists such that $DG_0(k) = k$, then $LG_1(i,j) = 0,0$: that is, no pair of logical indices is transferred into $LG_1(i,j)$. This happens if cell (i,j) is either faulty or an unused spare;

- no entry $LG_0(i,j)$ is duplicated into two entries of $LG_1(i,j)$ nor any pair of indices appearing in LG_0 disappears from the final logical matrix LG_1.

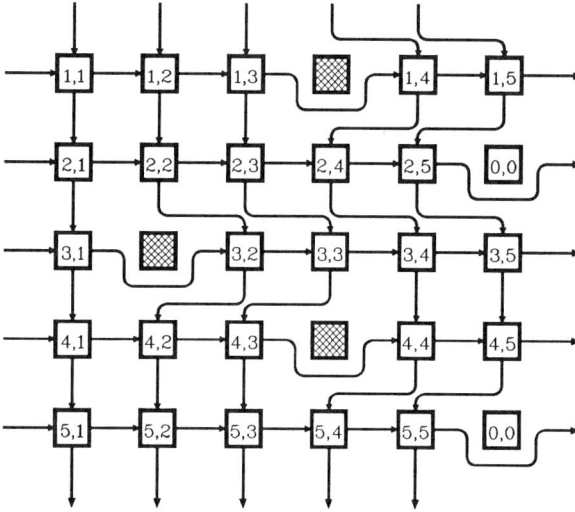

Figure 12.19

Example of single-operator reconfiguration (Algorithm 1).

For the example in figure 12.19, the various matrices are given in equations [12.2], [12.3] and [12.4].

$$LG = \begin{vmatrix} 1,1 & 1,2 & 1,3 & 1,4 & 1,5 & 0 \\ 2,1 & 2,2 & 2,3 & 2,4 & 2,5 & 0 \\ 3,1 & 3,2 & 3,3 & 3,4 & 3,5 & 0 \\ 4,1 & 4,2 & 4,3 & 4,4 & 4,5 & 0 \\ 5,1 & 5,2 & 5,3 & 5,4 & 5,5 & 0 \end{vmatrix} \qquad [12.2]$$

$$DG = \begin{vmatrix} 0 & 0 & 0 & 0 & 1 & 1 \\ 0 & 0 & 0 & 0 & 0 & 0 \\ 0 & 0 & 1 & 1 & 1 & 1 \\ 0 & 0 & 0 & 0 & 1 & 1 \\ 0 & 0 & 0 & 0 & 0 & 0 \end{vmatrix} \qquad [12.3]$$

$$LG_1 = \begin{vmatrix} 1,1 & 1,2 & 1,3 & 0 & 1,4 & 1,5 \\ 2,1 & 2,2 & 2,3 & 2,4 & 2,5 & 0 \\ 3,1 & 0 & 3,2 & 3,3 & 3,4 & 3,5 \\ 4,1 & 4,2 & 4,3 & 0 & 4,4 & 4,5 \\ 5,1 & 5,2 & 5,3 & 5,4 & 5,5 & 0 \end{vmatrix} \qquad [12.4]$$

The operation of the reconfiguration operator R_0 is made visible through a *deformation line* defined as an ordered sequence of index pairs $(i_1, i_1), ..., (i_n, j_n)$ such that:

(1) for any k, $1 \leq k \leq n$ $LG_0(i_k, j_k)$ is transferred into $LG_1(i_k + 1, j_k + 1)$;

(2) no pair of logical indices is transferred from LG_0 into $LG_1(i_1, j_1)$: rather, a null entry is written in that position: $LG_1(i_n, j_n) = 0, 0$

(3) $LG_0(i_n, j_n)$ *disappears* to allow correct reconfiguration; thus, $LG_0(i_n, j_n)$ must of necessity be a null entry (in our case, a spare).

In the above example, there are three horizontal deformation lines each starting on a fault and reaching the spare on the same row as the fault (in the last column to the right). Given the previous conditions for correct definition of deformation operator and deformation matrix, it follows that all deformation lines thus created will be disjoint (i.e., no physical cell may belong to two or more deformation lines).

This simple, single-operator algorithm, has rather low efficiency in terms of probability of survival (presence of two faults in the same row leads to fatal failure) and of utilization of spares (a 20×20 target array with 20 spare cells has a probability of 50 percent to survive up to 6.5 faults only, as shown in curve A in figure 12.25). There are two ways to increase performances:

(a) extend the *AD* for single-operator algorithms;

(b) adopt two-operator algorithms.

Let us consider alternative (a).

12.4.2. Algorithm 2

The *AD* is now defined as in figure 12.20; a spare row and a spare column are added to the basic array, so that a physical $(N + 1) \times (N + 1)$ array is available. Correct reconfiguration for any given set of f faults is achieved whenever a set of f deformation lines can be determined, characterized as follows:

- each line originates from a faulty PE and ends on a spare one;

- according to the *AD* used, each line consists only of rightward-oriented horizontal segments and of downward-oriented vertical ones;

- all lines are disjoint, due to the presence of one operator only.

The example in figure 12.21 shows, for the given fault pattern, the set of deformation lines built according to the above rules and the related index mapping. Deformation lines correspond to the *DG* matrix in equation [12.5].

$$DG = \begin{vmatrix} 0 & 0 & 0 & 0 & 0 & 0 & 0 \\ 0 & 0 & 0 & 0 & 1 & 1 & 0 \\ 0 & 1 & 0 & 1 & 1 & 0 & 1 \\ 0 & 2 & 1 & 1 & 2 & 1 & 1 \\ 0 & 0 & 0 & 2 & 1 & 1 & 0 \\ 0 & 0 & 2 & 1 & 0 & 2 & 1 \\ 0 & 0 & 0 & 2 & 0 & 1 & 0 \end{vmatrix} \qquad [12.5]$$

The final connections among fault-free operating cells are given in figure 12.22.

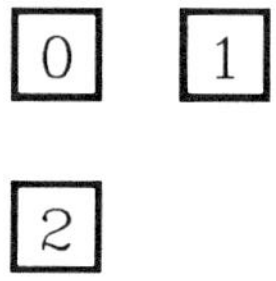

| 0 | 1 |

| 2 |

Figure 12.20
Adjacency domain coding for Algorithm 2.

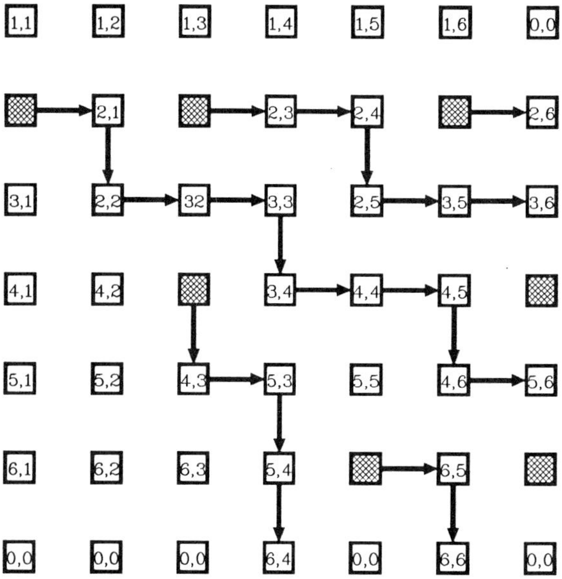

| 1,1 | 1,2 | 1,3 | 1,4 | 1,5 | 1,6 | 0,0 |

Figure 12.21
Index mapping by Algorithm 2.

A possible algorithm for the creation of deformation lines — which permits also to design relatively simple circuits for on-chip computation of the actual setting of connections among cells — can be defined as follows:

(1) The array is scanned to look for faulty cells, for increasing values of row index and, within a row, for increasing values of column index;

(2) for each faulty cell (i,j) thus found, an attempt is first made to create a horizontal deformation line ; if this line cannot be completed, because of interference with another faulty cell or with a fault-free cell already inserted in a previous deformation line, the line is continued with a vertical segment toward row $i + 1$;

(3) if the vertical continuation is also impossible, backtracking of one position is performed to explore an alternative possibility. If subsequent backtracking lead to (i,j), no successful deformation line can be built and fatal failure is declared;

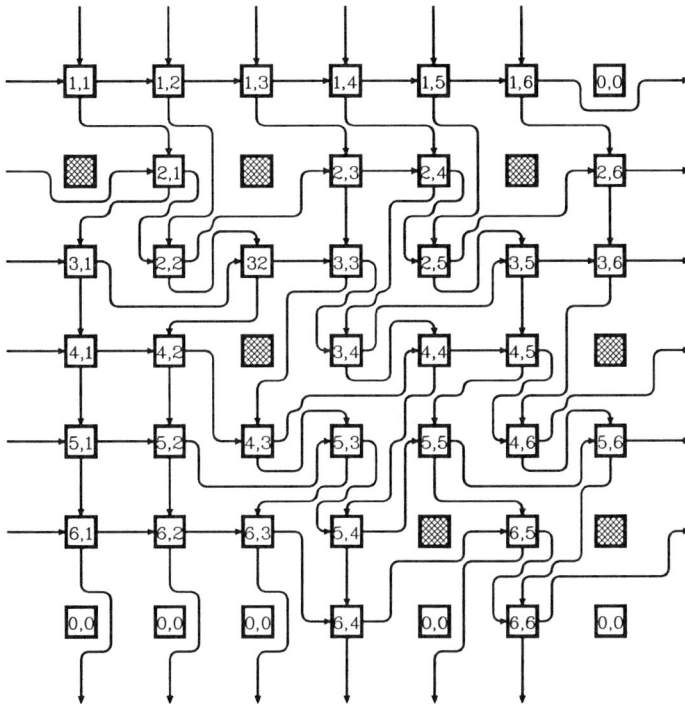

Figure 12.22
Complete reconfiguration by Algorithm 2.

(4) if the deformation line reaches a spare cell, it is completed and the procedure is repeated for another fault (if any such exists);

(5) the algorithm successfully terminates when f deformation lines have been completed for the f faults.

This algorithm has been adopted for the generation of figures 12.21 and 12.22. It is *not* an optimum algorithm, since it can be verified that for some fault patterns the solution may not be found, even though a solution exists. Still, probability of survival is quite good (see figure 12.23, curve a).

Algorithm 2 is one of a family of algorithms, each corresponding to a specific AD, of which algorithm 1 is the simplest one. A complete treatment of algorithm 2, as well as a description of a more complex algorithm with an adjacency domain consisting of (i,j), $(i,j+1)$, $(i+1,j)$, $(i+1,j+1)$ can be found in [SAM87]. A larger AD obviously leads to a more complex algorithm, and hence to a more area-consuming supporting hardware, as well as to better probability of survival (see curve b in figure 12.23).

The second alternative will now be considered, i.e., the use of two-operator sequences. Algorithms of increasing efficiency and complexity will be analyzed.

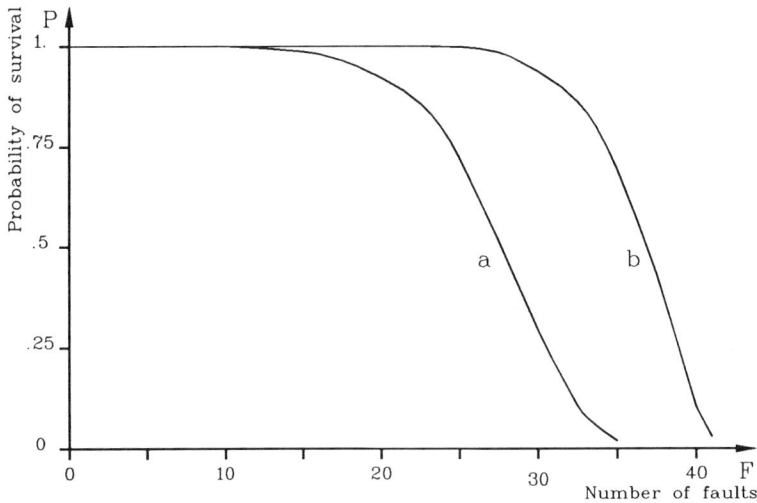

Figure 12.23

Probability of survival vs. number of faults for target 20×20 array. Curve a: Algorithm 2; curve b: single-operator algorithm with larger AD.

12.4.3. Algorithm 3: "simple fault-stealing, fixed choice"

This algorithm aims at overcoming the main drawback of algorithm 1. Still considering an array with just a spare column, we allow *stealing* of spare cells from adjacent rows whenever two (or more) faults are present in a single row. Let i be such a row: the rightmost fault invokes reconfiguration on the right, along the same row, following the previous rule: the other faults transfer their logical indices to cells in the same column, in row $i + 1$, *stealing* these positions that are marked as *pseudo-faults* since they will in turn require reconfiguration following the same rules.

The adjacency domain used are quite simple. For the horizontal deformation operator, it is $ADh : \{(i,j), (i, j+1)\}$ and the associated values are respectively $k = 0$ and $k = 1$. For the vertical deformation operator, it is $ADv : \{(i,j), (i+1,j)\}$ and the associated values are respectively $k = 0$ and $k = 1$. Referring to the fault distribution in figure 12.24, the two deformation matrices are given in equations [12.6] and [12.7].

$$DG_0 = \begin{vmatrix} 0 & 0 & 0 & 0 & 0 & 0 \\ 0 & 0 & 0 & 0 & 1 & 1 \\ 0 & 0 & 1 & 1 & 1 & 1 \\ 0 & 0 & 0 & 0 & 0 & 0 \\ 0 & 0 & 0 & 0 & 0 & 0 \end{vmatrix} \qquad [12.6]$$

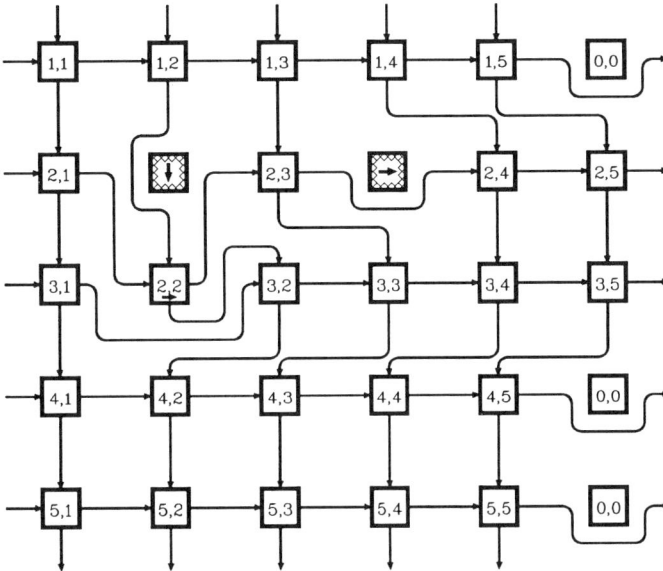

Figure 12.24
Example of reconfiguration by Algorithm 3 ("simple fault-stealing, fixed choice").

$$DG_1 = \begin{vmatrix} 0 & 0 & 0 & 0 & 0 & 0 \\ 0 & 0 & 0 & 0 & 0 & 0 \\ 0 & 1 & 0 & 0 & 0 & 0 \\ 0 & 0 & 0 & 0 & 0 & 0 \\ 0 & 0 & 0 & 0 & 0 & 0 \end{vmatrix} \qquad [12.7]$$

Deformation line segments created by the two operators are given in figure 12.24, representing the final reconfiguration (arrows mark the stealing and reconfiguration procedure). The cell with physical coordinates $(2,4)$ invokes reconfiguration along the row, using the spare at the right border; the cell with physical coordinates $(2,2)$, unable to invoke rowwise reconfiguration, *steals* the position of physical cell $(3,2)$, that in turn invokes rowwise reconfiguration.

Reconfiguration becomes impossible (*fatal failure* is reached) whenever one of the two following conditions is met:

(a) at least one cell requested for *stealing* is in turn faulty;

(b) two or more faulty or stolen cells are present in row N.

Restrictions created by this second condition are immediately overcome simply by adding one spare row (row $N+1$). Curve C in figure 12.25 gives the probability of survival in the second solution, obtained by simulation for a 20×20 target array (41 spare cells).

A more complex example of reconfiguration is given in figure 12.26.

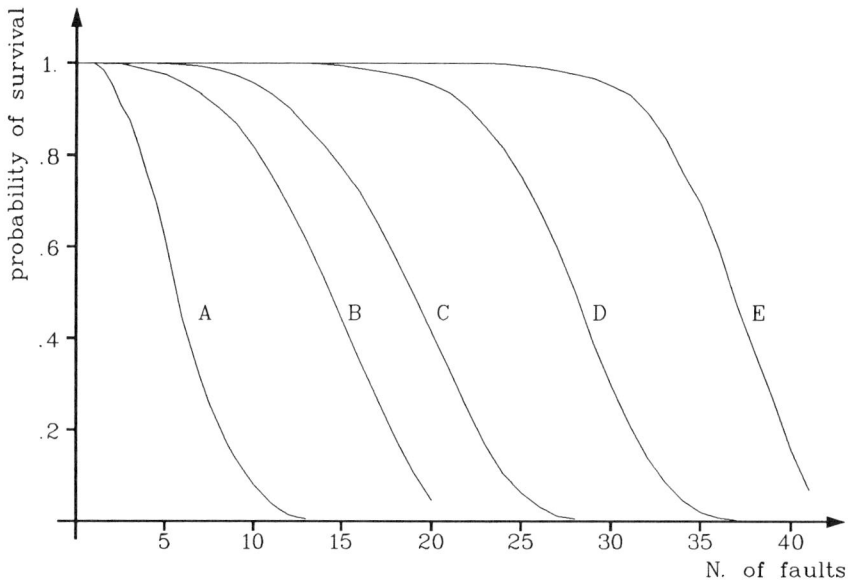

Figure 12.25

Probability of survival for 20×20 target array. A) Algorithm 1. B) Algorithm 3 one spare column.
C) Algorithm 3 one spare column and one spare row. D) Algorithm 4. E) Algorithm 5.

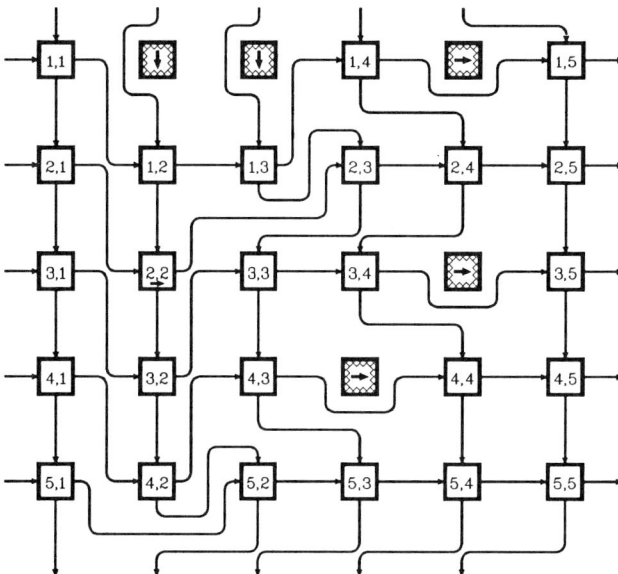

Figure 12.26

Further application of Algorithm 3.

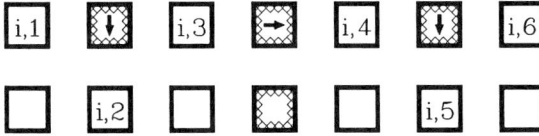

Figure 12.27
Algorithm 4 ("simple fault-stealing, variable choice"). Reconfiguration policy for a case not solvable by Algorithm 3.

12.4.4. Algorithm 4: "simple fault-stealing, variable choice"

Pre-determining which of the various faulty cells in one row is allowed to invoke reconfiguration along the row introduces one further restriction (appearing as a priority factor) that severely limits the possibility of reconfiguration. To overcome this limitation, we modify the previous algorithm as follows: reconfiguration along the row is invoked by the faulty or stolen cell (if any) that could not steal the cell in the lower row; otherwise, the rules of algorithm 3 are adopted.

The example in figure 12.27 shows an application of this modified fault-stealing. There, algorithm 3 would lead to fatal failure (caused by the cell with physical coordinates $(i, 4)$, that could not make use of stealing from row $i + 1$ since physical cell $(i + 1, 4)$ is faulty), while algorithm 4 allows reconfiguration by having physical cell $(i, 4)$ invoke reconfiguration along the row while all other faults in row i invoke stealing from row $i + 1$. Probability of survival is immediately higher (see curve D in figure 12.25). Creation of deformation matrices in this new instance follows the rules given below:

(1) rows are scanned for increasing row index. If in row i $(1 \leq i < N)$ there are s faulty or stolen cells such that:

  - considering any faulty or stolen cell in row i and considering cells in the same columns in row $i + 1$, at most one of these last-named ones is faulty;
  - j'' is the column index of such faulty cell in row $i + 1$;

  then:

  - cell (i, j'') invokes reconfiguration along the row, and $DG_0(i, j) = 1$ for all $j > j''$;
  - all other faulty or stolen cells in row i steal positions of cells in the same columns in row $i + 1$, and it is $DG_1(i + 1, j) = 1$ for all such column indices.

(2) Fatal failure is reached whenever at least two faulty or stolen cells in row i find faulty cells in the corresponding columns of row $i + 1$.

12.4.5. Algorithm 5: "complex fault-stealing, variable choice"

Algorithm 4 can be further modified to allow greater flexibility of choice as far as *stealing* is concerned. While maintaining the locality conditions imposed at

Figure 12.28
Algorithm 5 ("complex fault-stealing, variable choice"). Reconfiguration policy for a case not solvable by Algorithms 3 and 4.

the beginning, we can allow stealing not only in the *same* column, but also in the column immediately to the right of the given one. Thus, the horizontal deformation uses the same ADh as before, while the vertical deformation uses ADv : $\{(i,j),(i+1,j),(i+1,j+1)\}$, the associated values beeing, respectively, $k = 0$, $k = 1$, $k = 2$. By introducing this further alternative in algorithm 4, the following definition is arrived at (upon which more complex deformation matrices will be built):

(1) each row (for increasing row indices) is scanned from left to right: let (i,k) be the leftmost faulty or stolen cell;

(2) given any cell (i,h), $(h > k)$, faulty or stolen, it invokes stealing from $(i+1,h)$ if $(i+1,h)$ is not in turn faulty or stolen: otherwise, it attempts stealing from $(i+1,h+1)$. If this second attempt also fails, (i,h) invokes reconfiguration along the row, and all other faulty or stolen cells in row i (including (i,k)) invoke stealing from row $i+1$.

(3) Fatal failure occurs if and only if two (or more) cells in the same row invoke reconfiguration along the row.

See the example in figure 12.28. The fault in physical position $(i,2)$ cannot invoke reconfiguration on $(i+1,2)$, which is faulty; it therefore utilizes the physical cell $(i+1,3)$ for reconfiguration. The fault $(i,3)$ cannot in turn invoke reconfiguration on $(i+1,3)$ because it, although fault-free, has been "stolen," i.e., previously used for reconfiguration; thus, it utilizes $(i+1,4)$. This in turn induces $(i,4)$ to invoke horizontal reconfiguration (since $(i+1,5)$ is faulty) and $(i,5)$ to invoke reconfiguration on $(i+1,6)$ (All indices are *physical*).

Figure 12.29 gives an example of reconfiguration: the two deformation matrices are given by equations [12.11] and [12.12], while figure 12.30 gives the deformation lines (light segments relate to the first operator, bold ones to the second operator).

$$
DG_1 = \begin{vmatrix}
0 & 0 & 1 & 1 & 1 & 1 \\
0 & 0 & 0 & 1 & 1 & 1 \\
0 & 0 & 0 & 0 & 0 & 0 \\
0 & 0 & 0 & 0 & 0 & 1 \\
0 & 0 & 0 & 0 & 1 & 1 \\
0 & 0 & 0 & 0 & 0 & 0
\end{vmatrix}
\qquad [12.8]
$$

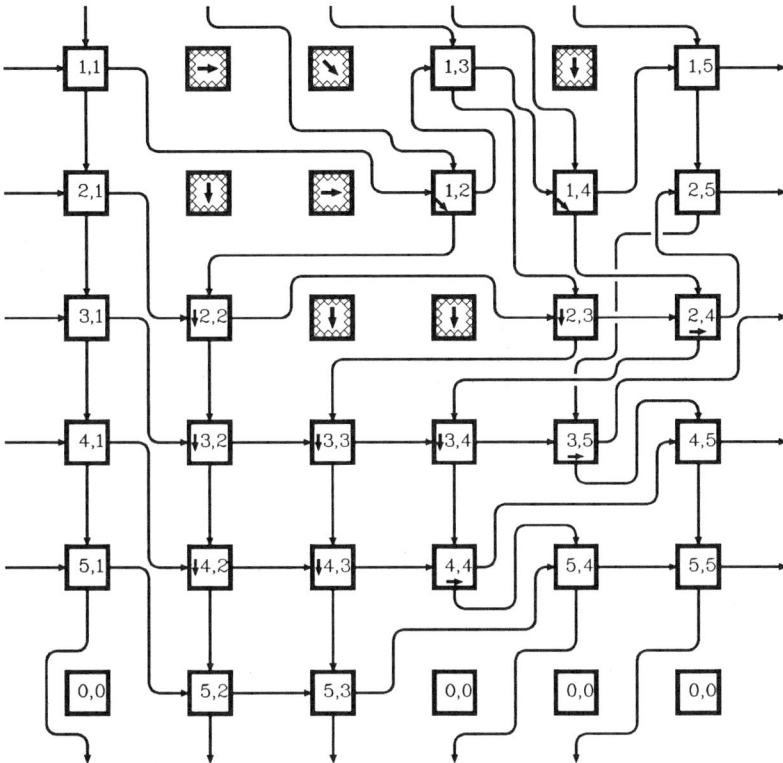

Figure 12.29
Example of reconfiguration by Algorithm 5.

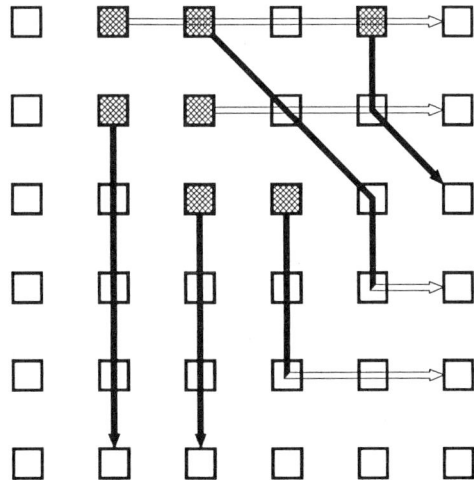

Figure 12.30
Deformation lines for the example in figure 12.29.

$$DG_2 = \begin{vmatrix} 0 & 0 & 0 & 0 & 0 & 0 \\ 0 & 0 & 0 & 2 & 1 & 0 \\ 0 & 1 & 0 & 0 & 2 & 2 \\ 0 & 1 & 1 & 1 & 1 & 0 \\ 0 & 1 & 1 & 1 & 0 & 0 \\ 0 & 1 & 1 & 0 & 0 & 0 \end{vmatrix} \qquad [12.9]$$

While it is quite complex, algorithm 5 gives a 50 percent probability of survival to a number of faults equivalent to 90 percent of available spares (see curve E in figure 12.25).

An interesting characteristic of all the algorithms represented by means of the deformation operator sequence technique is that they can be immediately mapped onto an *interconnection network* supporting reconfiguration, simply on the basis of adjacency domain and of operator sequence (that is, without referring to the actual algorithm as described by the deformation matrices). In fact, each single operator translates into a simple "shell" of multiplexers around the individual cell of the physical array, so that

- the cell may receive data from all its possible (alternative) "*input adjacents*";
- the cell may forward results to all its possible alternative "*output adjacents*."

Consider for example the single-shell structure in figure 12.31, corresponding to one-operator algorithms with a two-cells adjacency domain (algorithm 1): the large dots marked on this structure correspond to the *virtual* inputs and outputs of *logical* cell (i, j). The two input multiplexers to cell (i, j) provide it with the correct input data, while the results of operations performed by (i, j) are deviated to the correct output points (or excluded if cell (i, j) is faulty) through the two output demultiplexers. Data provided by the two alternative adjacent cells foreseen by the algorithm are *OR*ed: if DG_0 is correct, only one input to the OR will be active at any time, so that signals present at the terminals of the shell around a PE are actually the *correct data corresponding to the array location* (i.e., to the physical coordinates), whatever the deformation performed. In the example of figure 12.31 thick lines denote active links when $DG_0(i, j + 1) = 1$.

In the same way, figure 12.32 gives the structure for a single operator with AD consisting of (i, j), $(i, j + 1)$, $(i + 1, j)$; such a structure is particularly useful for algorithm 2. Similarly, when two operators are composed in a sequence the interconnection network can, in principle, be made up by composing two nested layers of *shells* around each PE, again with this same basic multiplexers structure, as is done in figure 12.33 where only the structure relating to the elementary cell is shown. The innermost layer becomes, to all external purposes, equivalent to a fault-free PE in the given physical position. A second layer of shell can therefore be superimposed around it with the same effect as the previous one, i.e., implementing the second operator of the sequence (the outer shell corresponds to the first deformation, the inner one to the second operator application).

For the structure in figure 12.33, the first operator (corresponding to the inner shell) was assumed to cause a horizontal deformation $(AD = (i, j), (i, j + 1))$;

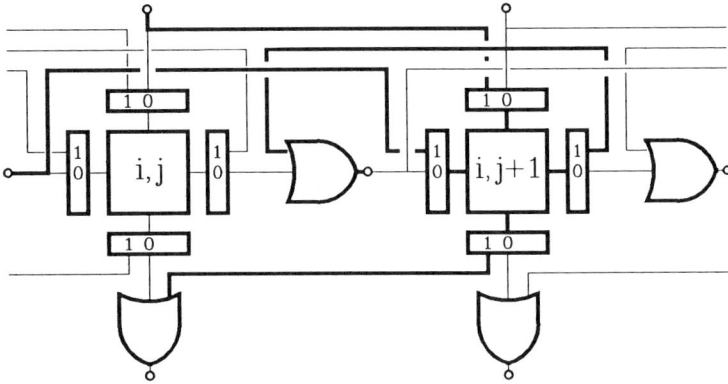

Figure 12.31
Single-shell structure, multiplexer-based solution for Algorithm 1.

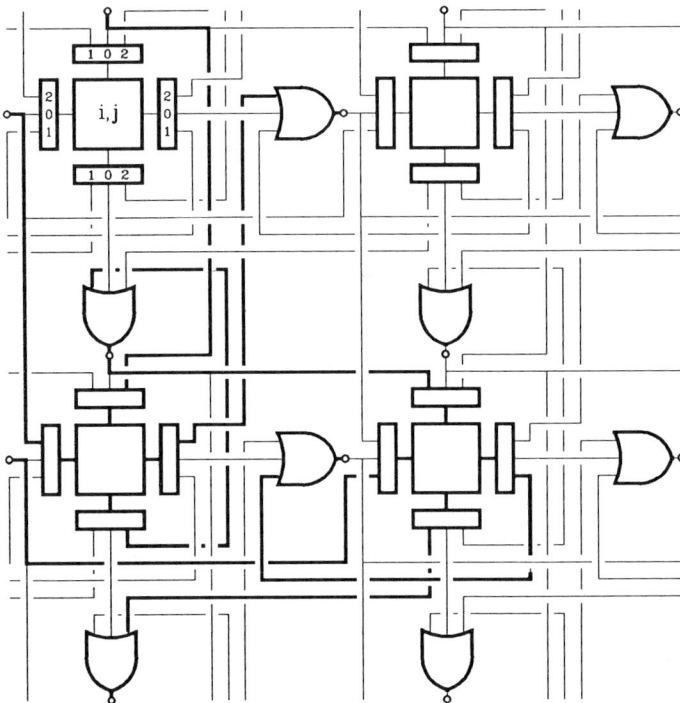

Figure 12.32
Shell structure for Algorithm 2.

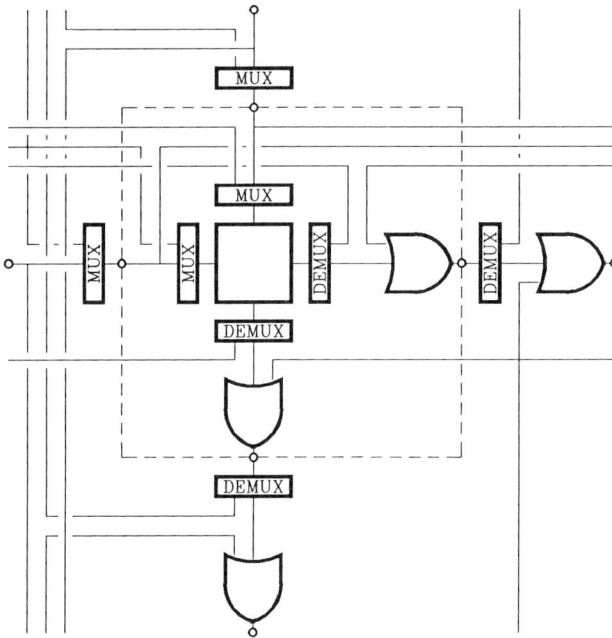

Figure 12.33
Two-layer shell for two-operators algorithms such as 3 and 4.

therefore, the structure can be used for algorithms such as the simple fault stealing (fixed or variable choice) as well as for algorithm 6.

The actual algorithm implemented by the operator sequence is mirrored by control signals for multiplexers: in figures 12.31 and 12.32 such signals correspond to the specific $DG_0(i,j)$, while in figure 12.33, inner and outer multiplexers are controlled, respectively, by the values in $DG_0(i,j)$ and $DG_1(i,j)$. Generation of such signals is the algorithm-specific part of the network. Circuits providing such signals, on the basis of the fault information present in the array, will have varying complexity. While that required for algorithm 1 is elementary, that corresponding to algorithm 4 will be quite complex, — so much so that if PEs are relatively small and simple it might be better to generate the control signals externally, e.g., by firmware on a specialized *host controller*.

It should be noted further that the *multiple layer* multiplexer structure illustrated above is simply derived for any given reconfiguration algorithm, but is far from optimized from the point of view of silicon area requirements (and also of communication delays, considering that four multiplexers are found for a two-operator algorithm on each PE-to-PE connection link).

A first compaction can be immediately obtained as follows: for the vertical (horizontal) input to cell (i,j) all alternative paths coming from output of different cells are followed. The *input shell* structure to (i,j) is substituted by one

Table 12.1
Connectivity table for algorithm 5

Horizontal	input	Vertical	input
$(i-1, j-3)$	$(i-1, j-2)$	$(i-2, j-2)$	$(i-2, j-1)$
$(i-1, j-1)$	$(i, j-3)$	$(i-2, j)$	$(i-2, j+1)$
$(i, j-2)$	$(i, j-1)$	$(i-1, j-2)$	$(i-1, j-1)$
$(i+1, j-2)$	$(i+1, j-1)$	$(i-1, j)$	$(i-1, j+1)$
$(i+1, j)$		$(i-1, j+2)$	$(i, j-1)$
		$(i, j+1)$	$(i, j+2)$

multiplexer having as many inputs as there are different paths, controlled by a function derived from the control signals for the multiplexers and demultiplexers found on the various original paths.

The backtracking procedure leads to creation of a *tree* of alternative interconnections. (See [SAM86a].) For algorithm 5, the set of cells whose outputs are connected to inputs of (i, j) is listed in table 12.1, and the corresponding schematic layout is given in figure 12.34. The list of all possible connections and the corresponding schematic layouts for all algorithms described here are given in [SAM87].

Restrictions introduced by specific algorithms on the set of possible reconfigurations (even within a predetermined adjacency domain) allow further simplifications of the interconnection network. An alternative solution — considered for many reconfiguration techniques — makes use of switched buses, i.e., a number of bidirectional buses and of 2×2 switches controlling them are substituted for the multiplexed direct links. In this case, it is not possible to identify a basic correspondence between operator sequence and elementary structure of interconnection network. Thus, the most complex algorithm previously discussed (algorithm 5) requires three buses in each interconnection channel and eight switches (see the structure of the interconnection network and the function of switches in figure 12.35), while the simpler algorithms require reduced sets of switches as well as reduced channel width.

As seen in figure 12.35, one vertical and two horizontal buses (together with four switches) support transmission of vertical signals, while the remaining interconnection network supports transmission of horizontal signals. Figures 12.36.a and 12.36.b show (for vertical and horizontal data) the sets of cells that might send data to (i, j) (as derived from table 12.1). Active bus segments and switch settings are also shown: to minimize the number of buses and at the same time avoid ambiguity in bus assignment, alternative paths had to be proposed for vertical signals.

Implementation of the example in figure 12.29 by this switched bus structure

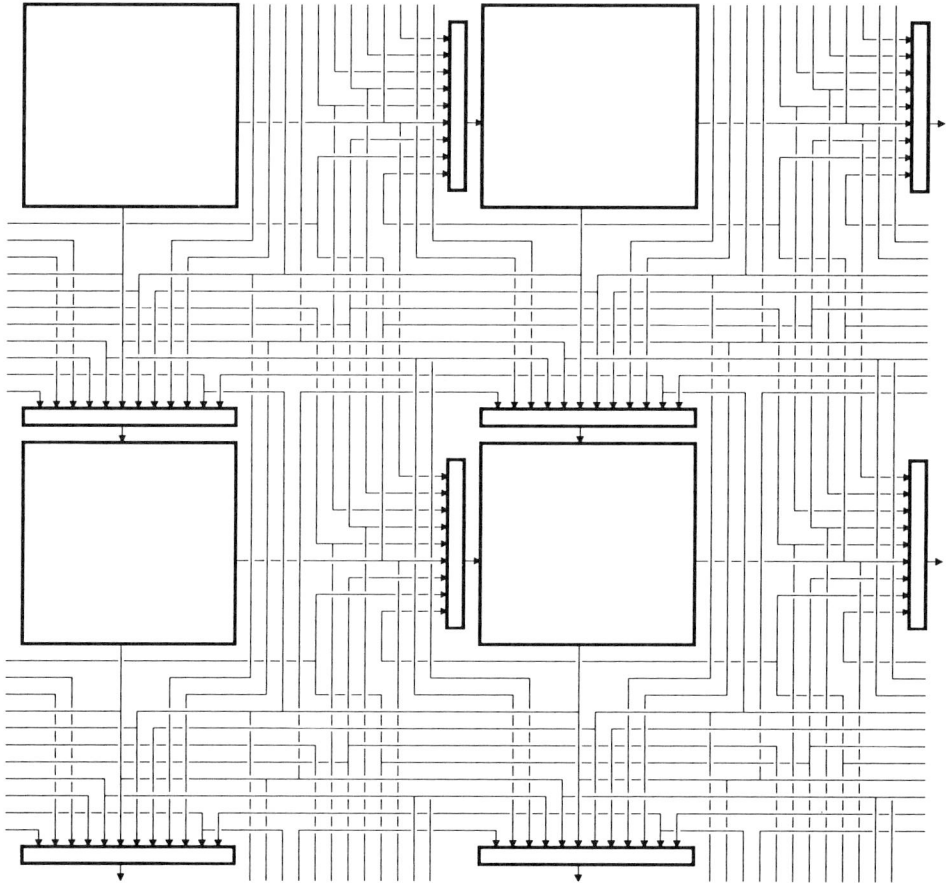

Figure 12.34
One-layer multiplexer structure for Algorithm 5.

leads to the activation of connection segments as shown in figure 12.37.

An even simpler switched-bus structure supports algorithms 3 and 4 (see figure 12.38). In this case, there is no need to devise alternative paths to avoid conflicts on buses. (An example of reconfiguration is given in figure 12.39.)

It would be interesting to compare the results granted by the algorithms described here — guaranteeing channel width and interconnection length bound by constants rather than by a function of array dimensions, but at the same time accepting a spares utilization lower than 100 percent — with other algorithms that are characterized by the same characteristics but that make use of *local* rather than *global* reconfiguration to achieve such results. The local technique described in [HED84] is particularly appropriate for comparison. In that study (chapter 10), the number of spares increases with N^2 (while in all techniques described in

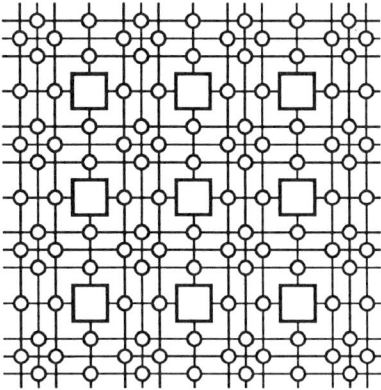

Figure 12.35
Switched-bus structure supporting Algorithm 5.

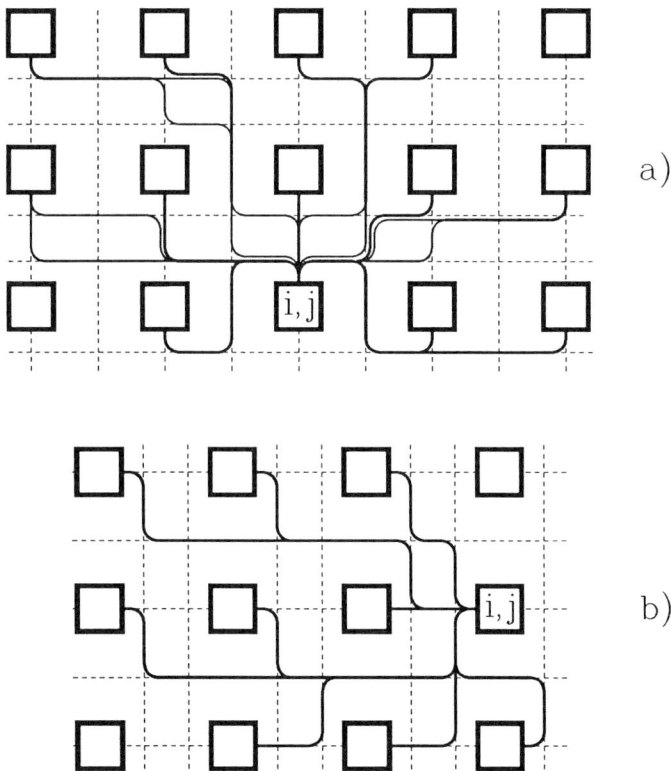

Figure 12.36
Set of possible connections between vertical (a) and horizontal (b) neighbors consequent to Algorithm 5: thin segments denote alternate paths.

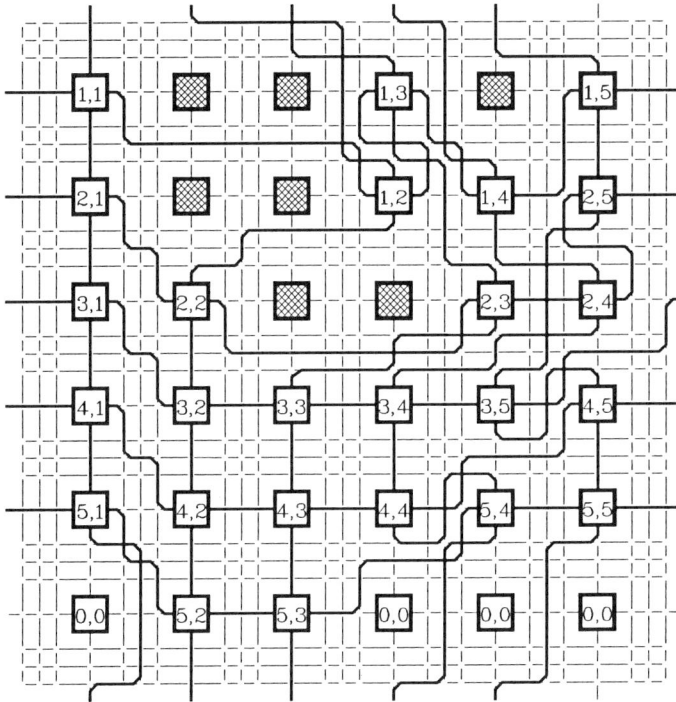

Figure 12.37
Implementation of the example in figure 12.29 by the switched-bus network.

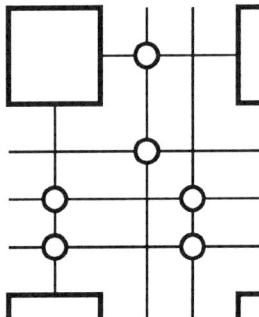

Figure 12.38
Switched-bus network supporting algorithms 3 and 4.

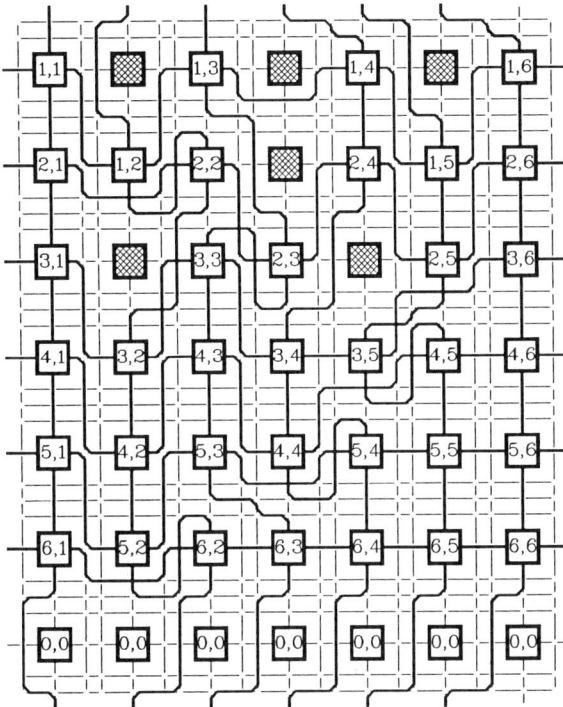

Figure 12.39
Example of reconfiguration by Algorithm 4 supported by switched-bus network.

this chapter it increases with N only), but at the same time the interconnection network is simpler than that required by algorithm 5. Probability of survival for the two alternatives, and for a 20×20 target array, is plotted in figure 12.40 (curve a for algorithm 5, curve b for Hedlund's local reconfiguration algorithm).

Thus, the reliability granted by algorithm 5 is higher than that granted by Hedlund's if the initial yield is high (below 35 faults). The attractiveness of the approach is even greater if normalization against silicon area is introduced — as might be expected, considering that greater area involves also higher probability of failures and lower overall reliability (see curve c in figure 12.40, where this normalization factor is taken into account). This factor includes the ratio between the number of physical cells required, in the two different instances, to implement the same $N \times N$ target array. Thus, curve c in figure 12.40 has been plotted against the normalized number of faults F', where:

$$F' = F \cdot (N + 1)^2 / 1.5N$$

This comparison is based on the assumption that the fault-stealing algorithms will also be host-driven rather than controlled on-chip, just as the local reconfiguration algorithm is, so as to make the evaluation for identical conditions. Otherwise, the

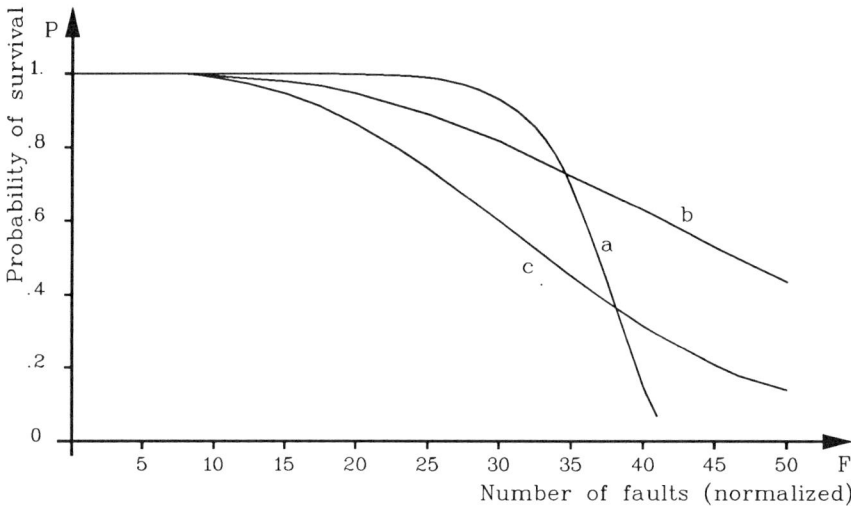

Figure 12.40

Probability of survival for a 20×20 target array. Curve a: Algorithm 5. Curve b: Hedlund's algorithm. Curve c: Hedlund's algorithm, results normalized to account for relative silicon requirements.

area required by on-chip circuits controlling reconfiguration would cause distorted results in the comparison.

12.4.6. Algorithm 6

Another algorithm, still based on the concept of *deformation line* and of matrix operators, and granting optimality as far as probability of survival and spares utilization, involves host-driven reconfiguration as a consequence of its complexity [LOM87]. This algorithm again requires one spare row and one spare column, and it uses two operators: one introducing a rightward displacement, with higher priority, and one introducing a downward displacement, with lower priority. Adjacency domains, for the two operators, are very restricted (identical to the ones defined for algorithms 3 and 4). For the horizontal operator, the adjacency domain of (i, j) consists of (i, j) and $(i, j + 1)$, while for the vertical operator the adjacency domain consist of (i, j) and $(i + 1, j)$. The same operators already used for algorithms 3 and 4 are adopted here, in the same order.

Deformation lines can create here *two* alternative associations between a faulty PE (i, j) and a spare:

- a column-wise association, by which (i, j) is associated with a spare $(N + 1, j)$;

- a row-wise association, by which (i, j) is associated with spare $(h, N + 1)$, $i \leq h \leq N$.

The choice of the association between a faulty PE and a spare completely defines the specific deformation line. While column-wise association creates one vertical deformation line from the faulty PE (i, j) to spare $(N + 1, j)$ on the same column, row-wise association leads to two possibilities:

- if the spare is $(i, N + 1)$, the deformation line consists of one horizontal segment;

- if the spare is $(h, N + 1)$, with $h > i$, the deformation line consists of a horizontal segment from (h, j) (a *virtual fault*) to $(h, N + 1)$, corresponding to the first operator, and of a vertical segment from (i, j) to (h, j) (now a *virtual spare*) corresponding to the second operator.

Note that such correspondence depends on the ADs chosen: it would not be valid if operators such as the ones used in algorithm 5 were adopted, where correspondence between fault and spare can be created by different distinct deformation lines.

Thus, we can refer to an *association matrix M*, and to its coverage (although with quite a different meaning from that of the coverage matrix Mg discussed in subsection 12.3). This new matrix has a row for each faulty PE and a column for each spare present in the physical matrix (thus, even if its dimensions vary at run time as new faults appear in the system). It describes all alternative possible associations, so that choice of a coverage (i.e., of associations between suitable spares and all faulty PEs) totally defines the whole system of deformation lines and therefore completes the index mapping. Dimensions of matrix M are never greater than $(2N + 1) \times (2N + 1)$: it is thus a much smaller matrix than the general one discussed in §12.3.

The only constraint introduced, besides that on adjacency domain and on priorities, is that the vertical segment of the deformation line will not contain any faulty PE except that which the line itself associates with a spare. Otherwise, deformation lines can intersect each other. (This is the same constraint previously introduced for two-operator reconfiguration sequences.)

Constraints are reflected in the organization of matrix M. Consider the example in figure 12.41, referring to a part of an array with two faults (a, b) and with spares $1, 2, ..., 9, A$. Fault a can be associated with any of spares $2, 3, 4$ by column-wise association (the possible deformation lines are shown in the figure) but *not* to spares $5, 6..., A$ since the vertical segment of the deformation line would then contain faulty cell b (contrary to the initial assumption). Fault b allows column-wise association with spares 5 or 6 and row-wise association with spare 9. For this example, matrix M has two rows (one for each fault) and ten columns (one for each spare); a mark is inserted in position $M(r, s)$ whenever fault r can be associated with spare s (table 12.2).

Matrix M is organized into two submatrices, by ordering columns so that submatrix M' corresponds to PEs present in the spare column (the first six columns in figure 12.41) and the subsequent submatrix M'' corresponds to PEs present in the spare row (the last four columns in figure 12.41). In submatrix M'', each column has (at most) one mark — as does each row segment contained in this submatrix. This corresponds to the constraint that the vertical segment of any deformation line (i.e., the segment exploiting a PE in the spare row) cannot contain any faulty

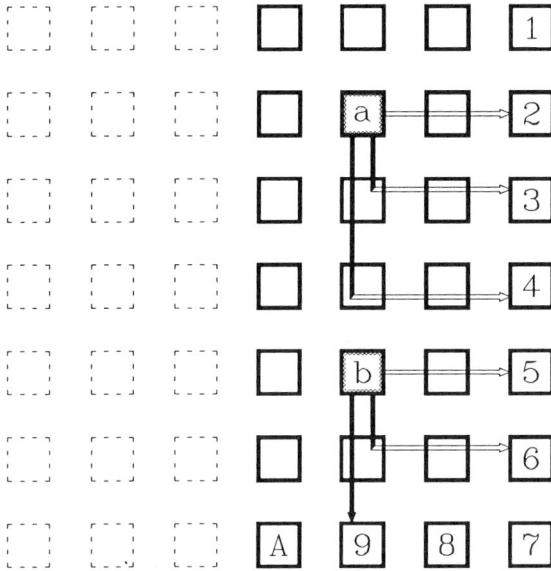

Figure 12.41
Possible associations of faults with spares with Algorithm 6.

Table 12.2
Association matrix M derived from figure 12.41

	1	2	3	4	5	6	7	8	9	A
a	x	x	x							
b				x	x				x	

PE except the one that is the "source," so to speak, of the deformation line. On the contrary, in M' each row segment contains a *string* of adjacent marks — corresponding to the various PEs in the spare column that can be adopted for the deformation line — and no constraint is introduced as regards the columns.

A more complex example, concerning an 8×8 physical array with 8 faults, coded $a, b, ..., h$, one spare column (cells coded $1, 2, ..., 7$) and one spare row (cells coded $8, 9, A, ..., F$) is represented in figure 12.42. Matrix M is given in table 12.3: matrix M' consists of the first seven columns, matrix M'' of the last eight ones.

The following basic rules can now be introduced to achieve correct coverage:

(1) In each row of M one and only one mark must be chosen (each faulty PE must be associated with one deformation line) in such a way that in each column one mark only (at most) will be present (no spare can be used to replace two or more faulty PEs);

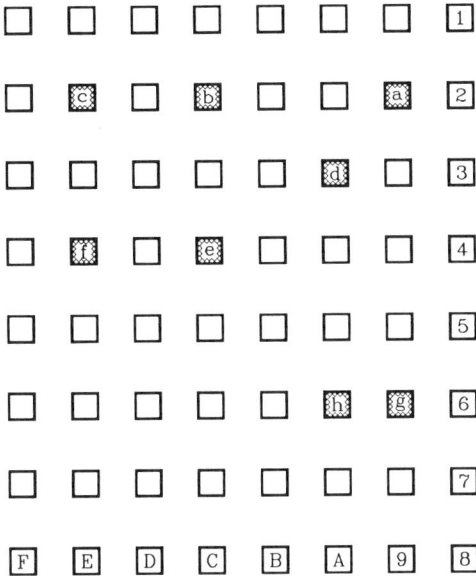

Figure 12.42
Sample array (a to h are faulty cells, 1 to F are spare cells).

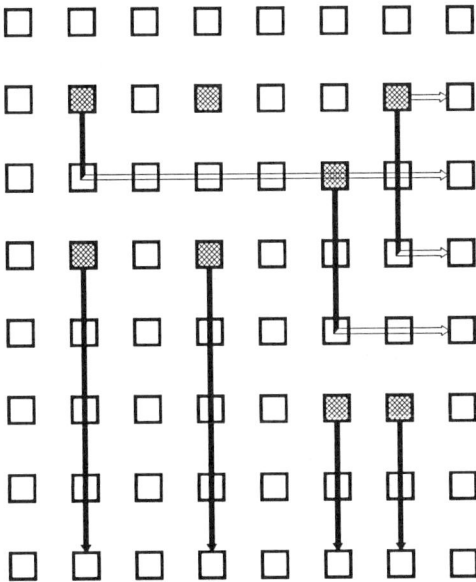

Figure 12.43
Association of faults with spares by algorithm 6, for the example in figure 12.42.

Table 12.3
Association matrix M derived from figure 12.42

	1	2	3	4	5	6	7	8	9	A	B	C	D	E	F
a	x	x	x	x											
b	x	x													
c	x	x													
d		x	x	x											
e			x	x	x	x					x				
f			x	x	x	x							x		
g				x	x		x								
h				x	x			x							

(2) If a row contains only one mark, the corresponding association is inevitable since it means that the given fault can be overcome only by using that particular spare;

(3) If a column contains only one mark, it can be proven that choosing it will not in any way endanger the possibility of reaching a solution;

The algorithm is then defined as follows:

(a) select all marks in M'' and correspondingly reduce M' (this corresponds to rule 3 above);

(b) verify conditions 1 and 2 above. Each time a mark is selected, row and column containing it are both deleted, thus simplifying the matrix;

(c) repeat step (b) until one of the following conditions is met:

- M' is empty and all rows have been selected in steps (a) and (b): this means that the reconfiguration has been successfully completed;

- M' does not contain any mark, but there is at least one row that has not been selected: this means that reconfiguration cannot be succesfully completed (the deformation line cannot be created);

- M' is not empty but it cannot be reduced by means of any of the previous steps: step d must then be attempted:

(d) among all rows having one mark in the first nonvoid column, choose the one with the shortest string of marks and select the first mark in it: then again apply the simplification procedure to M' and repeat all previous steps as applicable.

It can be proven that, if there is at least one solution, it will certainly be found by this algorithm: in this sense, the algorithm is optimum. No further cost factor is taken into account, so that no comparative evaluation is introduced among different solutions whenever such exist. On the other hand the following considerations hold:

- if by L we denote the length of the elementary cell (i.e., of the individual PE), maximum interconnection length is 2L;

- the array must be designed so as to grant correct synchronization even for the case of longest interconnections;

- given the above factor, as far as global delay is concerned, all solutions (even if they contain links of different lengths) will give the same performances.

In the above example, step (a) (concerning coverage of marks in M'') leads to the following associations:

$$\begin{cases} e \mapsto f \\ f \mapsto E \\ g \mapsto 9 \\ h \mapsto A \end{cases}$$

and the reduced matrix is given in table 12.4.

Table 12.4

	1	2	3	4	5	6	7	8
a		x	x	x	x			
b		x	x					
c		x	x					
d			x	x	x			

Further application of the algorithm leads to the following associations:

$$\begin{cases} b \mapsto 2 \\ c \mapsto 3 \\ a \mapsto 4 \\ d \mapsto 5 \end{cases}$$

(See, in figure 12.43, the deformation lines thus obtained and, in fig 12.44, the index mapping.)

Note that algorithms 3 and 4 would not have solved the above example, although the same ADs were used.

The reconfiguration algorithm can be supported by an interconnection network with two vertical buses and three horizontal ones (the same structure also capable of supporting algorithms 3 and 4, since possible paths between adjacent

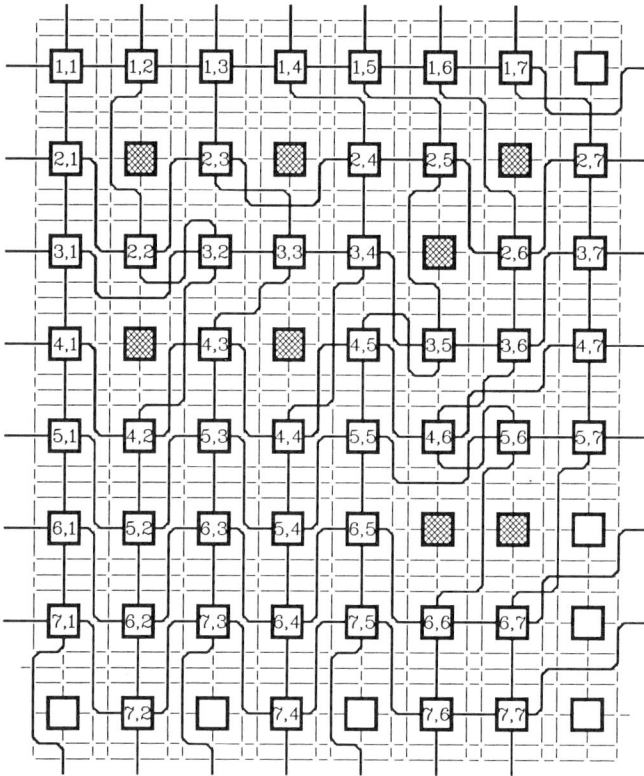

Figure 12.44

Final mapping and active interconnections for the example in figure 12.42.

cells, as defined by reconfiguration operators, are the same). The active bus segments implementing the reconfiguration example of figure 12.43 are given in figure 12.44.

The formal approach based upon the association matrix can be extended for any organization of adjacency domains and of spares. In [LOM88] the adjacency domains adopted are:

(a) for the horizontal deformation operator:

$$ADh : \{(i,j), (i,j-1), (i,j+1)\}$$

(b) for the vertical deformation operator:

$$ADv : \{(i,j), (i-1,j), (i+1,j)\}$$

A *total adjacency domain ADT* can be derived, as the set of coordinates upon which logical indices (i',j') can finally be mapped by application of both operators.

In this case it is

$$ADT : \{(i,j), (i,j-1), (i-1,j-1), (i-1,j), (i-1,j+1),$$
$$(i,j+1), (i+1,j+1), (i+1,j), (i+1,j-1)\}$$

The *ADT* is directly related to the distribution of spares that can be exploited; here, two spare rows and two spare columns can be added creating a *frame* around the nominal array (figure 12.45). It can be easily verified that larger numbers of spares would be useless, since the given *ADT* would not allow to reach them, while a more reduced distribution would forbid some of the reconfiguration patterns provided by the *ADT* itself.

The eight possible types of deformation lines created by a two operators sequence acting upon the *AD*s just defined and with a frame of spares are shown in figure 12.45. The solution of the association matrix is now more complex than in the previous case, since both operators can propagate a deformation line segment in any of two directions of the corresponding axis. The algorithm presented in [LOM88] proceeds in two phases; the first one operates following *simplified* rules (that would allow different deformation segments pertaining to the same operator to share common cells) while in the second phase the deformation lines thus obtained are modified so as to avoid such sharing. The algorithm has complexity $O(NlogN)$; an example of reconfiguration is given in figures 12.46 and 12.47. Probability of survival is given in figure 12.48 for arrays of varying dimensions: the greater freedom granted by bi-directional adjacency domains results in a much better exploitation of spares, and therefore in higher probability of survival.

12.5. Algorithms suited to single-bus networks

A reconfiguration algorithm that still belongs to the *index mapping* class and that can be supported by a very simple interconnection network has been described in [KUN86]. This low area overhead is paid for by lower probability of survival and higher complexity of the algorithm itself. Again, one spare row and one spare column are foreseen: the basic criterion (similar to that discussed also in [SAM83]) is that a given faulty PE (i,j) is associated either with spare $(N+1,j)$, i.e., along the same column (downwards deformation) or with spare $(i, N+1)$, i.e., along the same row (rightward deformation). If reconfiguration is envisioned once more in terms of deformation lines (as the algorithm allows), the constraints imposed in this case are:

(a) no vertical or horizontal deformation line will involve any faulty PE, excepting the *origin* one with which the line is associated;

(b) no two deformation lines can *cross*, i.e., share one (or more) PEs.

The second condition, while fairly restrictive, in the end allows the complexity of the augmented interconnection network to be reduced.

The algorithm is defined for a square $N \times N$ array, and it assumes a specific ordering of the faults for which reconfiguration is attempted (the *possibility* of

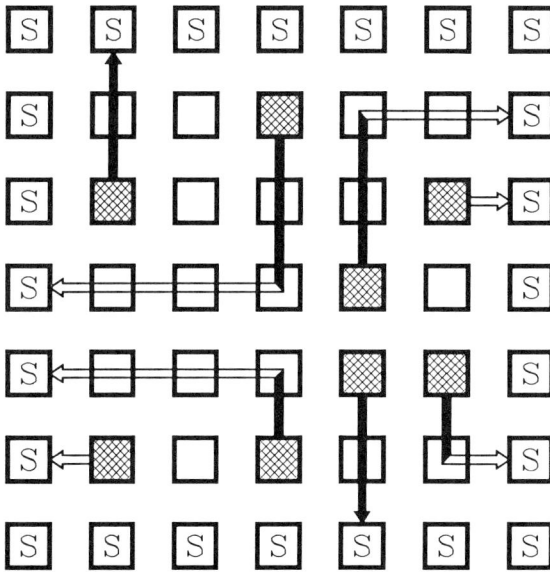

Figure 12.45
Frame of spares (cells S). Example of the eight possible types of deformation lines.

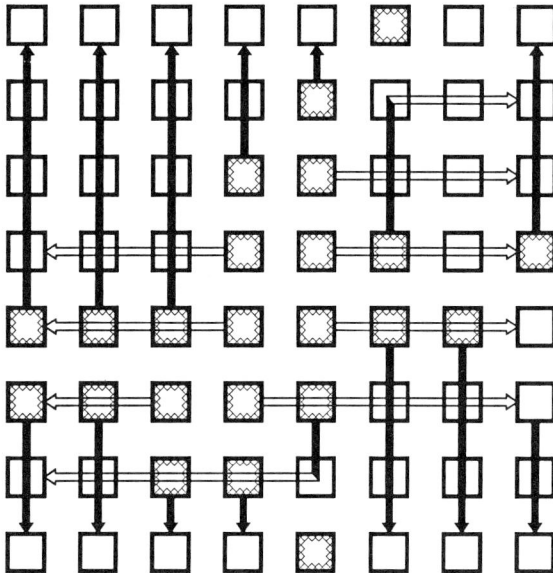

Figure 12.46
Frame of spares. Fault pattern and corresponding deformation lines.

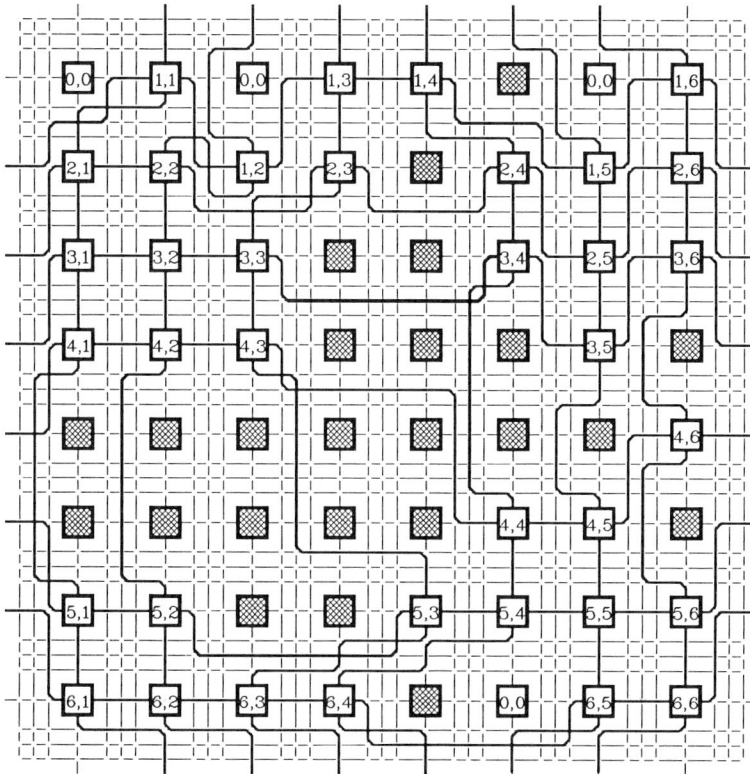

Figure 12.47

Frame of spares. Reconfiguration of the example in figure 12.46.

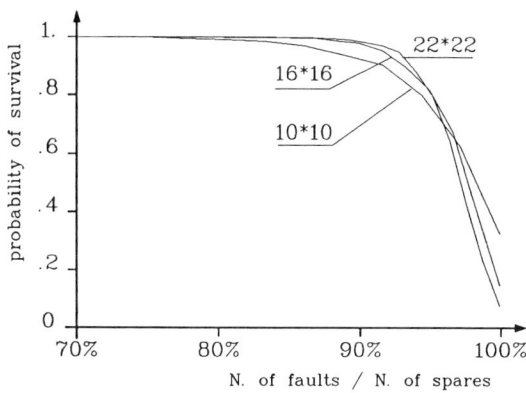

Figure 12.48

Frame of spares. Probability of survival for different array dimensions vs. number of faults normalized with respect to the number of spares.

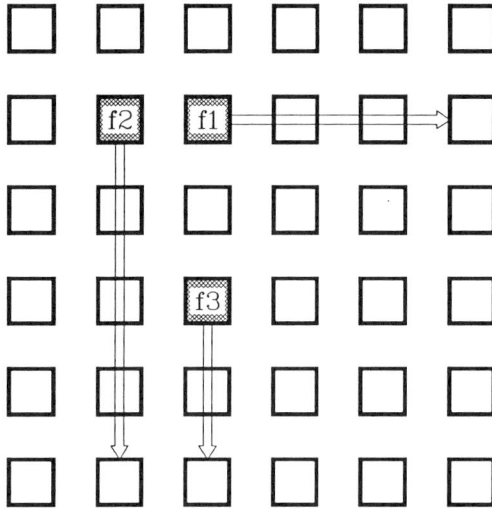

Figure 12.49

Example of reconfiguration by Kung, Chang and Jen's algorithm.

reconfiguration depends upon this ordering). Although the ordering criterion is not explicitly defined, it seems that *fault occurrence time* is implicitly assumed. The array is first subdivided into two triangular subarrays along the main diagonal (PEs on the diagonal itself are considered as belonging to the upper subarray). Then:

- a fault in the upper subarray produces a reconfiguration through a horizontal deformation line, unless such line intersects a vertical one. Otherwise, a reconfiguration through a vertical deformation line is adopted, unless such line intersects a horizontal one — if this second attempt also fails, fatal failure condition is reached;

- a fault in the lower subarray is treated in a symmetrical way (i.e., the vertical deformation is first attempted, and the horizontal one is considered as a second choice).

Consider the example in figure 12.49: fault 1 produces a horizontal reconfiguration, fault 2 (although, being on the main diagonal, it is also part of the upper subarray) therefore produces a vertical reconfiguration since otherwise the deformation line would comprise fault 1, and finally, fault 3 again requires vertical reconfiguration.

The interconnection network proposed by S.Y.Kung, Chang, and Jen is extremely simple, consisting as it does of one vertical bus and one horizontal bus per channel, with one switch associated with each bus (see figure 12.50.b). Nevertheless, it should be emphasized that the authors assume as possible (and, in fact, require as necessary) that an interconnection line will *cross over* faulty PEs (figure 12.50.a). This would involve very particular implementation techniques and is suited only to production-time static restructuring. To adapt the structure for

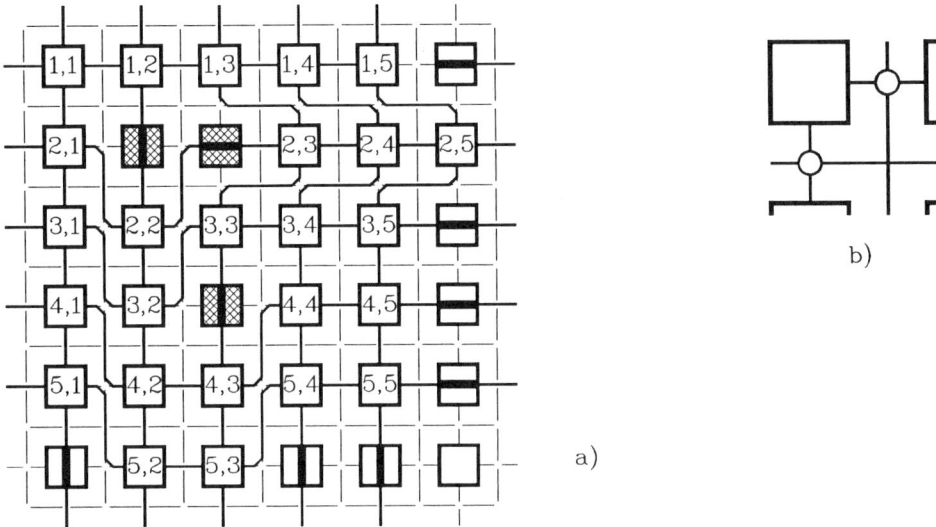

Figure 12.50

a) Example of reconfiguration by Kung, Chang and Jen's algorithm: buses are allowed to cross over faulty PEs. b) Structure of the interconnection network.

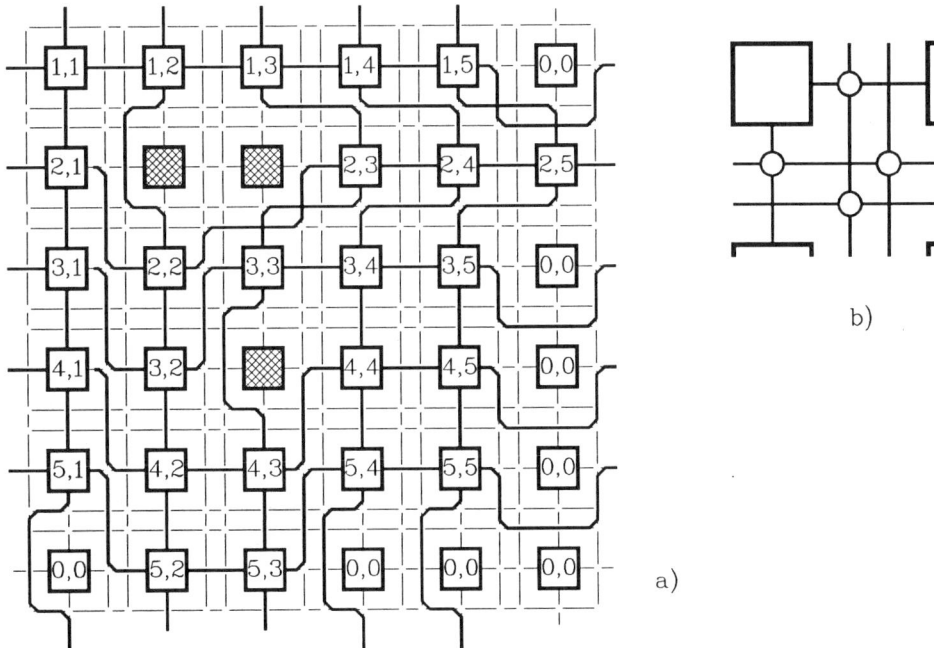

Figure 12.51

a) The example in figure 12.45 implemented on modified network b) that avoids crossing over faulty PEs.

run-time dynamic reconfiguration as well, and to use techniques coherent with the ones adopted for the other algorithms discussed here, it is necessary to introduce two horizontal and two vertical buses per channel, with four switches associated with each PE (figure 12.51). The same interconnection network also supports a simplified version of algorithm 3 (this simplified version is fully described as *simplified fault stealing* in [SAM87]) which gives a better probability of survival. In fact, simulations for a 20 × 20 target array (21 × 21 physical array) allow us to state that Kung's algorithm grants 50 percent probability of survival for up to 8.8 faults, while simplified fault-stealing grants the same probability of survival for up to 17.5 faults.

The algorithm could be modified to make it independent of the ordering among faults considered for reconfiguration. Thus, creation of a binary tree of all alternatives (vertical vs, horizontal reconfiguration) — while involving an NP-complete traversal algorithm — would grant that whenever the leaves are reached the solution is possible, and it is reached independently of any fault ordering.

12.6. References

[DIS88]   F.Distante, M.G.Sami, R.Stefanelli: *A general index mapping technique for array reconfiguration*, Proc. ISCAS-88, Tampere, June 1988.

[FRA87]   P.D.Franzon, S.K.Tewksbury: *"Chipframe" scheme for reconfigurable mesh-connected arrays*, Proc. 2nd IFIP Workshop on WSI, Brunel, Sept. 1987.

[KUN84]   H.T.Kung, M.S.Lam.: *Fault-tolerant VLSI systolic arrays and two-level pipelining*, Journ. Parall. and Distrib. Process., Aug. 1984, 32-63.

[KUN86]   S.Y.Kung, C.V.Chang, C.V.Jen: *Real-time fault-tolerant design for VLSI array processors*, Proc. Symposium on Realt-time systems, New Orleans, 1986, IEEE.

[HOP73]   Hopcroft, R.Karp: *An $n^{(5/2)}$ algorithm for maximum matching in bipartite graphs*, SIAM J. Computing, Vol. 2, Dec. 73, 225-231.

[LOM87]   F.Lombardi, R.Negrini, M.G.Sami, R.Stefanelli: *Reconfiguration of VLSI arrays: a covering approach*, Proc. FTCS-87, 1987, 251-256, IEEE.

[LOM88]   F.Lombardi, M.G.Sami, R.Stefanelli: *Optimum reconfiguration of arrays in a VLSI environment*, Proc. Supercomputing 88, Boston, May 1988, 272-280.

[MIC80]   S.Micali, V.V.Vazzirani: *An $O(\sqrt{|V|} \times |E|$ algorithm for finding maximum matching in general graphs*, Proc. 21st Annual Symposium on the Foundations of Computer Science, Long Beach, 1980, 17-27, IEEE.

[NEG85]   R.Negrini, M.G.Sami, R.Stefanelli: *Fault-tolerance approaches for VLSI/ WSI arrays, Proc. Phoenix Conference on Computers and Communications*, 1985, 460-468, IEEE.

[NEG86]   R.Negrini, M.G.Sami, R.Stefanelli: *Fault-tolerance techniques for array structures used in supercomputers*, IEEE Computer, Vol. 19, N. 2, 78-87.

[SAM83]   M.G.Sami, R.Stefanelli: *Reconfigurable architectures for VLSI processing arrays*, Proc. NCC 83, 1983, 565-577.

[SAM85]   M.G.Sami, R.Stefanelli: *Fault-stealing: an approach to fault-tolerance of VLSI array structures*, Proc. Int'l Conference on Circuits and Systems, Beijing, 1985, IEEE.

[SAM86a]   M.G.Sami, R.Stefanelli: *Reconfigurable architectures for VLSI processing arrays*, Proceedings of the IEEE, Vol. 74, N. 5, May 1985, 712-722.

[SAM86b]   M.G.Sami, R.Stefanelli: *Fault-tolerance and functional reconfiguration in VLSI processing arrays*, Proc. ISCAS 86, 1986, 643-648, IEEE.

[SAM87]   M.G.Sami, R.Stefanelli:*Fault-tolerant computing approaches*, in: *Systolic Signal Processing Systems*, (E.Swartzlander ed.), 1987, Marcel Dekker.

[SIN85]   A.D.Singh: *An area efficient redundancy scheme for wafer scale processor arrays*, Proc. ICCD 85, New York, Oct. 1985, 505-509, IEEE.

1 3   RECONFIGURATION BASED ON REQUEST-ACKNOWLEDGE LOCAL PROTOCOLS

In this chapter a class of reconfiguration algorithms will be considered, all derived by the same principle and characterized by fairly relevant simplicity. Such algorithms show a high efficiency in cases where other, more complex methods can be too costly to implement. They are primarily useful for *on-line* reconfiguration of *Ultra Large Scale* (e.g., Wafer Scale) integrated systems. *Self-reconfiguration* can also be foreseen, since the intrinsic simplicity of these algorithms leads in turn to simple circuits dedicated to control of reconfiguration.

This approach has been followed by researchers interested in architectures dedicated to digital signal processing. Many digital signal processing applications involve stringent reliability requirements; moreover, the overall complexity of such systems can be so high that on-line reconfiguration is mandatory to guarantee acceptable system life-time.

Typical methods and results have been presented in [NEG85] and, for specific application cases alone, in [EVA85] and in [GUP86]. As a case study, we will consider the method introduced in [NEG85], where it has been shown that this reconfiguration approach, characterized by perfect regularity and by a possibly complete dynamic self-reconfiguration, can give rise to a whole family of reconfiguration algorithms.

Although all approaches here discussed aim at *harvesting*, it is quite reasonable to accept them for run-time reconfiguration; in the cited signal processing applications, this philosophy means that the host controller will drive a processing array whose capacities may be gracefully degradable at operation time, as a consequence of faults.

The reconfiguration principle may be stated as follows. If the target structure is composed by lines of PEs (e.g., rows and columns in meshes) then it should be possible to identify each target line simply by building it, through an ordered connection of PEs, in an incremental fashion. Provided that some local simple conditions exist ensuring that the topology of the complete target structure is preserved (e.g., that crossings between rows and columns satisfy the basic ordering rules for a mesh), then it should be possible to *locally* identify which working PE can be chosen as the next in line. The PEs thus chosen must belong to an *allowable* set of PEs that can be linked to the one previously inserted in the line. If no such PE exists, e.g., because all the PEs that would allow connection to the last PE previously introduced are faulty, then this PE last inserted in the line should be marked as useless, even though fault-free; the line should be backtracked and another route chosen. Thus, large clustered distributions of faulty PEs, typical of WSI, can also be overcome. (A similar philosophy was adopted in some approaches to reconfiguration of linear arrays.)

A procedure that involves backtracking phases would lead to relevant algorith-

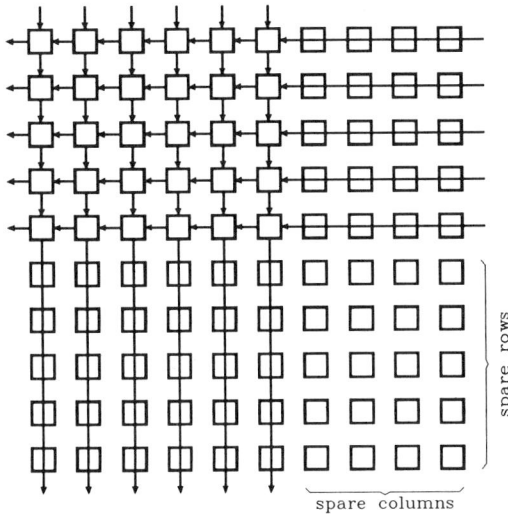

Figure 13.1
The physical array.

mic complexity when implemented sequentially on a host computer. This problem can be neglected, on the contrary, if the reconfiguration algorithm is implemented locally, in a distributed way, by circuits controlling reconfiguration locally for each PE. The intrinsic, very high parallelism of this implementation can eliminate the problems of algorithmic complexity (especially the ones deriving from time required for reconfiguration). The main figure of merit here becomes the simplicity of the reconfiguration controlling network, as compared to the intrinsic circuit complexity of the PE.

As for every on-line self-reconfiguration algorithm, this method can also be used for static end-of production restructuring.

A drawback of the incremental techniques here described will be found in the low harvesting obtained, in comparison to what can be obtained by means of more global approaches. Again, this suggests that the method is better tailored for ULSI or WSI systems consisting of a very large number of relatively small cells, where lower harvesting can be tolerated provided it is offset by more realistic fault models supported by the reconfiguration technique.

The method will be applied here only to *rectangular arrays* (see figure 13.1). The algorithm itself is not modified whether applied to an array of fixed dimensions provided with spare rows and columns or to the pure *harvesting* case, and can be applied for self-reconfiguration as well as for host-driven reconfiguration. In this chapter, we will deal with the algorithm alone, not with the control circuits necessary when self-reconfiguration is envisioned.

In all the examples considered, the communication lines among PEs are regarded as unidirectional and the data flow only in a direction between two PEs: if

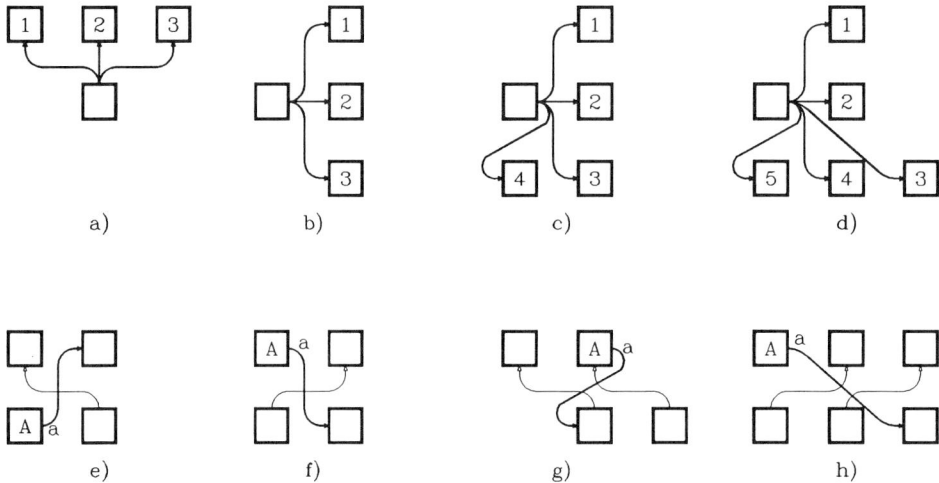

Figure 13.2

Horizontal and vertical neighboring sets: a) and b) for $/3*3/$, a) and c) for $/3*4/$, and a) and d) for $/3*5/$ algorithms. Link "a" in e), f), g), and h) must not be activated otherwise some columns are skipped.

paths in both directions are necessary, only the interconnection network must be modified, whereas all the reconfiguration controlling networks need not be changed. Similarly, structure and logic functions of the PEs will not influence the reconfiguration properties discussed below.

Given any PE i in the array, a small set of neighboring PEs such that the interconnection network allows to connect them with i can be identified: this is called the *neighboring set* of i. The reconfiguration procedure consists then in choosing inside the neighboring set of i a pair of *free* PEs, (i.e., not yet inserted in the array) that are capable of prolonging, respectively, the row and the column for which i is presently "head".

For example, suppose that rows are built linking PEs together from left to right, and that columns are built from bottom to top. If at each PE the row can be prolonged by choosing the right neighbor among y PEs, and the column can be prolonged by choosing the upper neighbor among x PEs, the two neighboring sets are identified by the symbol $/x*y/$. Figures 13.2.a-d show examples of different neighboring sets for rows and columns. When necessary, from now on, $/x*y/$ will also be used to identify a reconfiguration algorithm that uses the $/x*y/$ neighboring sets.

A few general properties required from all such algorithms can be identified. First, the reconfiguration control network should be very simple, otherwise too much circuit area would be required and there would be a high probability of failures in it (the need for arithmetic circuits should especially be avoided), and the control network of a PE should exchange signals only with neighboring PEs.

Second, the reconfiguration may waste some fault-free PEs (i.e., it is acceptable that fault-free PEs not be exploitable because of limitations in the spare paths); this has already been allowed in previous algorithm classes. The number of wasted PEs must of course be low, if related to the number of working PEs (this corresponds to the requirement of high harvesting).

Figure 13.3 shows an example of the kind of reconfiguration that is our goal: how columns can circumvent clusters of faulty PEs, trying to find a way on their right side, whereas rows try to pass below the faults. Inevitably, if the neighboring sets adopted are small, some large unused regions may well appear near large faulty zones. Fault-free PEs in such areas cannot in fact be reached, because columns (or rows) passing trough them would always eventually end against some faulty PE. On the other hand, if larger neighboring sets are adopted, these unused regions will be reduced, and additional columns or rows are possibly formed giving larger working arrays, but a larger area is wasted for the redundant interconnection networks required at each PE to connect the neighboring sets.

Depending upon the application, the neighboring set should be chosen so as to minimize the wasted regions (unused redundant interconnections *and* unused working PEs) as a function of the number of bits in data paths and of the PE area. In general, large PEs require large neighboring sets in order to maximize harvesting.

Figure 13.3 also shows that the unused regions can reach the edge of the array; thus, logically adjacent input signals on one edge (or outputs at the opposite one) can be noncontiguous, in the final structure.

This problem is common to algorithms seen in previous chapters (see, e.g., [KUN84]) and it basically involves connections to I/O paths. While it is fairly easy for end-of production restructuring, at run-time, circuits and buses of higher complexity are required on the arrays borders (e.g., implementing the *selectors* described in [GRE84]). Thus, contrasting figures of merit are obtained in the two instances: fast run-time reconfiguration involves difficulties for I/O connections, while algorithmic complexity for host-driven restructuring is balanced by ease of I/O interconnections.

13.1. The fault model

The fault model adopted in this chapter is quite different from that used earlier, being particularly designed to model the major characteristics of failures and defects inside ULSI and WSI; it will therefore be described prior to discussing the algorithms themselves and their implementation. While we assumed previously that all faults are permanent and confined inside the PEs (i.e., they do not affect control networks and communication paths), these restrictions must now be relaxed, at least partially.

While Poisson distribution can also be adopted here to model faults dynamically arising in ULSI or WSI arrays during field operational life, the modeling of ULSI or WSI production defects requires more complex assumptions. It can be

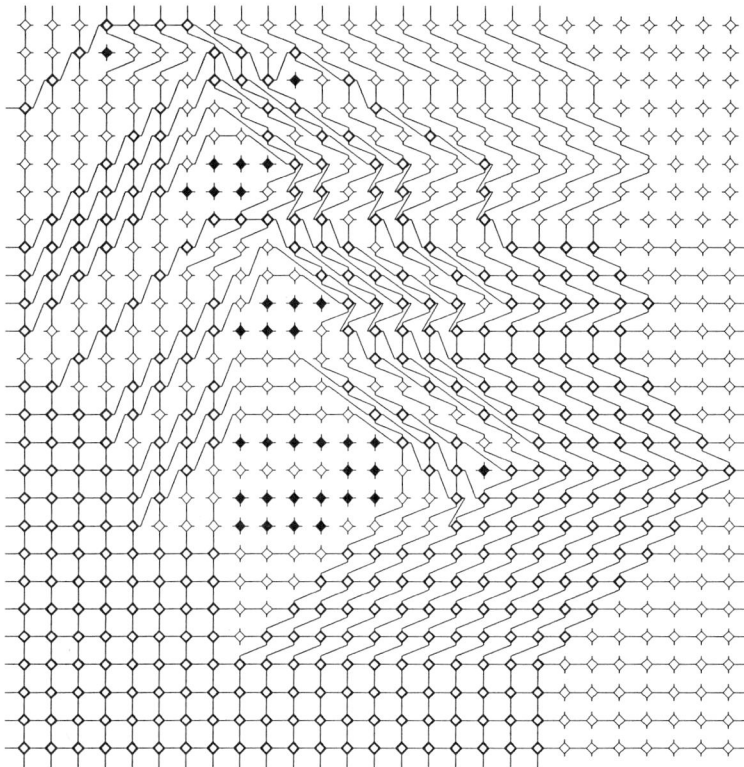

Figure 13.3

An example of reconfiguration. Filled PEs are faulty. PEs of the target array are connected both horizontally and vertically; some PEs, working as relay elements only, are connected either horizontally or vertically; not connected PEs are unused.

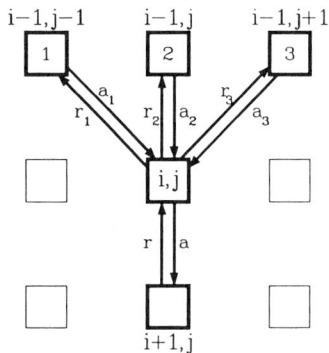

Figure 13.4

Request signals (r, r_1, r_2, r_3) and Acknowledge signals (a, a_1, a_2, a_3) that can implement a simple reconfiguration protocol.

supposed that clustered defects may arise from large fabrication defects involving many PEs as well as segments of interconnections positioned among them and the related reconfiguration control circuits. In this case, faults and defects hit irregular groups of nearby PEs, and group shape and size are random variables. Thus, it is safer, when reconfiguring the array:

(a) not to use connections passing through clusters of faulty PEs, and

(b) to exclude reconfiguration control circuits of faulty PEs from the determination of the reconfigured array.

The algorithms considered here are based upon mechanisms that can satisfy both (a) and (b).

13.2. The algorithms

Given the above fault model, approaches such as those described in chapters 9 to 12 cannot be immediately adopted. All these techniques assumed that:

(a) interconnections were fault free (as were all related control circuitry, whenever on-chip self-reconfiguration was used);

(b) as a consequence of point (a), in the case of on-chip self-reconfiguration some kind of *global* signal could be propagated through the array in order to specify conditions for renewed mapping of logical onto physical indices (see chapters 10 and 11).

Any *global* information must now be avoided: reconfiguration algorithms must reconfigure the array by *locally* identifying for each PE how to connect it with further devices in the target architecture by choosing — inside the two neighboring sets for rows and columns — the two PEs that must be connected to prolong the row and the column. We will therefore consider *incremental techniques*. Furthermore, such techniques must also be *managed* locally, by control circuits that will be considered as part of the PEs (and therefore subject to the same fault assumptions, at least when run-time reconfiguration is envisioned). Reference will be made here to one possible physical implementation of this local process, based upon a simple protocol with *Request and Acknowledge signals* exchanged among working PEs.

Algorithms foresee incremental creation of rows and columns; columns have a priority over rows, meaning that modification of a column may also lead to modification of rows, while the converse is never true. Should host-driven implementation be foreseen, columns would be created first, and rows would be completed subsequently.

Locally, each working PE manages its Request and Acknowledge signals: the procedure starts from two edges of the physical array, and requests to prolong rows and columns are then propagated inside the array. This philosophy is consistent with the fault assumptions, since a fault inside a PE *or* in its protocol management circuits *or* even in related interconnection links will lead quite simply to the corresponding silicon region being considered as useless and therefore avoided during reconfiguration.

The basic principle of all the algorithms can be presented in figure 13.4, which refers to the simple case of creation of columns, when the neighboring set of figure 13.2.a is adopted for this phase.

If PE (i, j) receives an active request signal r coming, e.g., from a PE in the lower row, e.g., PE $(i + 1, j)$, it tries to propagate upwards this request to create a column.

It first activates request r_1, trying to connect PE $(i - 1, j - 1)$. If this PE is fault-free and not yet requested, it generates its acknowledge signal a_1, and PE (i, j) in turn activates generation of acknowledge signal a. Otherwise, PE (i, j) activates request r_2; again, if acknowledge a_2 is not received, r_3 is activated. If a_3 is also not received, then acknowledge a is negated to PE $(i + 1, j)$, which will in turn forward the request to its next possible neighbor PE $(i, j + 1)$, and a similar process of request/acknowledge propagation will start again from this last PE.

Fixed priorities are associated with these signals: values inside PEs in figure 13.4 define priorities both for acknowledge and for request signals.

In actual implementation, request signals r_1, r_2 and r_3 are simultaneously sent to PEs 1, 2, and 3. When an acknowledge signal comes back from PE i, it deletes all requests associated with PE values higher than i. Thus, only the accepted request associated with the lower value is maintained.

Simultaneous request signals might arrive to a PE coming from different neighbors: in this case, the same values of figure 13.4 can be associated with requests (i.e., 1 associated with requests coming from the down-right neighbor, 2 associated with request coming from the lower neighbor, and 3 with requests coming from the down-left neighbor). Only requests marked with the highest value will then be accepted.

As a consequence of the priorities chosen when examining requests and acknowledges, columns tend to bend to the left, but are hindered by the left border of the array, by columns previously built at the left, and by faulty PEs: faults are circumvented on the right.

This basic mechanism of request/acknowledge propagation must be simultaneously implemented in each fault-free PE of the array. Similar principles work with different neighboring sets, and can also be applied to the creation of rows. Columns never intersect each other, i.e., the ordering of PEs on the bottom of the array is the same as for those at the top. A similar property holds for rows.

Figure 13.5 shows an example of reconfiguration based upon the /3 ∗ 3/ neighboring sets. Each PE is marked with a pair of indices denoting its logical position in the working array. Note that some fault-free PEs cannot be reached and are thus left unused. Other PEs can be connected for building rows but not for columns, or vice-versa. They are therefore used simply as *relay elements*, i.e., as segments of interconnection paths, without performing any processing on the data they transmit (a procedure otherwise used only for a few algorithms in the reconfiguration of linear arrays, as was seen in chapter 8). PEs that operate as relays both for columns and for rows are easily identifiable because are connected to the *same* PE

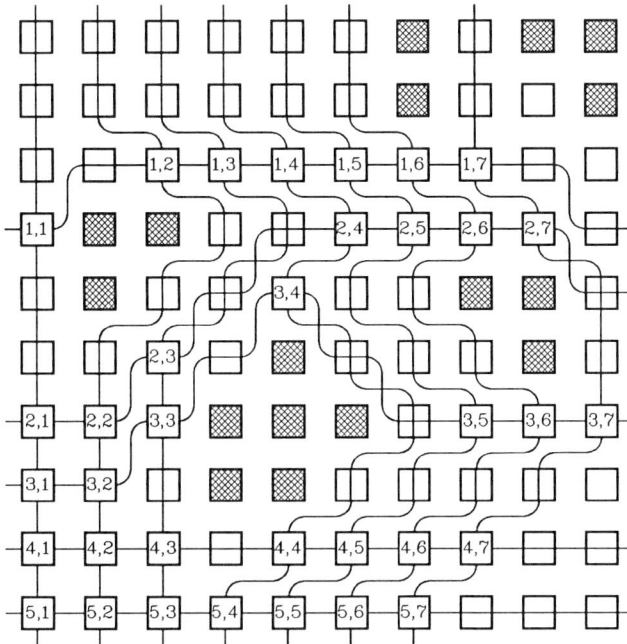

Figure 13.5

An example of reconfiguration with the $/3 * 3/$ algorithm.

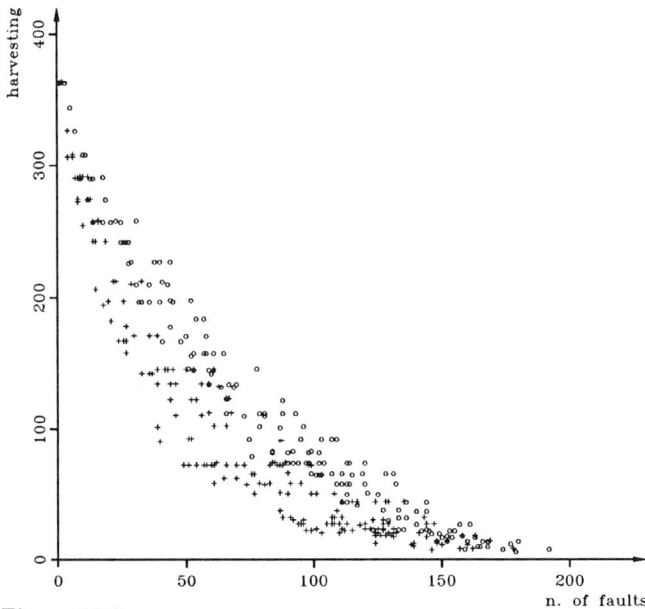

Figure 13.6

Harvesting of $/3 * 3/$ in a 20×20 physical array.

both along the row and the column.

Some algorithms and the structure of a possible control network that implements them will be considered in detail.

The algorithm can be decomposed into:

(1) propagation of vertical Request/Acknowledge from the bottom to the top of the array (i.e., creation of columns);

(2) propagation of horizontal Request/Acknowledge from left to right (i.e., creation of rows).

These two actions can be executed at the same time, but for clarity will be presented here as if they occurred successively, i.e., as if the creation of columns happened first, and then — only after columns were completely determined — rows were built. As will be seen, phases (1) and (2) of the reconfiguration algorithm are in fact linked by a logical priority ordering, since the creation of columns is *independent* from the creation of rows, whereas the configuration of rows can change if columns are changed.

The creation of a new column can change both preexisting columns and rows, whereas creation of a new row can induce rearrangement only on rows. For example if, in case of on-line self-reconfiguration, a new fault arises (i.e., a PE fails), the reconfiguration control network of this PE ceases to produce Acknowledge signals both for columns and for rows. A new propagation of Request/Acknowledge signals starts, and, simultaneously, new rows and columns are built. During the transient phase, columns and rows are tentatively built, until the new configuration is reached.

These rearrangements do not affect columns at the left of the original column where the newly faulty PE was, because these columns have a higher priority compared to the columns that must be rearranged. On the contrary, rows are repeatedly modified, following possibilities from moment to moment that are dictated by the creation of columns. Only when the columns are stable, can the rows reach a stable reconfiguration. Note that this algorithm builds the largest possible array, compatible with the neighboring set chosen.

13.2.1. Configuration of columns

A formal definition of this algorithm can be given by adopting two matrices to represent request and acknowledge signals (R and A) and a matrix U, denoting availability of the various PEs (a PE is considered to be permanently unavailable, i.e., useless, whenever it is unreachable, even if it is fault-free).

A, R, and U have the same dimensions as the physical array, e.g., $N \times M$. Using the numeric marks shown inside the PEs in figure 13.4 as entries for matrices A and R, the basic rules are given in table 13.1. A matrix F identifies the faulty PEs (see table 13.1).

In this off-line implementation, request and acknowledge signals are simulated

Table 13.1

Basic step of the propagation algorithm.

;exam of PE (i, j)
;$F(i, j) > 0$ means that PE is faulty
;$U(i, j) = 1$ means that PE is fault-free but cannot be connected
;$R(i, j) = 0$ means that PE is not requested
;$R(i, j) = 1, 2,$ or 3 are the request marks as in figure 13.4
;$A(i, j) = 0$ means that PE does not acknowledge requests
;$A(i, j) = 1, 2,$ or 3: request 1, 2, or 3 is acknowledged

;if PE is faulty/permanently unavailable/not requested:
;then cancel all requests and acknowledges generated by itself
if $F(i, j) > 0$ or $U(i, j) > 0$ or $R(i, j) = 0$ then
        if $R(i - 1, j - 1) = 1$ then $R(i - 1, j - 1) = 0$
        if $R(i - 1, j) = 2$ then $R(i - 1, j) = 0$
        if $R(i - 1, j + 1) = 3$ then $R(i - 1, j + 1) = 0$
        $A(i, j) = 0$
        EXIT

;if PE is fault-free and requested then propagate incoming requests
if $R(i, j) > 0$ then
    $A(i, j) = R(i, j)$;anticipated acknowledge
    ;
    ;request upper left PE, if free and fault-free
    if $F(i - 1, j - 1) = 0$ and $U(i - 1, j - 1) = 0$ and
        $R(i - 1, j - 1) \leq 3$ and $R(i - 1, j - 1) \leq 2$
        then $R(i - 1, j - 1) = 1$ and EXIT
        ;
    ;request upper PE, if free and fault-free
      if $F(i - 1, j) = 0$ and $U(i - 1, j) = 0$ and $R(i - 1, j) \leq 3$
        then $R(i - 1, j) = 2$ and EXIT
      ;
    ;request upper right PE, if fault-free
      if $F(i - 1, j + 1) = 0$ and $U(i - 1, j + 1) = 0$
        then $R(i - 1, j + 1) = 3$ and EXIT
      ;

    ;all propagations are impossible, clear acknowledges
    ;and mark PE as permanently unavailable
    $A(i, j) = 0$
    $U(i, j) = 1$
    EXIT

---

by means of numeric marks. In other words, PE (i, j) tries to mark all its upper neighbors, and the values of marks represent priorities of request. For example, referring to PE $(i - 1, j - 1)$, if in the first attempt it is $R(i - 1, j - 1) = 2$, this means that this PE has already been requested by the higher-priority PE $(i, j - 1)$, because this is the only PE that can leave this mark. This procedure must be repeated by scanning the columns iteratively from left to right, until a stable configuration is reached. (Whenever an operation upon indices, e.g., $i - 1$, $j + 1$ and so on, brings them out of the array boundaries, they are then considered as corresponding to *dummy* fault-free PEs.)

13.2.2. Configuration of rows

The same method previously seen for columns can be applied to rows also, through the proper A, R, and U matrices. For example, the case of the row neighboring set of figure 13.2.d, can be treated considering five marks, one for each horizontal redundant path, as depicted inside the PEs in the same figure. Thus, the algorithm will require five attempts to mark the five horizontal neighbors in an orderly way. Again, the values of the marks represent both the order and priorities of the request.

In [NEG85] quite simple reconfiguration control networks were defined, capable of managing the Request/Acknowledge signals previously defined without any arithmetic processing.

Configuration of rows must also eliminate the possibility that reconfiguration errors similar to those of figures 13.2.e-h arise. Some otherwise possible request signals cannot therefore be propagated: PEs marked as A cannot request paths marked as a. These few requirements can be satisfied by conditioning the horizontal control network through the acknowledge signals generated by the vertical control network. A PE determines if it has to act as a relay along a row or a column, or if it has to remain inert by sensing both its horizontal and vertical Acknowledge signals.

13.2.3. Harvesting

In [NEG86], harvesting for the cases $/3 * 3/$, $/3 * 4/$ and $/3 * 5/$ has been discussed. These cases all have the same reconfiguration properties for columns, but rows are built with different capacity. The corresponding neighboring sets are shown in figure 13.2.

These three algorithms have been evaluated by simulating a high number of reconfiguration problems for the case of an array of 20×20 PEs, injecting faults in the PEs.

The reconfiguration capacity is measured by inspecting the dimensions of the largest working array that is obtained by applying the reconfiguration algorithm under test. In this way, harvesting is defined as the number of PEs that constitute the final working array (therefore excluding pass-through or *relay* elements, unused fault-free PEs, and faulty PEs).

In figure 13.6, harvesting is shown as a function of the total number of injected faults for the case of /3 * 3/ neighboring sets. A "+" identifies the harvesting for an array hit by clustered faults, a "o" does the same for random faults.

Only faults inside PEs have been simulated, not those in interconnection paths or in reconfiguration control circuits; thus, reconfiguration properties obtained by simulation do not depend on the choice of various possible implementations for interconnection networks. The possibility that links or control circuits inside a cluster of faulty PEs are, in turn, faulty is both automatically accounted for and solved by the algorithm, so that relevance of the simulation is not affected.

Faults have been injected through two fault generators. The first generates multiple random (single) faults; the second generates multiple random clusters (each defined by an ellipse with random position and random axes). Random faults and random clusters have not been injected simultaneously, so as to separate reconfiguration capacities for these two distributions.

Figures 13.6, 13.7, and 13.8 depict the results of many reconfiguration problems, as solved by /3 * 3/, /3 * 4/ and /3 * 5/ algorithms, respectively. Harvesting (number of nodes of the maximum reconfigured array) is shown as a function of the total number of faults injected.

For clustered faults it can be seen that algorithms /3 * 3/ and /3 * 4/ present very similar reconfiguration capacities, and that /3 * 4/ is slightly better than /3*3/ against random faults. As predictable, /3*5/ shows a higher reconfiguration capacity. Algorithms /3 * 4/ and /3 * 5/ require similar amounts of interconnection circuits. Thus, /3 * 3/ and /3 * 5/ seem to be the only practical algorithms.

Harvesting drops rapidly when the number of faulty PEs increases; this is a result of the small dimensions adopted for the neighboring sets. On the other hand this choice guarantees low silicon area overheads for spare paths. As suggested in [MAN82] when faults affecting added silicon areas are accounted for, they rapidly increase with added area and reduce the practical advantages given by the higher reconfiguration possibilities granted by the presence of more interconnection paths.

13.3 Interconnection networks

When aiming at on-line self-reconfiguration — the most interesting case for the algorithms discussed here — the interconnection networks necessary to achieve dynamic reconfiguration can be implemented in different ways.

Two solutions have been investigated.

A *first* implementation (suggested in [NEG85]) is based upon spare links among every PE and its neighbors, and upon input multiplexers at each PE. Outputs coming, through the spare links, from all the PEs of the neighboring set are linked to an input multiplexer, and a suitable control network manages the

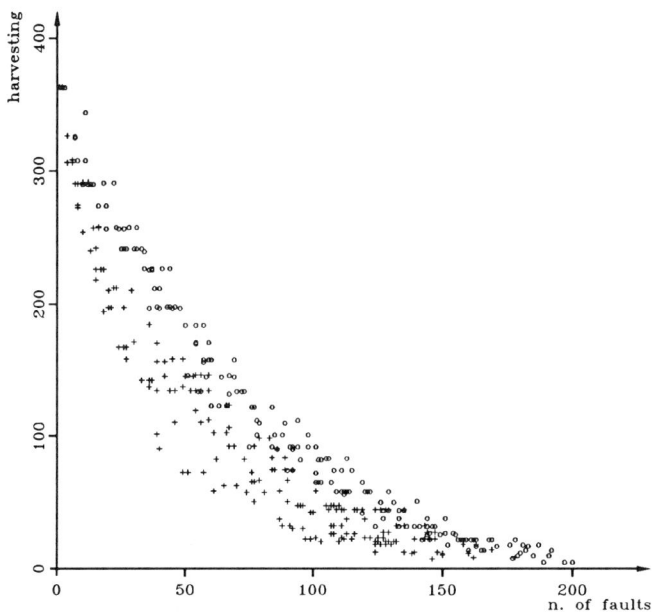

Figure 13.7

Harvesting of $/3 * 4/$ in a 20×20 physical array.

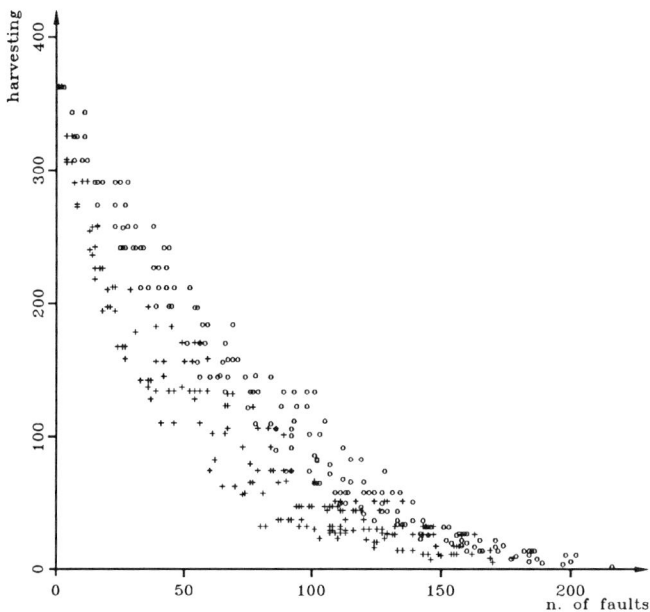

Figure 13.8

Harvesting of $/3 * 5/$ in a 20×20 physical array.

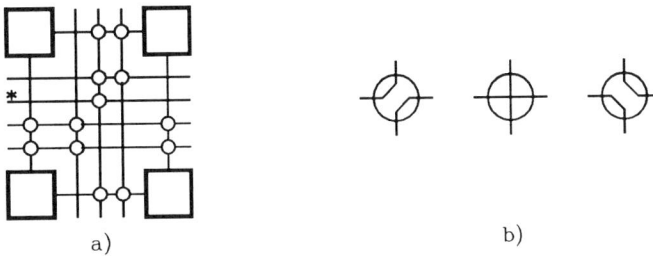

Figure 13.9

a) Structure capable of implementing $/3*3/$, $/3*4/$ and $/3*5/$ through switched buses; the bus marked by a $*$ is not necessary for $/3*3/$. b) Possible settings of switches.

control signals of the multiplexer, choosing the correct output (i.e., the correct link).

The main drawback of this solution is the large amount of area required for spare links; control networks are simple.

At each PE, two multiplexers are used for reconfiguration of paths (i.e., for the choice, as horizontal and vertical inputs, of the correct signals among those coming as outputs from the neighboring sets of PEs). Two other multiplexers are used to bypass the PE, if this acts as a relay element only.

When the $/3*4/$ and $/3*5/$ algorithms are considered, more spare links and bigger multiplexers have to be introduced for rows, thus requiring more silicon area. It is apparent, for example, that rows in $/3*5/$ require five-inputs multiplexers.

A *second* implementation has been suggested in [NEG86], based upon switched buses. Four vertical and three horizontal buses are inserted between PEs as in figure 13.9.a (the bus marked with a star is required only for $/3*4/$ and $/3*5/$). This figure also shows the switches that are needed at bus crossings; figure 13.9.b shows the possible settings of these switches.

Reconfiguration is actuated by correctly setting all the switches on the buses linking a PE to its four logic neighbors. Test design has proven that switches are reasonably compact, introduce very small delays, and that simple reconfiguration control networks can be designed that set the switches, .

While the first method of implementation (adopting multiplexers) minimizes circuits that control reconfiguration, this second method minimizes the number of necessary interconnection links. Switched bus structures seem to have a small advantage compared to structures based on spare links and multiplexers; the advantage grows when wide choices of possible neighbors are allowed for. Switched structures require less area for spare links because of bus multiplexing and because bus segments can be shared both for horizontal and for vertical links.

Row paths that can be built by the switches are given in figure 13.10 for algorithm $/3*5/$; paths for algorithms $/3*4/$ and $/3*3/$ are subsets of the ones in figure 13.10. Column paths for all the algorithms may be obtained by

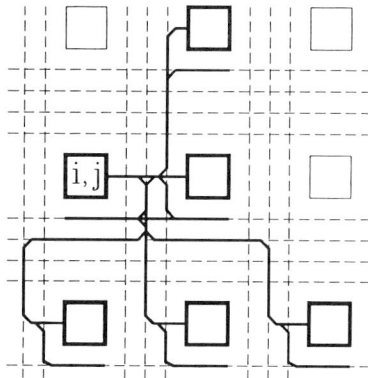

Figure 13.10
Paths that can be built by setting the switches.

rotating figure 13.10. When a path corresponding to an acknowledge signal has to be built between two PEs, the acknowledge signal itself can be used to command the switches upon the path.

Figure 13.11 and 13.12 show how algorithms /3 * 4/ and /3 * 5/ reconfigure the same problem seen in figure 13.5.

13.4. Area and time overheads

Costs of a given on-line self-reconfiguration algorithm might be determined in terms of additional circuit area (for self-checking and self reconfiguration circuits and for spares), and of increased processing time due to additional delays of data propagation through the interconnection network. Other negative figures of merit are related to time wasted during reconfiguration after diagnosis of a new fault, or to working PE area wasted for limits set by the interconnection network structure, allowing for incomplete harvesting only.

With regard to the *positive* factors of merit, the main focus is on yield and probability of survival to given classes of fault distributions, as previously discussed. These parameters are strongly related to each other, and can be often determined only through statistical evaluations in a manner similar to that adopted for obtaining harvesting in figures 13.6, 13.7 and 13.8.

These parameters, moreover, acquire complete and real significance only when an *application frame* is given, attributing, for example, an order of magnitude to costs of reconfiguration failures and of reconfiguration time.

Thus, it was decided that additional area and processing time required also be evaluated. In fact, reconfiguration time is of secondary importance in the anticipated applications. Furthermore, these reconfiguration algorithms adopt a brute force attack to the problem of determining the reconfiguration: they fully exploit parallelism in its determination, and thus allow reduced reconfiguration

Figure 13.11

An example of reconfiguration with $/3 * 5/$ and switched buses.

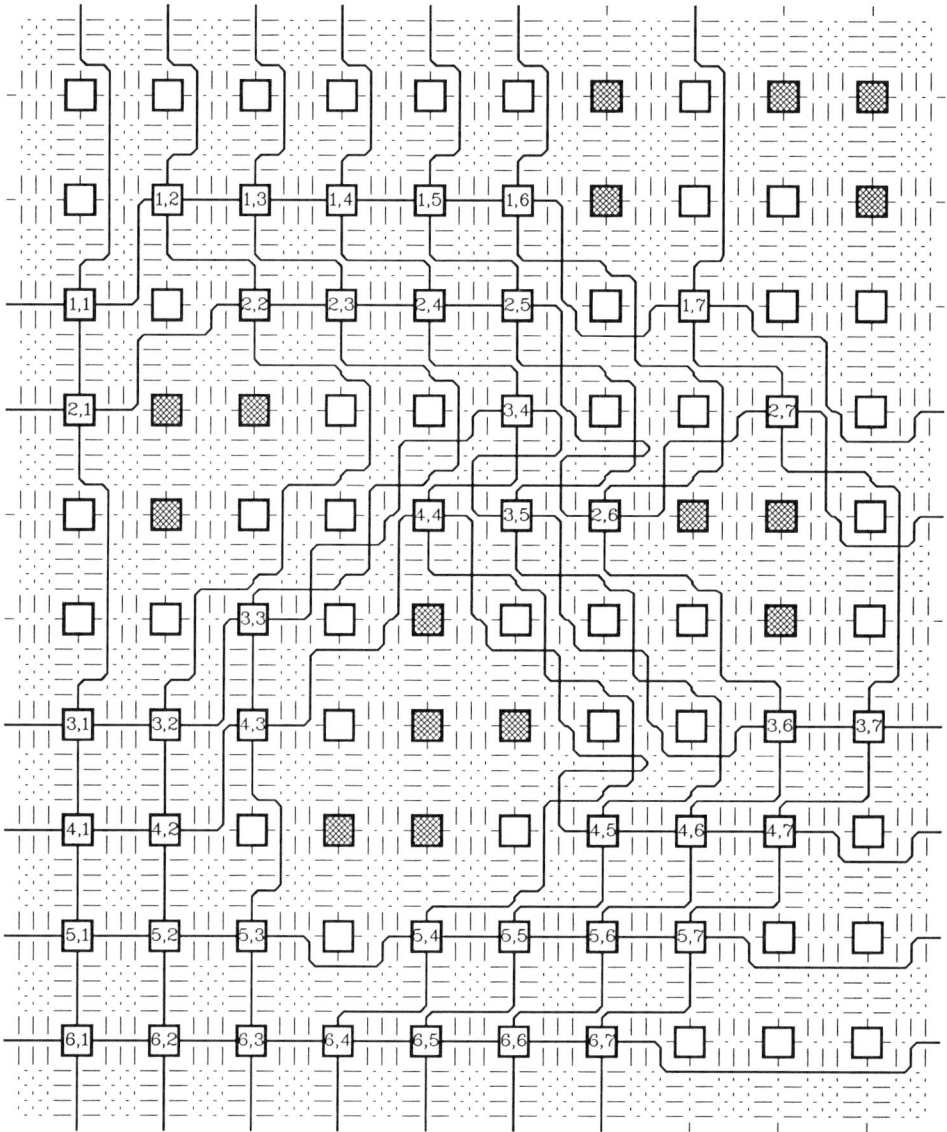

Figure 13.12

The same problem of figure 13.11 solved by $/3 * 4/$.

time. Area requirements have been modeled only for rectangular PEs, by means of the following parameters:

B = base length of nominal PE

H = height of nominal PE

e = fraction of area increment of PE due to self-testing

c = fraction of area increment of PE due to internal registers multiplexing

n = percentage of spare PEs introduced, referred to one PE

h = number of vertical buses

v =number of horizontal buses

i =bus width (number of wires)

The nominal PE area BH is approximately incremented by the following terms:

$$BH(n + e + ne) + K_{1s}Hvi + K_{2s}Bhi + K_{1s}K_{2s}vhi^2$$

where K_{xy} are appropriate coefficients. This shows how heavily areas depend on the value of i (bits/bus). Note that BH also heavily increases with i. This formula can be used to approximately determine production yield decrement (before reconfiguration is made) as a function of area increment, given the usual assumptions about the probability of faults, similar to that followed in [MAN82].

If the notation is adopted:

S = maximum number of switches along a path

T_s = maximum delay of a switch

T_L = maximum delay of a bus segment

M = maximum number of bus segments in a path

D = maximum between delay for internal multiplexing and multiplexing of I/O signals of the PE,

propagation delays can be approximately summarized by

$$ST_s + K_sT_LM$$

13.5. Application to different arrays

This analysis refers mainly to planar rectangular arrays of PEs having two inputs and two outputs. As seen in chapter 1, other planar arrays have a wider application: this is particularly true for four-neighbors mesh structures in which data transfers are bidirectional (applications are found, e.g., in image processing).

In this case, reconfiguration methods discussed previously can again be easily applied, changing the structure of hardware connections.

A more complex problem is the application of these methods to the case of three-dimensional (e.g., pyramidal) arrays. Extension to pyramids is made difficult by the complex relationship and communications existing among PEs of the array, and loss of communications locality when pyramids are projected onto flat surfaces. It would seem that complex arrays, like pyramids, can better be treated by end-of production static routing algorithms, based upon a regular structure composed of uniformly distributed PEs that are separated by wide routing channels.

On the contrary, *uniform multiple-pipeline* structures can be easily and efficiently reconfigured through the algorithms considered here. This particular case will be examined in chapter 14.

13.6. Other case studies

The approach based upon local Request/Acknowledge protocols is a quite natural one, and as such it has been proposed independently by several authors each of whom has given different emphasis to the degrees of generality granted by the approach itself.

A typical case is that of [EVA85], where two algorithms are presented for on-line self-reconfiguration of two-dimensional meshes. These two algorithms can be seen to correspond to cases /3 ∗ 1/ and /3 ∗ 3/. The suggested implementation is based upon Request and Availability signals that correspond to Request and Acknowledge signals.

Instead of determining harvesting, the authors investigate the *overhead factor* as a function of the production yield of the PE. For a given target array size, the overhead factor indicates the factor by which the number of cells in the target array must be multiplied in order to determine the size of the physical array that, on average, makes it possible to form the target array. The overhead factor is shown in figure 13.13. This figure is also practically valid for the /3 ∗ 3/ implementation discussed in the previous pages.

A second case is presented in [GUP86]: an approach to reconfiguration of two-dimensional meshes called *fan-out scheme*. Even if the algorithm is presented in a more complex way that suggests adoption only for off-line restructuring, it produces exactly the same configurations of /3 ∗ 1/, being an alternative formulation of it. The suggested structure corresponds to that of [EVA85]. The algorithm will be dealt with in greater detail in chapter 14.

A third case, again for two-dimensions meshes, is presented in [GRE85], as a method for building with high probability a number of $(1 - e)\sqrt{N}$ tentative horizontal chains of blocks. This method can be described by means of a neighboring set characterized by

$$/1 * 2C_2\sqrt{lgN}/$$

Essentially the same procedure is then adopted for identifying the vertically running chains; the blocks are subsequently expanded (see chapter 9, where this method has been discussed in detail).

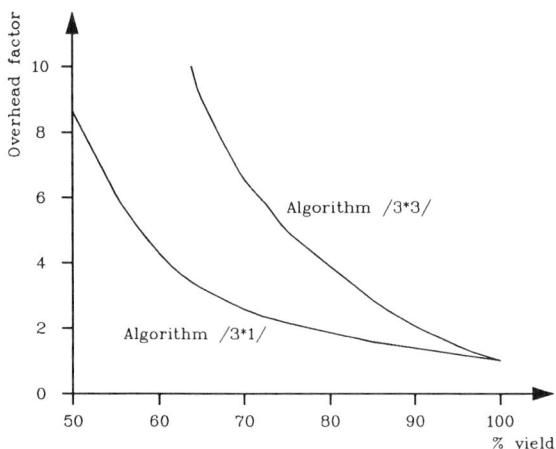

Figure 13.13
Performances of the algorithms from [EVA85].

In conclusion, it can be noted that the hard point of the algorithms (the *only* hard point, actually) is the determination of the rules that must be followed for identification of those fault-free PEs that cannot be used during the second phase of the algorithms, to satisfy the constraints given by the global array topology. This point requires the identification of all the possible topological errors, as was shown in figures 13.2.e-h. The $/3*1/$ algorithm of [EVA85] and the $/1*N/$ of [GUP86], as any $/1*x/$ or $/x*1/$ algorithm, do not face this hard point, because in any $/1*x/$ a target row cannot skip any physical column, and, for the same reasons, a target column cannot skip any physical row in $/x*1/$ algorithms.

In the [GRE84] case, the problem is *implicitly* solved by the fact that the array is first regularly partitioned into redundant blocks that are marked good and used globally only if they contain at least the necessary number of t PEs where t depends on the dimensions of the target array. The problem is also *explicitly* solved by the fact that the horizontal chains (after having built the vertical chains) are again modified in order to disconnect the PEs that are not used in the vertical chains.

13.7. References

[EVA85]   R.A.Evans, J.V.McCanny, K.W.Wood: *Wafer-scale integration based on self-organization*, in: *Wafer Scale Integration*, (W.Moore, C.Jess-hope eds.), Proc. Int'l Workshop, Southampton, July 10-12 1985, Adam Hilger.

[GUP86]   R.Gupta, A.Zorat, V.Ramakrishnan: *A fault-tolerant multipipeline architecture*, Proc. FTCS-16, 1986, 350-355, IEEE.

[GRE84]   J.W.Greene, A.El Gamal: *Configuration of VLSI arrays in the presence of defects*, JACM, Vol. 31, N. 4, Oct. 1984, 694-717, IEEE.

[KUN84]   H.T.Kung, M.S.Lam.: *Fault-tolerant VLSI systolic arrays and two-level pipelining*, Journ. Parall. and Distrib. Process., Aug. 1984, 32-63.

[MAN82]   T.E.Mangir, A.Avizienis: *Fault-tolerant design for VLSI; effect of interconnect requirements on yield improvement of VLSI designs*, IEEE TC, Vol. C-31, N. 7, July 1982, 609-615.

[NEG85]   R.Negrini, R.Stefanelli: *Algorithms for self-reconfiguration of wafer-scale regular arrays*, Proc. Int'l Conference on Circuits and Systems, Beijing, 1985, IEEE.

[NEG86]   R.Negrini, R.Stefanelli: *Comparative evaluation of space- and time-redundancy approaches for WSI processing arrays*, in: *Wafer Scale Integration*, (G.Saucier, J.Trilhe eds.), Proc. IFIP WG 10.5 Workshop, Grenoble, Mar. 17-19 1986, North-Holland.

13.8. Further readings

A. Zorat: *Construction of a Fault-Tolerant Grid of Processors for Wafer-Scale Integration* Circuits Systems and Signal Processing, Vol. 6, N. 2, 1987 (special issue on VLSI Technology and Computer System Design). A paper that contains, among others, a /3 * 1/ algorithm.

14 RECONFIGURATION OF
MULTIPLE PIPELINE STRUCTURES

A peculiar type of architecture — which can be considered as a subset of the rectangular array class — is comprised of *multipipelines*, i.e., sets of identical pipelines each consisting of several stages. Structures of this type are often found in supercomputers for performing vector operations [KOG81, HWA84]. Some dedicated architectures in the area of Digital Signal Processing belong to this same class, e.g., bit-serial DFT structures [BRU86], and pipelined FFT devices such as the ones discussed in chapter 1. While the individual pipe is obviously a linear array, the complete architecture can be seen as a rectangular array with a simplified interconnection structure.

Different instances correspond to the various applications noted above. In the case of vector processors, the different stages of a single pipe often perform different operations and the corresponding processing elements are therefore different. Thus, the multipipeline array is not completely homogeneous (as in the cases examined earlier), but homogeneity is found only *column-wise* (figure 14.1). In the case of *pipelined FFT*, the processing (butterfly) cells proper are identical, but associated memories (shift registers) have different lengths. Homogeneity can, of course, be reached at the expense of additional area by requiring all registers to have the maximum length. Finally, in bit-serial devices such as the DFT architectures in figure 1.14 all stages are perfectly identical, and total homogeneity of the rectangular array is reached. Reconfiguration policies will vary depending upon the particular instance.

14.1. Column-wise reconfiguration for non-homogeneous structures

The first class of architectures inevitably creates strong restrictions on the reconfiguration policies that can be adopted. Unless relevant redundancy is introduced (either by adopting *functionally reconfigurable* processing cells whose specific function is reprogrammed whenever reconfiguration is performed, or by duplicating each single column), reconfiguration can be performed only column-wise (see the physical organization in figure 14.1). A reconfiguration procedure explicitly dedicated to this instance is that presented in [GUP86] (examined in chapter 13 as an example of $/1 * N/$ and analyzed here from a different point of view): it is envisioned for run-time, dynamic reconfiguration, and its simplicity allows it to be implemented by on-chip switch-programming logic. The goal of the algorithm, on the other hand, is recovering the largest number of working pipes from a given array - i.e., optimizing *harvesting*, a factor of merit that is rather adopted in production-time restructuring. This seeming conflict can be overlooked if it is assumed that the host controller of the multi-pipeline system has the capacity of adapting to an array of variable dimensions and performances.

There is a basic drawback in the column-wise reconfiguration technique, and

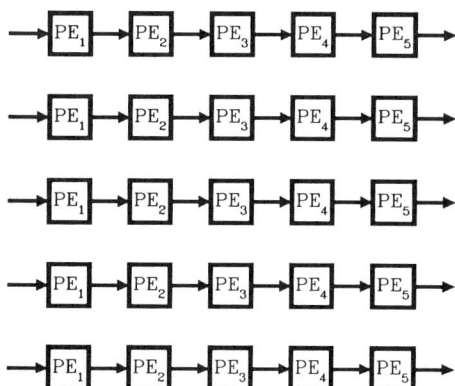

Figure 14.1

Example of multipipeline structure in a vector processor: PEs inside each column are identical.

the authors of [GUP86] try to overcome it. If during reconfiguration the modified
interconnections created have the sole function of avoiding the faulty elements in
the individual columns, connecting a cell in row i of column j with a cell in a row
$i + x$ of column $j + 1$ (the value of x depending on the presence of faulty cells
in column $j + 1$), either a restriction on the number of adjacent faulty cells in
any individual column has to be accepted (and this, of course, leads to premature
fatal failure occurrence) or else a large channel width (actually depending on the
array dimensions) must be provided. The solution chosen by the authors is to
accept that some fault-free cells be considered as *pseudo-faults* and to bypass them
so as to create fault (and pseudo-fault) patterns that are better circumvented
by the interconnection network. Taking this philosophy into account (which is
adopted in other instances as well, i.e., as discussed in chapter 12, with the identical
purpose of limiting channel width and interconnection length) the basic algorithm
proposed in [GUP86] can be summarized (see figure 14.2 as an example of the final
organization):

14.1.1. Algorithm 14.1 ($/1 * N/$)

- denote by i the current physical row index and by i' the index of the target
  pipeline being built: construction starts with $i = 1$, $i' = 1$. The algorithm
  configures the i'-th pipeline by selecting the first usable (i.e., neither faulty
  nor a pseudo-fault) stage from the top, in every column.

- As soon as all necessary PEs have been assigned to target pipe i', switches
  are set to connect them in an ordered way and pseudo-faults are determined
  as follows. Consider two consecutive stages j, $j + 1$ and assume that PE (e, j)
  and PE $(f, j + 1)$ have been assigned to the target pipe i', then:
    - if $e = f$, the two cells are directly connected (the intermediate switch is
      set to the *through* mode): see all the PEs in the first target pipe in figure
      14.2;

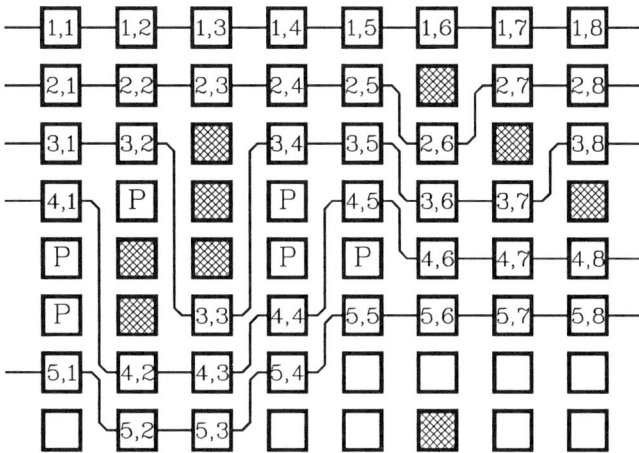

Figure 14.2
Column-wise reconfiguration by Gupta's algorithm.

- if $e < f$, a *downwards* connection must be created from stage j to stage $j+1$. To this end, the switches on outputs of PE (e, j) and of PE (f, j) will be set on the *down-connection* mode, while every PE (k, j), $e < k < f$, will be assumed to be a *pseudo-fault* and as such will become unusable in the construction of any target pipeline (see PEs $(4, 1)$ and $(3, 4)$ in figure 14.2);

- if $e > f$, the above procedure is simply reversed and every PE $(k, j + 1)$, $f < k < e$, is declared to be a *pseudo-fault* (see target PEs $(3, 3)$ and $(3, 4)$ in figure 14.2). Note that in this way a completely symmetrical structure is achieved.

Thus, when building pipeline i', if for each stage j there is at least one unused fault-free (or pseudo-fault-free) PE, the pipeline can be completed.

The algorithm can be supported by a simple on-chip structure of switched buses (one bus in each vertical channel) and by switch-controlling logic: control signals propagated through the array to such control logic are few, and thus added silicon area is kept quite low. With regard to I/O connections, the authors do not discuss the problem of creating such connections between extremes of the target pipes and the device pins, although, of course, such connections could be created by making use of the *selector* model presented in [GRE84].

Yield provided by the above algorithm is not, generally, 100 percent, because of the appearance of pseudo-faults. With regard to delays, while worst-case fault distribution leads to delays $O(N)$, N being the number of pipes, the authors prove that the probability of limiting delays to $O(\log N)$ is actually high; larger yield can be reached for larger channel widths (related to $\log N$).

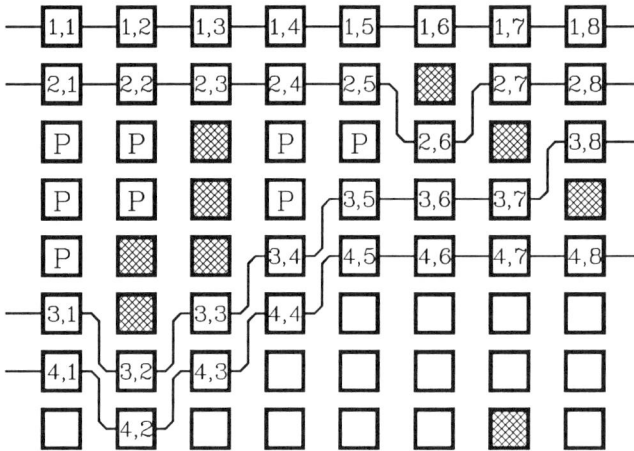

Figure 14.3
Column-wise reconfiguration by algorithm 14.2.

14.1.2. Algorithm 14.2 (/1 * 3/)

Dynamic reconfiguration algorithms based on the propagation of protocol signals such as the ones discussed for meshes in chapter 13 can be used also in multiple pipeline arrays: they allow circumvention of large defects affecting clusters of faulty PEs and the segments of the interconnection network inside the clusters. The physical architecture has as many columns as the target array, and a number of spare rows. The algorithm is identical to that in [NEG85] and in [EVA85] (chapter 13) as far as line propagation is concerned; they all can be seen as belonging to the /1 * x/ class, All algorithms of this class can be proposed for dynamic, run-time reconfiguration driven by on-chip circuits as well. The augmented interconnection network consists of one vertical bus per channel with one switch associated with each PE; an example of reconfiguration is given in figure 14.3.

As might be expected, given the type of algorithm chosen, harvesting does not reach 100 percent. In fact (as also occurs in [GUP86]), a number of fault-free PEs cannot be inserted in the target array and should therefore be considered as *pseudo-faults* (such pseudo-faults P can be seen in figure 14.3).

14.2. Homogeneous multipipelines

In considering the second, totally homogeneous architecture class, an immediate solution would be to also apply here the algorithms developed for rectangular arrays in chapters 12 and 13, suitably simplified as far as the interconnection network is concerned. This approach can be adopted both with regard to dynamic *reconfiguration* assuming an array with a predetermined number of spares, and with regard to *restructuring*.

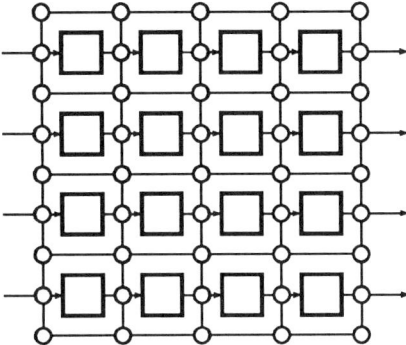

Figure 14.4
Switched-bus structure supporting algorithms 12.3 and 12.4 for a multipipeline structure.

14.2.1. Algorithm 14.3

Index-mapping techniques described in chapter 12 (in particular, algorithms 3, 4 and 6) can be extended to the multipipeline instance. The interconnection network required here is much simpler than the one necessary for the rectangular array. Consider, for example, figure 12.36 in chapter 12, which shows that horizontal and vertical signals require different numbers of buses and switches. A reduced silicon area will therefore be required here, by keeping horizontal signals and deleting the vertical ones. Thus, extension of the above algorithms, together with introduction of one spare row and one spare column, will require one horizontal and one vertical bus, two switches being associated with each PE (see figure 14.4).

In figure 14.5, a sample 7×7 physical array comprising one spare row and one spare column (inserted as the lowest and rightmost ones, respectively) is shown: a random fault distribution has been superimposed onto it, and the ensuing deformation lines are shown. Mapping of logical onto physical indices and final interconnection paths are shown in figure 14.6.

With such techniques, probability of survival is not 100 percent, but on the other hand, the algorithms guarantee an exact upper limit for the interconnection links and therefore for the communication delays. Moreover, interconnections to the I/O pins are effected by the same algorithm implementing reconfiguration, so that reconfiguration will actually be transparent to an external machine interacting with the array.

Another approach — closer to that previously described in subsection 14.1 — takes into account the specific characteristics of the multipipeline arrays. In essence, the aim is to extract linear arrays of length N from a given rectangular distribution of cells whose direct interconnections follow only one direction (the structure of the augmented interconnection network will be deduced from the particular algorithm adopted). Consider first dynamic reconfiguration in the presence of a given distribution of spares. A very simple solution can be described

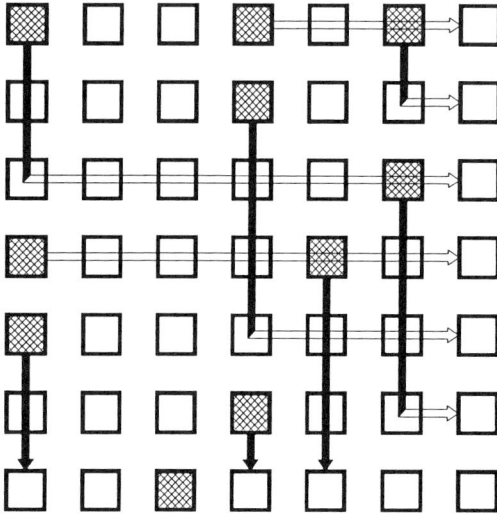

Figure 14.5
Reconfiguration rules for multipipeline by algorithm 12.4.

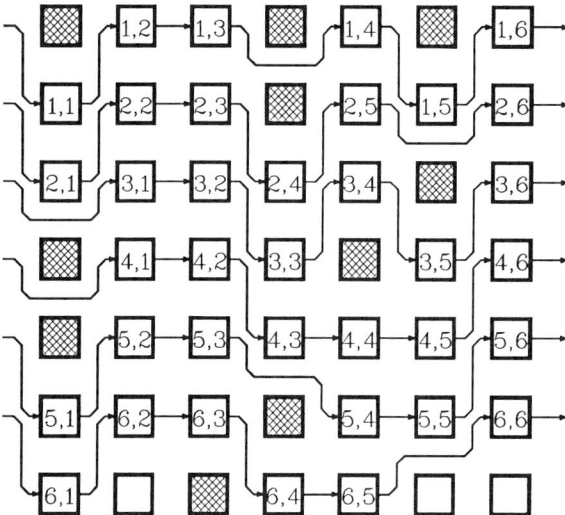

Figure 14.6
Index mapping and interconnections for the example in figure 14.5.

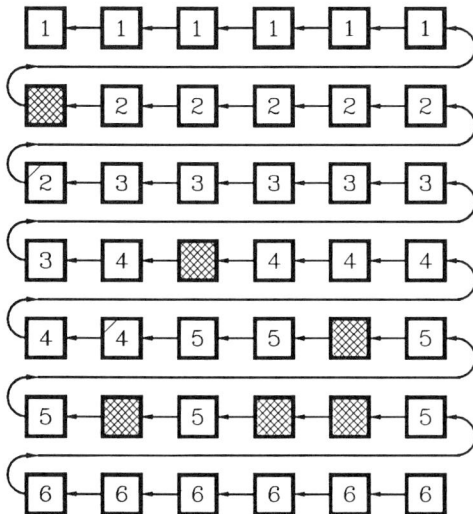

Figure 14.7
Initial pipeline assignment by algorithm 14.4.

as follows:

14.2.2. Algorithm 14.4

Assume a basic array consisting of M pipelines of length N and a single row of spares added to it: the goal is to again obtain M pipelines in the presence of any distribution of up to N faulty cells.

(1) Starting with cell $(M + 1, N)$ and with target pipe $i' = M$, begin counting fault-free cells with physical indices (i, j) for decreasing values of j and — in order — of i, with the purpose of building pipe i'. As soon as a set of N fault-free cells has been found, they are assigned to pipe i' which is then accepted as completed; i' is decreased (unless, obviously, the first target pipe had already been reached) and the procedure is iterated (see figure 14.7).

(2) At the end of step (1), all target pipes have been identified: column indices must now be associated with the individual cells. Given the distribution of spares, cells attributed to pipe i' may belong to physical rows $i+1$ $(i = i')$ and — possibly — i, as shown in figure 14.8.a. To assign the column index, for each pipe i' start with the leftmost cell assigned to pipe i' in row $i+1$, denote it with column index $j = 1$, then proceed rightward on row $i + 1$, increasing index j, for all cells belonging to pipe i'. When the last cell belonging to i' has been reached, if $j < N$, then cells belonging to pipe i' in row i are inserted starting again (as far as column index is concerned) from the leftmost one, as shown in figure 14.8.b.

The length of connections with I/O pins is immediately minimized; it is easily

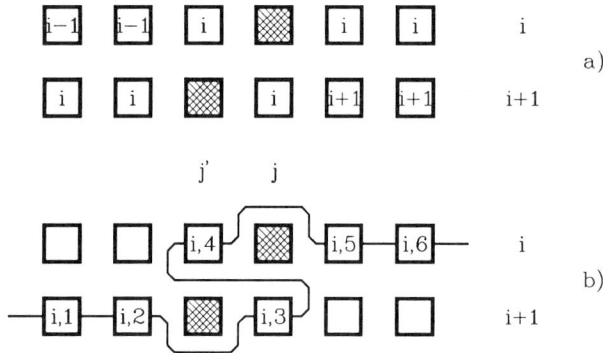

Figure 14.8

a) Possible assignment of cells to pipes in adjacent rows. b) Column index mapping for case a).

proven that the leftmost fault-free PE in row $i + 1$ is the *input* PE of pipe i', as the rightmost fault-free PE in either row $i + 1$ or (if it is also used for pipe i') in row i is the *output* PE of the same pipe. I/O connections, moreover, again require one vertical bus and one switch per pin, and are effected by the same technique used for internal connections; this allows easy run-time reconfiguration.

The reconfiguration procedure is very simple (although in principle it is foreseen to be host-driven, it actually requires low computing capacity). As for interconnection paths between two *logically adjacent* cells, four instances may occur:

- direct link between physical adjacents: no buses or switches are involved;

- interconnection between two cells $(i + 1, k)$, $(i + 1, h)$ with $h > k + 1$ (this is a result of the presence of faults in cells $(i + 1, k + 1)...(i + 1, h - 1)$: the connection is made by exploiting the horizontal bus segment *below* the faulty cells (and, obviously, vertical bus segments downward at the output of $(i+1, k)$ and upward at the input of $(i + 1, h)$.

- interconnection between two cells of row i with intermediate faulty cells (defined as above): the connection is made by using the segment of horizontal bus *above* the faulty cells;

- connection from the last fault-free cell of row $(i + 1)$, $(i + 1, k)$, belonging to pipe i' to the first fault- free cell of row i, (i, h), belonging to pipe i': in that case, the horizontal bus segment between the two rows from column h to column k is exploited.

An example of reconfiguration is shown in figure 14.9. It can be easily proven that no conflicts can arise either on the horizontal or on the vertical bus.

The above technique guarantees 100 percent probability of survival; against this advantage, intercell connection length can reach $N + 1$ (see figure 14.10).

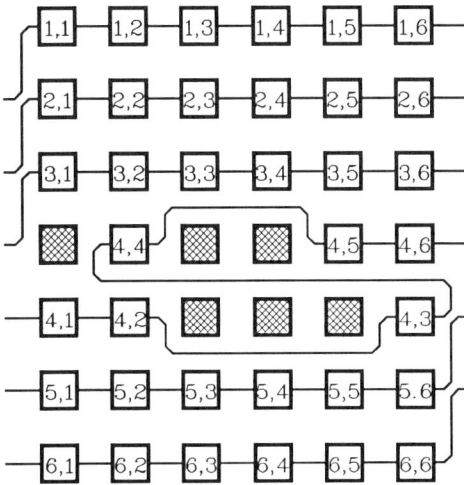

Figure 14.9
Example of reconfiguration by algorithm 14.4.

14.2.3. Algorithm 14.5

The same criterion can be extended to production-time restructuring, with the aim of reaching 100 percent harvesting. In that case, a target pipeline can be made up of PEs belonging to more than two physical rows, with the provision that the leftmost cell in the lowest row and the rightmost cell in the uppermost row contributing to i' are to be connected with I/O pins (see the example in figure 14.11, where target pipe $i' = 3$ extends over four physical rows). I/O connection lengths are not bound, so that the technique is suitable only for production-time restructuring. Moreover, the augmented interconnection network here requires *two* horizontal buses and one vertical bus per channel.

For large fault distributions the total pipe-level delay introduced by the above criterion can become excessive. An algorithm described in [NEG87] allows the goal of 100 percent harvesting to be preserved while obtaining lower mean values of pipe-level delays. It can be summarized as follows.

14.2.4. Algorithm 14.6

Given a physical array of $M \times N$ cells, with any given distribution of F faulty cells, the aim is to extract from it K pipelines of length N, with $K = \lfloor (NM - F)/N \rfloor$;

- denote by Fi the number of faults in row i. Construction starts with pipe $i' = 1$, and from cell $(1, 1)$. If there are no faults in row 1, the whole pipe $1'$ will be created with row i; otherwise, $F1$ fault-free cells will be borrowed from row 2 and (if necessary) from the following rows as well, and these will

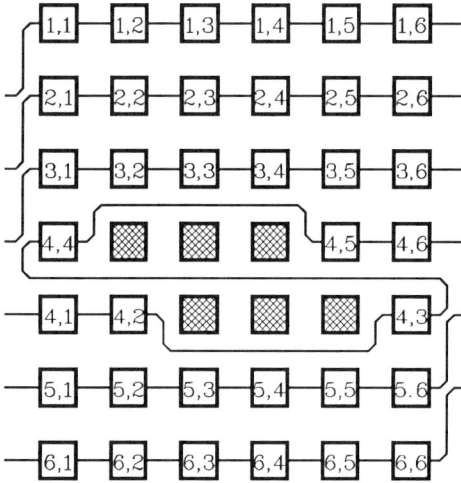

Figure 14.10

An example of reconfiguration solved by algorithm 14.4 showing a worst-case interconnection length.

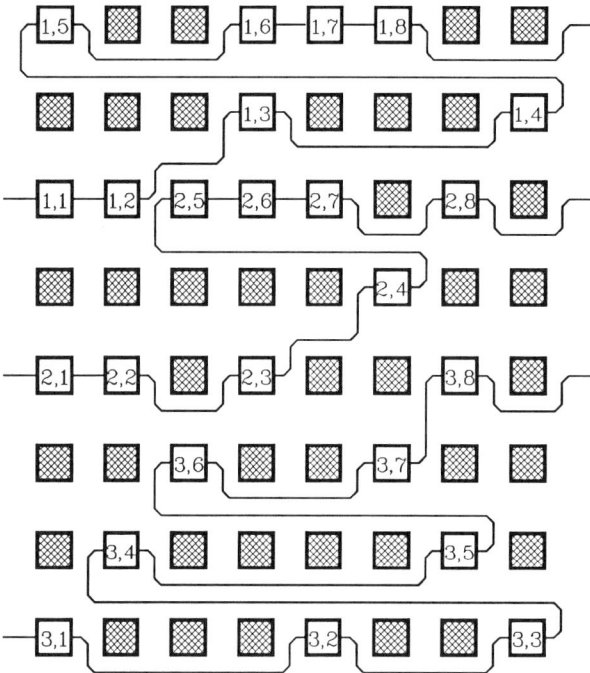

Figure 14.11

Example of restructurating by algorithm 14.5.

then be considered, while constructing subsequent pipelines, as *stolen cells*, and as such unavailable.

- Let i be the row to which the first cell of target pipe i' belongs: pipe i' will in general have to borrow $Fi + Pi$ cells from the following rows (Pi being the number of stolen cells in row i). Cells are borrowed starting from the right and going to the left in each row: if row $i + 1$ contains $Fi + Pi$ fault-free cells, these will all be blocked as *borrowed* by row i: if not enough fault-free cells are available in row $i + 1$ (in which case, of course, *all* fault-free cells in i will be borrowed) the missing ones will be borrowed from rows $i + 2$... etc., always with the same rule (from the right to the left).

- denote by (h, k) the physical cell reached while building pipe i'; let R be the number of cells that must still be inserted into pipe i' in order to reach its nominal length N.

If $h = i$, then:

  - if (h, k) is fault-free, it is directly inserted in pipe i';

  - if (h, k) is faulty, column index p of the first available cell $(h + 1, p)$ reserved for pipe i' and still unused is checked: if p is lower than the column index of the first fault-free cell in row h, $(h + 1, p)$ is inserted into the pipe and pipe i' is continued on row $h + 1$ (see rules below);

- if $h > i$, then check whether available cells in rows $h - 1...i$ are still unused and fault-free (or pseudo-fault free) and in that case switch to the highest row in which there is such a row. Otherwise, continue the construction of pipe i' on row h (and on other possible rows with indices $g > h$ reserved for pipe i') with the rules already presented.

By the above rules (as for the previous algorithms), a pipe always starts with the cell nearest to the left border in the highest row reserved to the pipe itself, and terminates with the cell nearest to the right border in the terminal row of the pipe.

The example in figure 14.12 refers to the same fault pattern adopted in figure 14.11. Comparison of this algorithm with the previous one enables us to see that much lower delays can be reached for a given fault distribution. In fact, while the three pipes in figure 14.11 have lengths, respectively, of 21, 13, 32, (length computation is made by the usual conventions), lengths of the same pipes in figure 14.12 are (in the same order) of 11, 12, 16. Silicon requirements are largely the same, since this second algorithm requires two buses in the vertical channels and one in the horizontal ones.

This same algorithm can also be adopted to restructure pipes of length $Q > N$, without any modification: for lengths lower than N, further horizontal buses would be needed to accommodate multiple I/O connections from the same row but the need for *larger* numbers of I/O pins than originally provided would make restructuring unreasonable!

It is interesting to compare the various possible algorithms for multipipeline reconfiguration, considering a number of alternative figures of merit. An initial

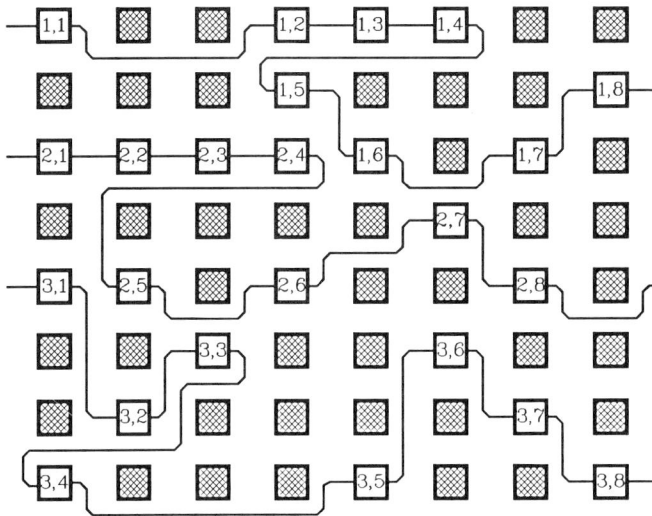

Figure 14.12
Example of figure 14.11 solved by algorithm 14.6.

evaluation may take into account the following factors:

- length of intercell connections fixed or depending on array dimensions;
- number of spares (where applicable, i.e., when the goal is probability of survival rather than harvesting);
- time of restructuring/reconfiguration (production time *vs.* run time);
- efficiency of harvesting/spares utilization (100 percent *vs.* variable with fault distribution);
- number of buses/cell.

Table 14.1 permits a comparative evaluation of the various algorithms discussed in this chapter (either application-specific or extensions of techniques for rectangular arrays). Further evaluations can be made on a statistical basis:

14.3. Harvesting

Algorithms 14.1 ($/1 * N/$) and 14.2 ($/1 * 3/$) are compared by means of simulation on a 10×10 array; results are given in figure 14.13.a for algorithm 14.1 and in figure 14.13.b for algorithm 14.2 [NEG85, EVA85]. P is the probability of obtaining the maximum number K of pipes (of 10 cells each) in the presence of a given number of faults (variable on the horizontal axis).

The restrictions of algorithm 14.2 on allowable interconnection lengths with respect to algorithm 14.1 leads to lower performances, because of the introduction of higher numbers of pseudo-faults.

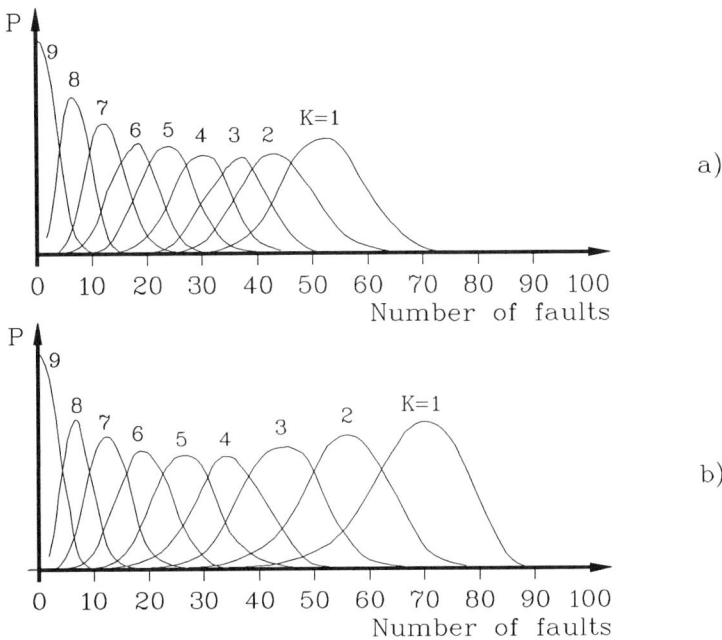

Figure 14.13

Probability of harvesting against increasing numbers of faults a) by Gupta's algorithm 14.1 (i.e., $/1 * N/$), and b) by algorithm 14.2 (i.e., $/1 * 3/$).

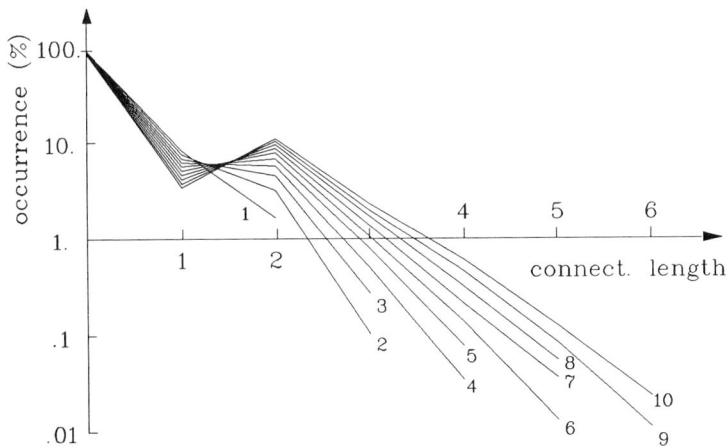

Figure 14.14

Probability of occurrence of interconnections vs. interconnection length for varying numbers of faults, Algorithm 14.1.

Algorithms 14.5 and 14.6 always grant 100 percent harvesting, by extracting $H = \lfloor (M \cdot N - F)/N \rfloor$ pipes for any number F of faults.

14.4. Length of interconnections between logically adjacent PEs

As was done earlier, the unit adopted is the edge of an ideal square cell containing the PE: channel width is not taken into account. Connections with I/O pins will also be evaluated (for run-time reconfiguration, I/O pins are in fixed positions; for restructuring, they are chosen as the nearest pads to extreme cells in each pipe).

(a) Algorithm 14.2 gives links of length 0 or 1 only (and no pass-through cells are allowed);

(b) For algorithm 14.4, plots in figure 14.14 (referring to a 9×10 target array from a 10×10 physical one) give the probability that an interconnection link assumes a given length (length is the variable on the horizontal axis) in correspondence of different numbers of faults. In figure 14.15, probability of occurrence for each given length is plotted *vs.* numbers of faults. A physical $N \times N$ array contains $N \times (N-1)$ internal links and $2 \times N$ connections to I/O pins; therefore the probable number of interconnections with a given length can be obtained by multiplying the probabilities in fig 14.15 by $N \times (N+1)$. The representation used in figure 14.15 will also be adopted for the remaining algorithms.

(c) For algorithm [GUP86], results are plotted in figure 14.16. Maximum length, i.e., $N-1$ (in our case, 9) has probability higher than one per thousand only when the number of faults is higher than 56 percent.

(d) In figure 14.17 results referring to algorithm 14.5 are plotted.

(e) For algorithm 14.6, results are plotted in figure 14.18. Connections of length $N-1$ occur with probability higher than one per thousand only when the number of faults exceeds 78 percent.

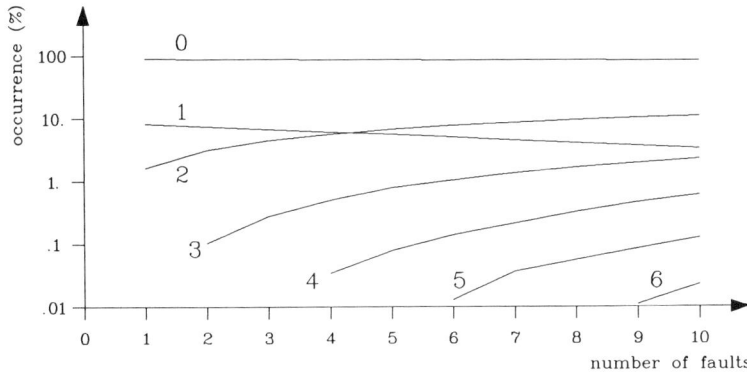

Figure 14.15

Probability of occurrence of interconnections vs. numbers of faults, for varying interconnection lengths.

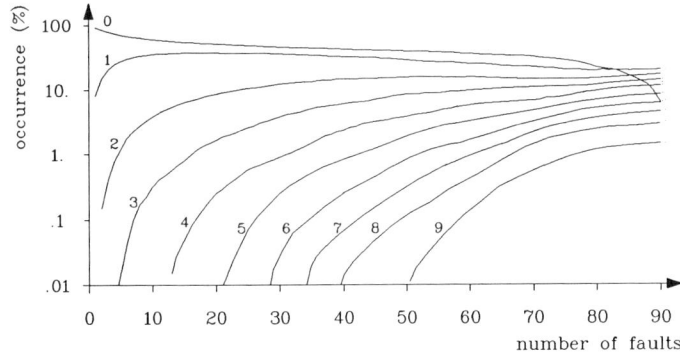

Figure 14.16

As in figure 14.15, Gupta's algorithm.

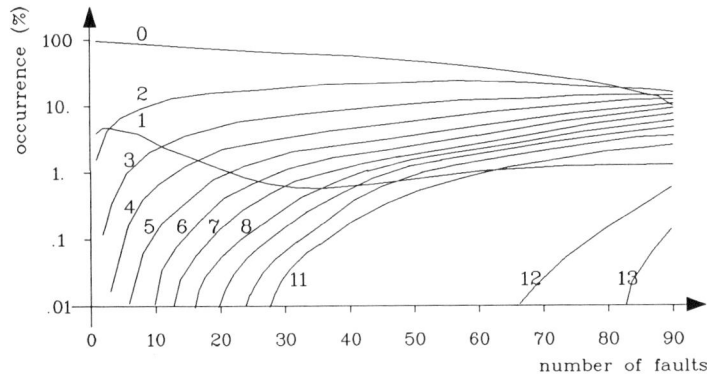

Figure 14.17

As in figure 14.15, Algorithm 14.1, for more than N faults.

Table 14.1

Performance comparison of the algorithms described in this Chapter, applied to a $N \times N$ physical array.

LI = Max length of interconnection links

NS = Number of spare PE: may be fixed (one row and/or column) or variable when harvesting is considered

HE = Harvesting efficiency. That of algorithm 14.1 can be considered high, and that of algorithm 14.2 can be considered medium (see figure 14.13).

NB = Total number of buses in horizontal and vertical channels

SW = number of switches per PE

NU = Non uniform array: capability of reconfiguring *column-wise uniform* arrays

TL = Variable target length: capability of obtaining target pipes with length greater than the physical one (functional configuration)

PP = Position of I/O pins: may be *near* or *far* from the cell to be connected

algorithm	14.1	14.2	14.3	14.4	14.5	14.6
LI	N-1	1	2	N+1	2N+1	2N+1
NS	var	var	2N-1	N	var	var
HE	high	med	-	-	100%	100%
NB	1	1	2	1	3	3
SW	1	1	2	2	3	3
NU	yes	yes	no	no	no	no
TL	no	no	no	no	yes	yes
PP	far	far	near	near	far	far

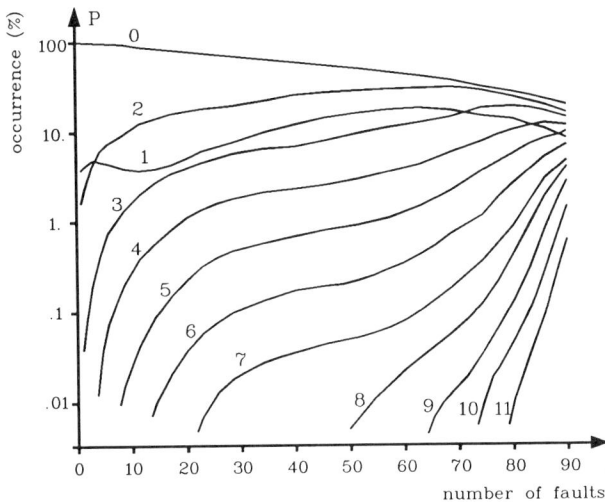

Figure 14.18
As in figure 14.15, for algorithm 14.2.

14.5 References

[BRU86] O.Bruschi, R.Negrini, S.Ravaglia: *Systolic arrays for serial signal processing*, Microprocessing and Microprogramming, Vol. 20, N. 1-3, 1987, 133-140.

[EVA85] R.A.Evans: *A self-organizing fault-tolerant 2-d array*, Proc. VLSI-85, Tokyo, 233-242, IEEE.

[GRE84] J.W.Greene, A.El Gamal: *Configuration of VLSI in the presence of defects*, JACM, Vol. 31, N.4, 694-717, Oct. 1984

[GUP86] R.Gupta, A.Zorat, V.Ramakrishnan: *A fault-tolerant multipipeline architecture*, Proc. FTCS-16, 1986, 350-355, IEEE.

[HWA84] K.Hwang, F.A.Biggs: *Computer architecture and parallel processing*, New York, 1984, McGraw Hill.

[KOG81] P.Kogge: *The architecture of pipelined computers*, New York, 1981, McGraw Hill.

[NEG85] R.Negrini, R.Stefanelli: *Algorithms for self-reconfiguration of wafer-scale regular arrays*, Proc. Int'l Conference on Circuits and Systems, Beijing, 1985, IEEE.

[NEG87] R.Negrini. M.G.Sami, R.Stefanelli: *Restructuring and reconfiguring DSP multipipeline arrays*, Proc. MTNS87, Phoenix, June 1987.

15 SOME EXTENSIONS TOWARDS TIME-REDUNDANCY

In the reconfiguration techniques analyzed in the previous chapters, array-level processing speed was kept basically unchanged, and maintaining added delays as low as possible was one of the most important figures of merit. As an alternative, the possibility of accepting degradation with regard to operation speed can be explored, thus introducing the concept of a *time redundancy*, i.e., that the nominal speed should be higher than the one related to strictest requirements (*redundant with respect to design specifications*) and that reconfiguration involves lowering of such speed.

Most proposals involving time redundancy are based upon a close examination of the specific *algorithm* mapped onto the array. Variations of such mapping due to introduction of faulty PEs are examined and the corresponding variations in overall processing speed are evaluated. A basic example can be derived in the case of linear arrays and of operations such as convolution: whenever the number of PEs in the array (independent of faults) does not coincide with that of elementary operation modules present in the systolic definition of the array, a *folding* of the algorithm upon the array becomes necessary and a corresponding operation time is evaluated. Introduction of faults (and a simple corresponding bypass of faulty PEs) immediately leads to a different folding of the same algorithm upon a different set of PEs and thus to a new evaluation of a (in general, reduced) operation speed.

This type of approach is outside the scope of this book, since it is better derived from an *algorithmic* study of systolic and wavefront-computation arrays (see, for example, [KUN84] and [MAJ88]). We will discuss here a few reconfiguration proposals that simply transfer reconfiguration approaches adopted in the usual *area redundancy* solutions to the *time redundancy* domain. In such techniques no mention is made of the algorithm implemented on the array, but rather the modified mapping of array functions upon fault-free PEs (and the modified interconnections activated among these same PEs) are considered; the same process will apply with regard to the introduction of *spare processing phases* (i.e., a form of time redundancy) instead of *spare processing PEs*.

Two different proposals, each referring to a different fault model, will be examined in detail: the first accepts random faults located in PEs only, while the second accepts clustered faults affecting the segments of the interconnection network as well as PEs inside the clusters.

15.1. Index-mapping in the time domain

The basic concept of index mapping has been discussed in chapter 12; its extension to the time domain can be derived from the following: whenever a PE is declared faulty, its functions (represented by its *logical indices*) are taken over by another fault-free PE. If such a PE had non-null logical indices (i.e., it was already assigned

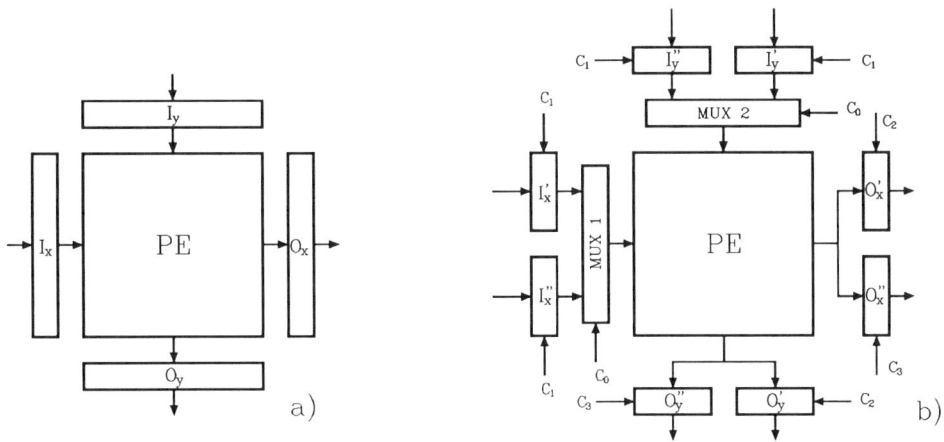

Figure 15.1

a) Basic structure of the individual PE with I/O latches. b) Modified structure supporting time redundancy: each latch is substituted by a pair of latches, controlled by a staggered clock.

a function in the working array) the new logical indices will cause such a PE to operate in a *spare* processing phase (either nonexistent in the nominal, fault-free mode, or a phase during which the PE was otherwise inactive).

It is self-evident that this type of approach — whatever its specific implementation — requires strict timing to grant correct operation even after reconfiguration. Synchronous arrays are then considered, and the basic (nontime reconfiguring) structure for the individual PE is that given in figure 15.1.a [SAM84]. Each PE has two input latches I_x, I_y, two output latches O_x, O_y, and two staggered clocks C_1 and C_2 (with identical frequencies) are required for operation. Clock C_1 controls operation of the PE, which processes data available in its input latches and loads results in output latches; clock C_2 controls transfers, respectively, from $O_x(i,j)$ to $I_x(i,j+1)$ and from $O_y(i,j)$ to $I_y(i+1,j)$ in order to set up conditions for the subsequent processing step. Together, the two clocks contribute to one single processing phase.

To achieve reconfiguration, the single phase is split into two subphases, now controlled by three clocks C_1, C_2, C_3 — again characterized by the same frequency but staggered with respect to each other, and by a fourth clock C_0 identifying the operation subphase. Each I/O latch, in turn, is substituted by two latches (see figure 15.1.b), each of them used each in one of the two subphases. As a consequence of reconfiguration, any PE (i,j) can be associated with two different pairs of logical indices, (i',j') during subphase 1 and (i'',j'') during subphase 2. Horizontal and vertical input signals are forwarded through input multiplexers that select input data for each alternating phase: in the same way, control signals drive the loading of output latches. The three staggered clock signals control array

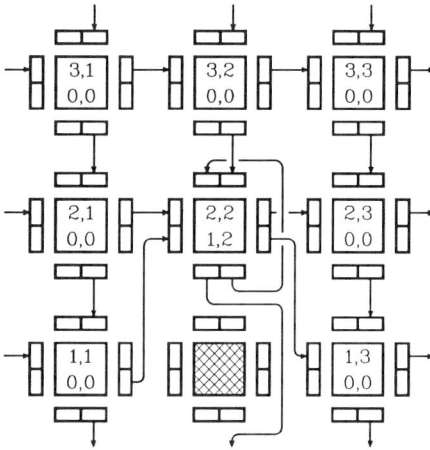

Figure 15.2
Example of reconfiguration for a single fault.

operation as follows:

(1) Clock C_1 controls loading of latches I'_x, I'_y from the output latches of first-phase logical adjacents and of I''_x, I''_y from output latches of second-phase logical adjacents (denote such adjacents, in order, as $(i'-1, j')$, $(i', j'-1)$, $(i''-1, j'')$, $(i'', j''-1)$);

(2) C_2 controls first-subphase processing and loading of output latches O'_x, O'_y;

(3) C_3 controls second-subphase processing and loading of latches O''_x, O''_y.

If no faults are present, only the first subphase involves actual processing; in the presence of faults, and as a consequence of reconfiguration, some fault-free PEs will be operating in both subphases while faulty PEs — and, in general, some fault-free PEs as well — will be inactive in both subphases. To reduce the complexity of the augmented interconnection network (which obviously introduces a *structure redundancy*), the reconfiguration algorithm is very simple, involving only cells along the same columns as the faulty ones and based upon a set of redundant links connecting each PE with neighbors only at unit distance from it and in the same columns as those of the *nominal* neighbors.

To describe the reconfiguration algorithm in its simplest terms, first consider one single fault (see figure 15.2, where row indices increase from bottom to top). In the example, cell $(3,2)$ is faulty and cell $(2,2)$ substitutes for it, performing its functions during phase 2 and acting as a *time spare*. The detailed operation is then as follows:

(1) Clock C_1:

$$\text{load } I'_x(i, j+1) \text{ from } O''_x(i+1, j),$$
$$\text{load } I'_y(i-1, j) \text{ from } O''_y(i+1, j)$$

load $I_x''(i+1,j)$ from $O_x'(i,j-1)$

load $I_y''(i+1,j)$ from $O_y'(i+1,j)$

In this way, during the two different subphases of one phase, cell $(i+1,j)$ actually becomes *adjacent to itself* on the vertical axis;

(2) Clock C_2:

control the first subphase processing and the loading of O' latches for all operating PEs, on the basis of *first-subphase* logical indices (upper pairs of indices in figure 15.2);

(3) Clock C_3:

control the second subphase processing *limited to* those PEs that are acting as time-spares (in our particular instance, limited to PE $(2,2)$) on the basis of second-subphase logical indices. Such indices are set to $(0,0)$ for all PEs inactive during this phase.

Reconfiguration can be performed only in the interval between signals C_3 and C_1. Thus, worst-case error latency will correspond to one full processing cycle and faulty results *might* appear at the external pins (as a consequence of error propagation) but it can be proven that after a transient of N cycles (at most) correct results will again be provided.

When more extended fault distributions are present, the reconfiguration algorithm becomes somewhat more complex, since first of all it involves the computation of the *displacement*, i.e., the number of rows in each column that will be active during both subphases. Since each cell active in two subphases will provide not only its second-subphase processing capacity, but also a horizontal link *borrowed* for information transfers by lower rows, it becomes necessary to create paths that avoid any possible conflict for access to horizontal links. To this end, some fault-free PEs may have to be declared as *pseudo-faults* and thus allowed to be inactive during both phases (the philosophy of *pseudo-faults* has been already introduced in chapters 13 and 14). We do not deal in detail here with the evaluation of displacements and with the definition of pseudo-faults. An example of reconfiguration for a relatively complex fault distribution (figure 15.3) illustrates the choice of pseudo-faults, aimed basically at granting that active cells belonging to the same logical row should never be at a relative distance higher than unit (along the columns) from each other.

Probability of survival to faults is plotted as curve A in figure 15.4 (for a sample 12×12 array). As might be expected, it is not very satisfactory, since the presence of even a single fault in row 1 will lead to fatal failure as no *upper* row from which a time-spare could be requested is available (and, under the assumption of random fault distribution, probability of finding such a fault is $1/N$ for the $N \times N$ physical array). Performances are improved by introducing a measure of structure redundancy (in addition to that already resulting from the augmented interconnection network), consisting simply of an added row as *row 0* at the top of the array. The following facts are then verified:

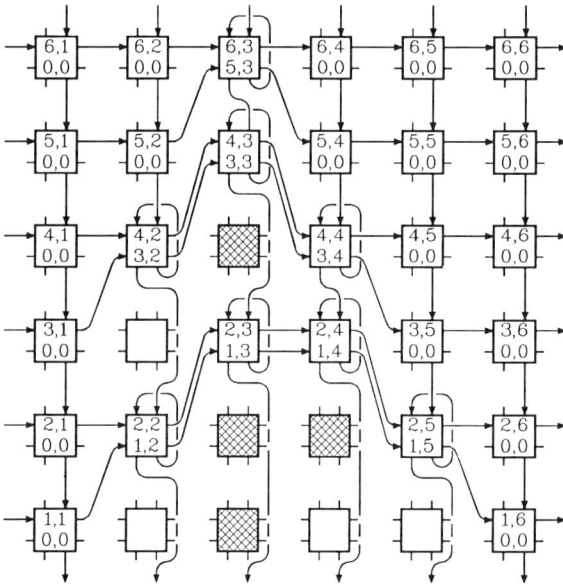

Figure 15.3

Reconfiguration through time-redundancy for a complex fault distribution.

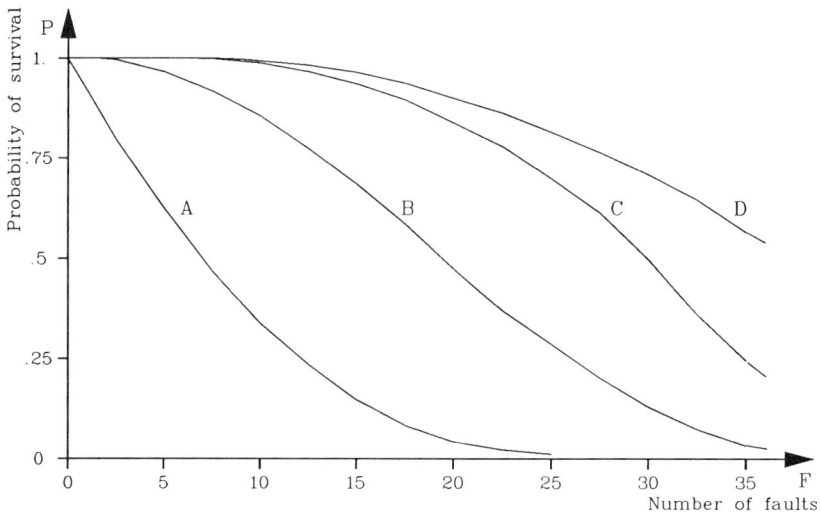

Figure 15.4

Probability of survival for a 12×12 array: curve A: basic algorithm; curve B: results after addition of one spare row $(13 \times 12$ array); curve C: pre-configured $12 X 12$ array; curve D: preconfigured array with spare row.

- single faults can always be corrected;

- double faults lead to fatal failure only if faulty cells are adjacent in a column and if they belong to column 1 or to column N or else, again, if one of the cells belongs to row $N+1$ (the last fact is due to the constraint that requires connections to I/O pins to reach only length 0 or 1). Probability of such an occurrence is inversely proportional to N^3.

Thus, curve B in figure 15.4 gives probability of survival for this new 13×12 array, in which row 0 is a spare one. Such performance is far better than for any *structural redundancy* approach with just N spares but, on the other hand, speed has evidently been decreased. The whole solution might be envisioned, with regard to its external behavior, as a conventional, structural redundancy one in which PEs half as fast as the nominal ones are used and the number of available spares equals in fact $N \cdot N + 2N$. This interpretation derives from the fact that each *nominal* cell could act as a time spare, and that each cell in the spare row, moreover, may act as a time spare twice. This first technique in fact is based upon the attempt to maintain nominal speed as long as there are no faults, while degraded performances are accepted only after the appearance of the first fault. If speed decrease is accepted during the whole lifetime of the reconfigurable system (a reasonable assumption, seeing that it leads to simpler control structures and I/O synchronization solutions), better performances can actually be obtained. Such a solution leads to a *pre-configured* structure. Again, the basic idea can be that of keeping as many cells in the lower rows working during *both* subphases — as long as no faults are present — and of using cells in the upper rows only as actual spares. Thus, a 6×6 array in the absence of faults will involve the index mapping seen in figure 15.5, while a sample fault distribution will lead to the index mapping in figure 15.6. Probability of survival is now given by curve C in figure 15.4 (12×12 array) and by curve D (addition of a spare row).

15.2. Protocol-driven reconfiguration in the time domain

In chapter 13, protocol-driven reconfiguration techniques were discussed with particular reference to restructuring of wafer-scale arrays, with an underlying fault assumption by which a whole area might become defective, thus involving not only PEs but also interconnection links and related circuits. In [NEG86] an approach was presented by which time redundancy was also applied to that type of fault model, thus extending protocol-driven techniques to the time domain.

Here too, *request* and *acknowledge* signals constitute the basis of the reconfiguration procedure (as for all other protocol-driven techniques): these signals are managed by fault-free cells only, and control signals do not flow through defective areas. Each physical cell is made to correspond to two *virtual cells* — one for each of two operating subphases, as in the case discussed in subsection 15.1 (see figure 15.1).

With these conditions, reconfiguration is performed just as in the structure-redundancy case, by transposing the protocol philosophy into the time domain,

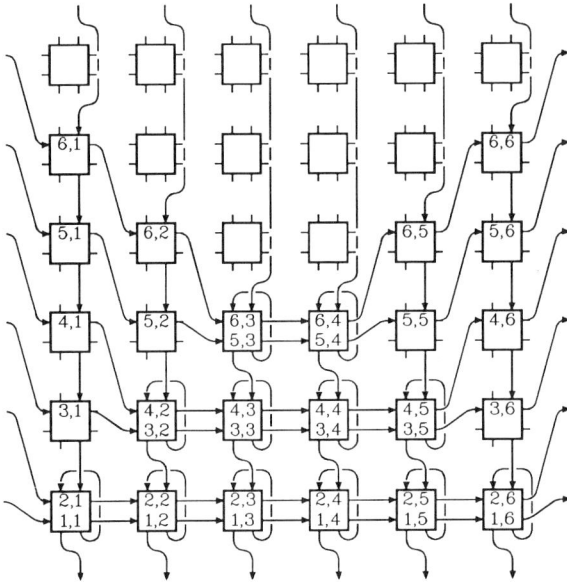

Figure 15.5

Modified organization of index mapping.

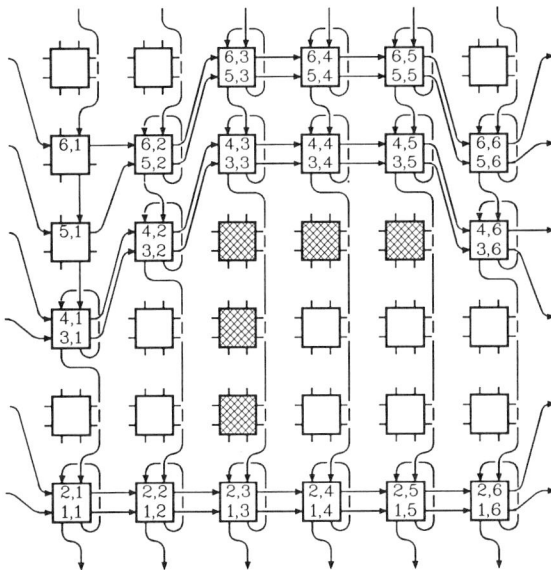

Figure 15.6

Reconfiguration on structure of figure 15.5.

through exploitation of the twin subphases. In each subphase, any given working cell acts as a *virtual* cell and attempts to prolong a row or a column by requesting operation of neighboring virtual working cells. Choice of the particular protocol will obviously determine the reconfiguration obtained (and, implicitly, the augmented interconnection network as well). Consider, for example a case in which each physical cell is linked with three alternative physical cells (i.e., with six alternative *virtual cells*) in order to attempt to prolong a row, and with three alternative physical cells to prolong a column. We accept the possibility that two adjacent rows pass through the same physical cell, because of suitable mapping of two different virtual cells upon it: both virtual cells may then belong to the same column. An additional condition requires that if a physical cell hosts two different columns it will not accept two different rows, and vice-versa.

By creating a protocol based on the above, reconfiguration shown in figure 15.7 can be obtained; adopting the symbols of chapter 13, it can be defined as a /3∗6/ algorithm. The interconnection network supporting it requires four vertical and four horizontal buses. As in the structure-redundancy case, faulty areas are completely bypassed and, as in the case of time-redundancy transposition of index-mapping, a number of cells are left inactive in both subphases (*pseudo-faults*).

While the main goal remains that of configuring the maximum number of rows and columns, two alternative approaches can be identified:

(1) creation of all possible rows is attempted first, by exploiting to the maximum the possibility of having two adjacent rows passing through one physical cell: thus, two rows are tentatively started at each fault-free physical cell on the lefthand border. Arrays are then configured as in the case of protocol-driven structure redundancy using the same protocol (i.e., the same set of neighbors), but a number of logical rows double that obtained in structure redundancy is made available. This solution corresponds to the *preconfigured philosophy* used in subsection 15.1.

(2) For each row of physical cells, a row is created: it must start and end on the first (and, respectively, last) cell of the physical row but, if necessay, passing from one physical row to another one is allowed. The example in figure 15.7 was solved using this second technique.

Direct simulation of the second approach led to the harvesting distribution shown in figure 15.8.

Since protocol-driven techniques are specifically envisioned for very large arrays — typically, wafer-scale arrays of relatively small PEs — evaluation of this class of approaches must take into account the typical factors related to such devices. Time redundancy involves a measure of area redundancy (greater than in §15.1, given the greater complexity of the augmented interconnection network), and its relevance strongly depends on bus width. On the other hand, the actual time overhead introduced in computations is not so easily computed since in the case of space-redundancy, protocol-driven techniques may also involve fairly relevant delays. The main point, in any event, relates to the *mode* of operation required to use time redundancy, i.e, by the strict synchronization involved (a syn-

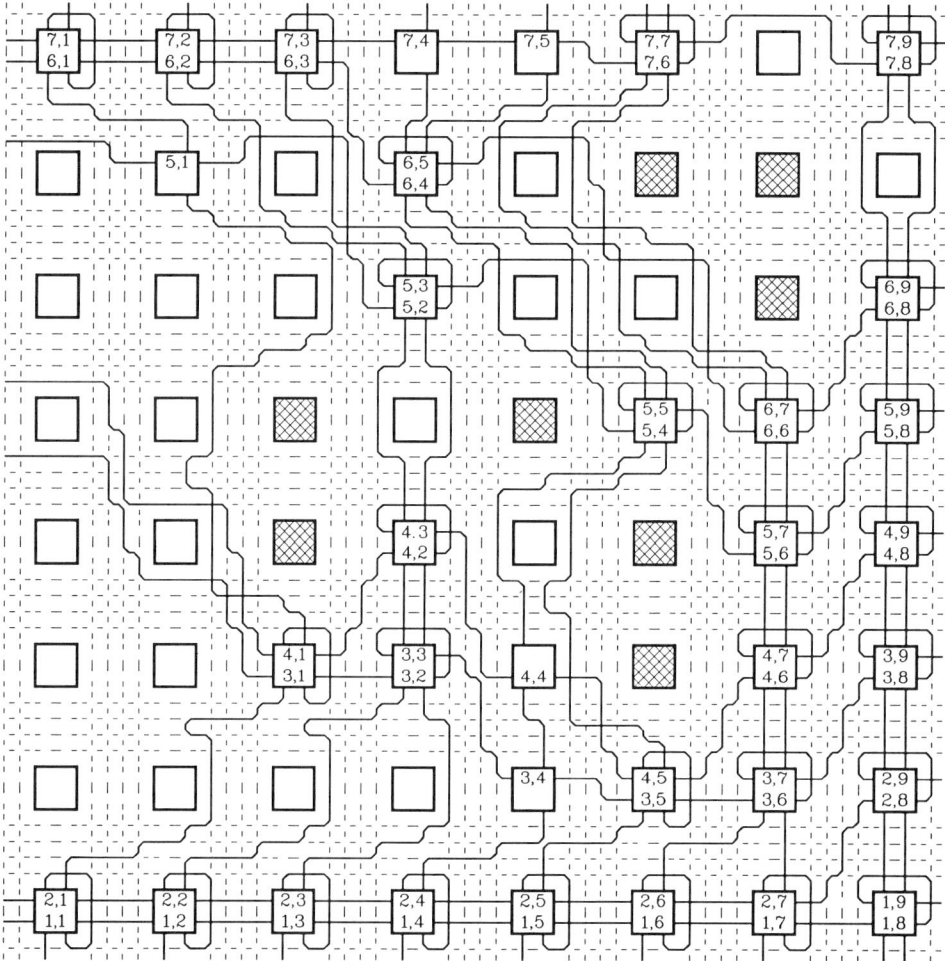

Figure 15.7
Reconfiguration by protocol-driven technique.

chronization that extends here to multiple clocks with well-defined relative delays).
Clock distribution is always an important problem in wafer-scale structures, where
phenomena such as *clock skew* easily affect the good performance of the whole system. It is self-evident that extending the complexity of a global synchronization
system, and even requiring switches commuting from subphase to subphase with
strict timing rules, adds to the problem in a way that may easily become unacceptable. Thus, time-redundancy appears to be better suited for relatively smaller
arrays, where synchronization with global clocks is better achieved.

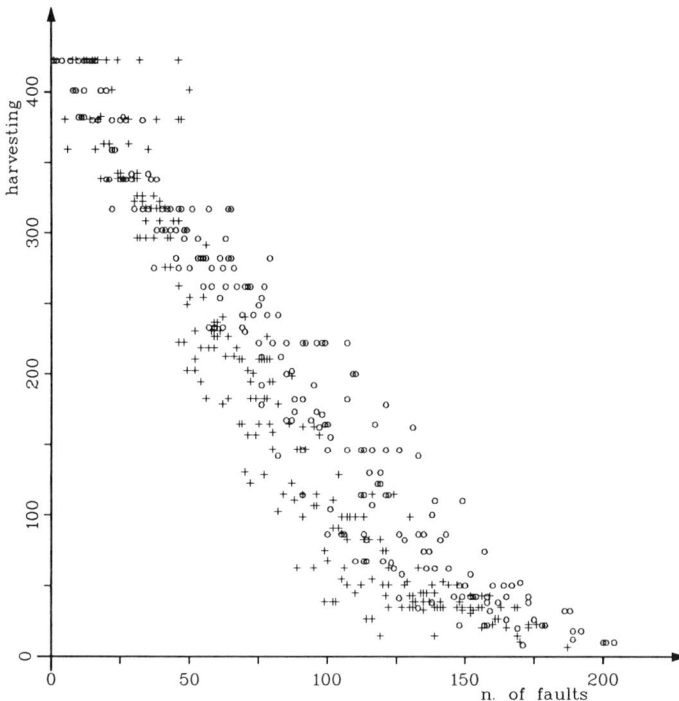

Figure 15.8

Harvesting for protocol-driven technique $(20 \times 20$ physical array).

15.3 References

[GUP86]   R.Gupta, A.Zorat, V.Ramakrishnan: *A fault-tolerant multipipeline architecture*, Proc. FTCS-16, 1986, 350-355, IEEE.

[KUN84]   H.T.Kung, M.S.Lam: *Fault tolerance and two-level pipelining in VLSI systolic arrays*, Proc MIT Conf. on Adv. Res. in VLSI, Jan. 1984.

[MAJ88]   A.Majumdar, C.S.Raghavendra, M.A.Breuer: *Fault-tolerance in linear systolic arrays using time redundancy*, Proc. 21st Annual Hawaii Int'l Conference on Systems Sciences, Kailua-Kona, Jan. 1988, IEEE.

[NEG86]   R.Negrini, R.Stefanelli: *Comparative evaluation of space- and time-redundancy approaches for WSI processing arrays*, in: *Wafer Scale Integration*, (G.Saucier, J.Trilhe eds.), Proc. IFIP WG 10.5 Workshop, Grenoble, Mar. 17-19 1986, North-Holland, 207-222.

[SAM84]   M.G.Sami, R.Stefanelli: *Fault tolerance of VLSI processing arrays: the time-redundant approach*, Proc. Real-time systems symposium, 1984, 200-207, IEEE.

16 SOME TECHNIQUES FOR RECONFIGURATION OF BINARY TREES

The last basic class of array architectures that should be examined is that of tree structures. Actually, tree structures have been widely studied since many years. A well-known paper by Hayes [HAY76] describes a tree organization capable of surviving to any single fault; other papers examined in particular survival to *link* failures, assuming such failures to be more probable than node failures, or alternatively (in the case of node failures) aimed at achieving graceful degradation rather than actual fault tolerance. In this second case, fault-tolerance policies concern mainly the message routing algorithm; thus, in [HOR81], a message routing algorithm achieving fault-tolerance in a leaf-ringed tree is presented, such that — whenever correct routing between sorce and destination of a message cannot be completed — the message is routed back to the source.

Initially considered for complex access modes to large memories, trees have more recently been advocated to support so-called "functional programming" or to execute "reduction languages" (see [BAC78, MAG80]). In such applications, nodes are relatively simple PEs and it is quite reasonable to envision VLSI implementation of even large trees. Thus, papers dealing with fault models and figures of merit better suited to such implementations have been presented. Here, we examine some such solutions, taking into account problems such as complexity of the augmented interconnection network and ease of mapping the reconfigurable architecture on silicon. The specific architecture considered is the *binary tree*, by far the most usual one; extension to other classes of tree is straightforward.

As it happens for other classes of arrays examined in previous chapters, in this context most authors assume probability of failure to be much higher for nodes than for links: thus, reconfiguration involves use of suitable spare nodes, and spare links are mainly introduced to create an augmented interconnection network supporting reconfiguration.

A first VLSI oriented solution, concerning binary trees also, is embedded in the Diogenes approach discussed in chapter 7.

In a similar way, Koren [KOR81] examines the possibility of extending the reconfiguration technique proposed mainly for linear arrays (chapter 8) to the case of binary trees. His solution balances relatively low spares utilization with relevant algorithmic simplicity as it was done for linear arrays. The approach starts from a rectangular grid of PEs in which a preliminary testing phase identifies all faulty cells; all fault free elements in rows and columns that contain at least one faulty PE are used simply as *Connecting Elements* (CEs), so that in practice the grid is *shrinked* to a smaller operational one. Upon this grid the usual *H-tree layout* is mapped to obtain a binary tree [MEA80]. An example is given in figure 16.1, where a 5-level binary tree is extracted from a grid containing one faulty PE; PEs inserted in the tree are marked with the ordering number of the level in which they are inserted. Besides the CEs in the row and column of the faulty cell, a number

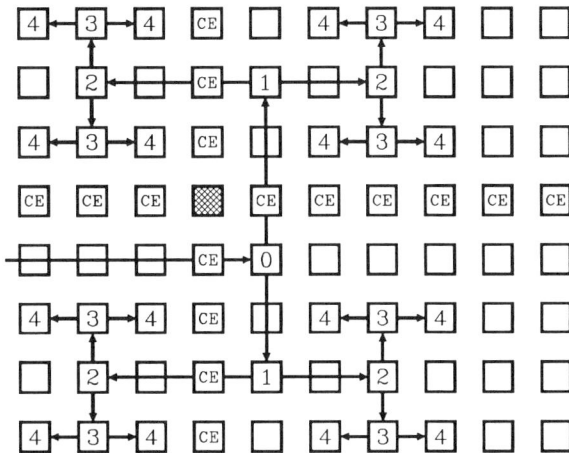

Figure 16.1

Example of reconfiguration for a binary tree mapped onto a rectangular grid: Koren's solution.

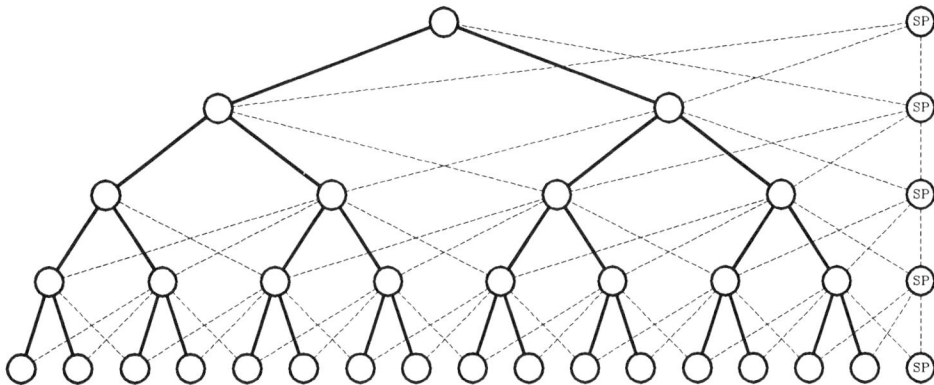

Figure 16.2

Basic redundancy scheme proposed by Raghavendra, Avizienis and Ercegovac. Thick lines denote nominal links; dashed lines denote links introduced to achieve fault tolerance.

of other fault free cells are used as connecting elements only, or left unused, so that the final area efficiency is rather low.

We will consider now only approaches directly related to the specific tree architecture.

In [RAG84] a fault-tolerance scheme adding spares at each level of a tree and introducing an augmented interconnection network implemented by decoupling networks is described. The general tree scheme is given in figure 16.2; a spare node is added for each level of the tree, and redundant links interconnecting the spare node of each level L with suitably chosen nodes of level $L+1$. Redundant links are also added linking *nominal* nodes of each level L with suitable nodes of

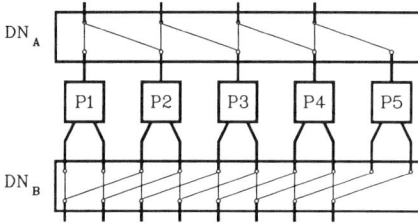

Figure 16.3
Decoupling networks inserted between pairs of levels to achieve reconfiguration.

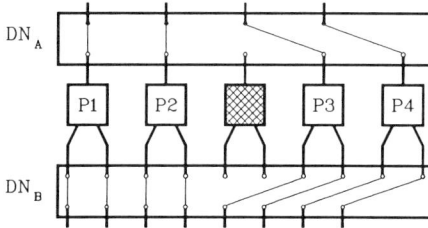

Figure 16.4
Modified interconnections effected by the decoupling network in presence of a fault.

level $L + 1$ (besides the physical adjacents). Thus, each node has between three and five redundant links. The reconfiguration upon node failure involves selection of a spare node and appropriation of links for communications.

To simplify the physical implementation of the interconnections, decoupling networks are inserted between pairs of levels as illustrated by figure 16.3. When a processor in position i of level L fails, all the links of processors in positions $i + k$ ($k \geq 1$) are readjusted to right neighbors, as shown in figure 16.4.

This solution has some affinity with the "row-wise" reconfiguration techniques discussed for rectangular arrays (see chapter 12). Similarity extends even to distributions of faults that can be overcome; the tree tolerates the failure of any single PE, and it can tolerate even multiple failures, provided they are in different levels of the tree (just as row-wise reconfiguration with one column of spares required that at most one fault per row be present). Such a fault assumption is anyway more stringent for tree architectures, in which the number of nodes per level increases with the cardinality of the level itself. As noted in [RAG84], assuming equal probability of failure of a single node separately for each level, i.e., independent of the number of nominal nodes in the level itself, may become unrealistic for large trees.

An alternative solution is consequently proposed, by which variable numbers of spares are introduced at the different levels of the tree, based upon the concept of *modular sparing*, i.e., the introduction of one spare per group of k nodes. An example of such a philosophy is given in figure 16.5. The problem then arises of finding the best possible distribution of spare modules with a fixed number of

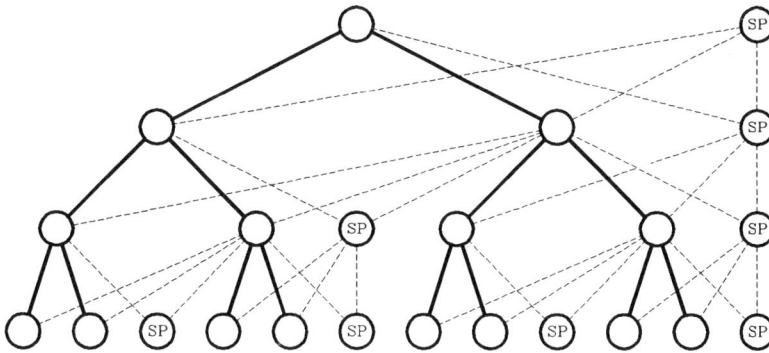

Figure 16.5

Example of modular sparing for $k = 2$.

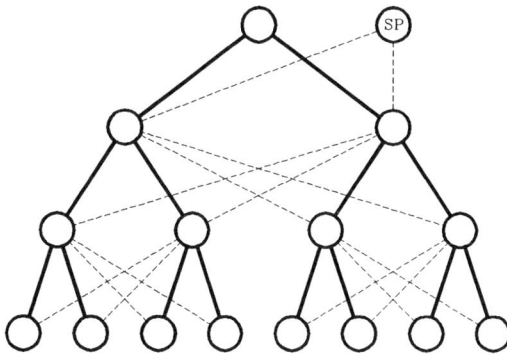

Figure 16.6

Augmented interconnection network for a binary tree allowing performance degradation in the presence of faults.

spares. The solution proposed in [RAG84] for very large trees is to introduce one spare from level 0 up to a level k characterized by 2^k nominal nodes, and then provide one spare for every 2^k nodes. This would double the number of spares per level after the k-th level.

Finally, Raghavendra, Avizienis and Ercegovac examine the alternative of a scheme with performance degradation, requiring one spare for the root only, and extra links to provide for routing in the presence of faults (figure 16.6). This last solution assumes that in presence of a failure, the neigboring node at the same level of the faulty one will perform the tasks of the faulty node in addition to its nominal ones. Therefore, a measure of speed degradation (or, in other words, a *time redundancy* factor) must be taken into account.

A more specifically VLSI-oriented approach is described in [HAS85]. There, the specific layout of the binary tree is taken into account from the beginning,

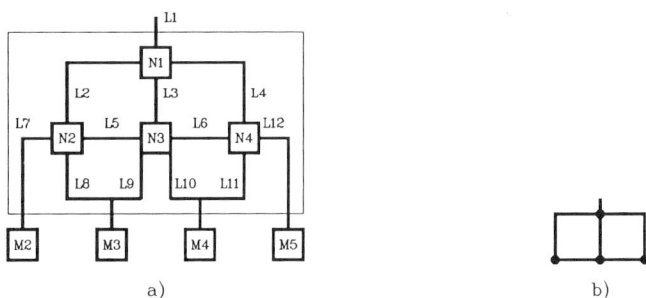

Figure 16.7

a) Basic redundant four-node module proposed by Hassan and Agarwal. b) Schematic diagram.

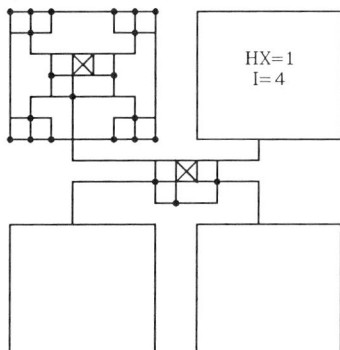

Figure 16.8

The HX-tree layout obtained by composition of basic modules. Five modules are connected in a larger module, four at the corners and one at the center, in an X pattern; such a larger module is then iterated.

and the well-known *H-tree* mapping (known to grant optimum area efficiency) is adopted as the basis of the fault-tolerance policy, so that area efficiency remains an essential figure of merit of the final scheme.

In [HAS85] also, as in [RAG84], reference is made to a *modular approach;* the philosophy is anyway quite different in the two cases. Here, *modularity* implies an approach that is similar to the *local reconfiguration* techniques discussed in chapter 10; an elementary tree module consisting of four nodes and of a redundant interconnection network (see figure 16.7.a) allows local reconfiguration for any single fault, while composition rules for larger trees grant possibility of survival to multiple faults — provided, of course, that the allowable limit of one fault per module is not exceeded. A larger tree is then built by composition of such modules, following a scheme that Hassan and Agarwal call *HX-tree* that can be recursively applied just like the basic H-tree layout. If the basic four-node module is schematically represented as in figure 16.7.b, a five-module pattern can be created and then recursively repeated as in figure 16.8. The most interesting result in [HAS85] grants that the silicon area thus required by a tree of n nodes has complexity $O(n)$, just as it happens for the non-redundant tree, while the Diogenes approach involves

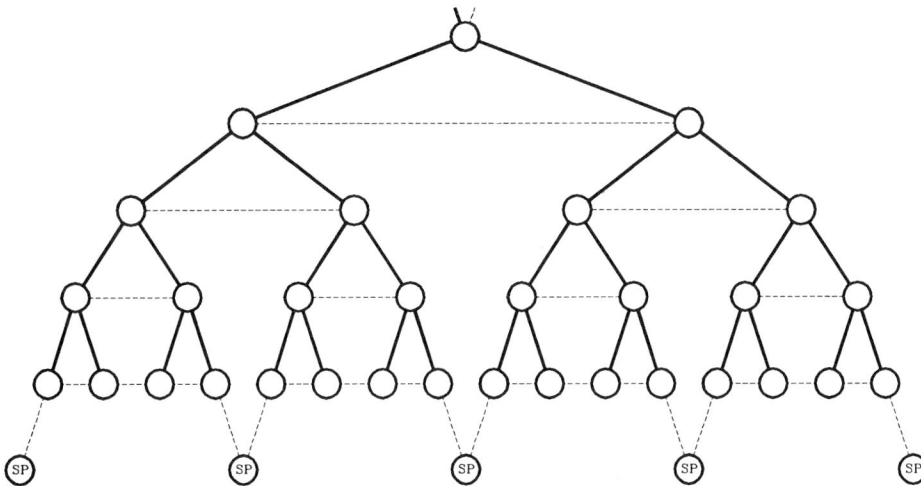

Figure 16.9

Basic SOFT architecture; spare links are represented by thin lines; spare nodes are denoted by a circle with a dot.

silicon requirements $O(n \log n)$ and the approach discussed in [RAG84] involves area complexity $O(n^3 \log n)$.

While an immediate comparison with [RAG84], with regard to probability of survival, is far from trivial, since here restrictions to fault distributions are made with reference to twin-level modules rather than to individual levels, in this approach also redundancy increases with the level ordering. Thus, it can be stated that in large binary trees distribution of redundant modules will be directly related to the number of nodes in a level and, therefore, to the corresponding probability of failure.

Finally, a global approach that makes use of *deformation lines* to achieve reconfiguration has been presented in [LOW87], where the *SOFT* approach is described (SOFT standing for Subtree Oriented Fault Tolerance). The basic philosophy consists in adding a number of spare PEs as nodes of level $L+1$ (for an L level tree), as well as a number of spare links connecting, inside each level, each pair of "brother" nodes and, with regard to spare node, spares with suitable level-L nodes. An example of such a redundant structure is given in figure 16.9; actually, there is a relevant degree of freedom in choosing the amount of redundant nodes to be added, as long as this number does not exceed 2^{L+1}.

The actual opening or closing of a (nominal or redundant) link is effected by a simple switched-link structure schematized in figure 16.10. The formal description of the reconfiguration algorithm, as given in [LOW87], is rather complex: we will summarize it here intuitively in terms of deformation lines. Refer to figure 16.11. Consider fault of node $N_{0,3}$; the node itself is bypassed, and its functions (i.e., its logical indices) are given to its right-hand son, $N_{1,4}$ (a *displacement* of the node is effected). Node $N_{0,4}$ retains as logical indices its physical ones, and the spare

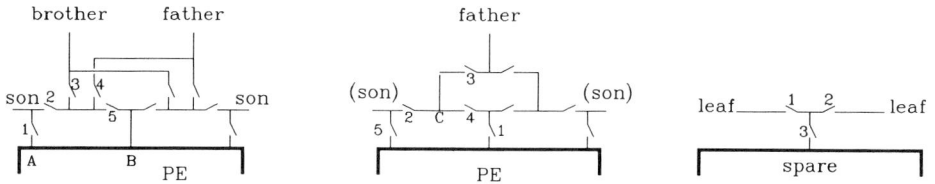

Figure 16.10

SOFT switching in binary trees. a) Upper level switching. b) Leaf switching. c) Spare switching.

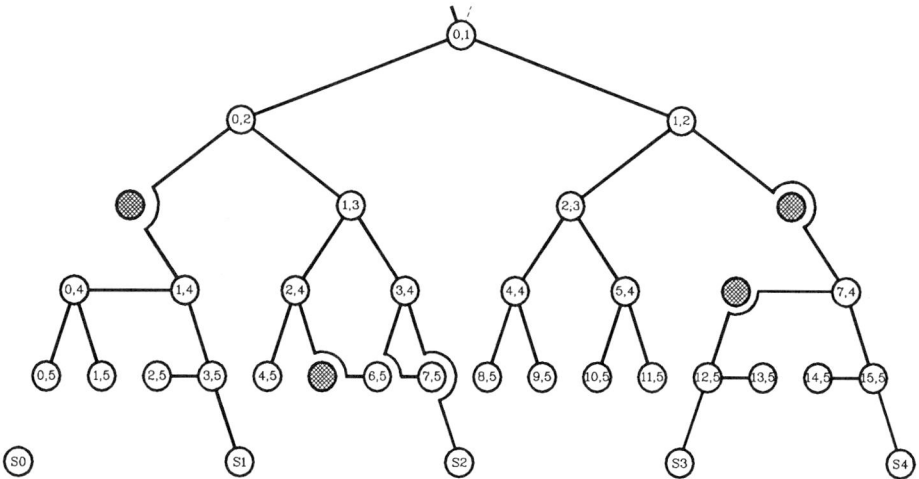

Figure 16.11

Example of reconfiguration in presence of faults.

link between it and $N_{1,4}$ is activated. In turn, the functions previously retained by $N_{1,4}$ are taken over by its own right-hand son, and the procedure of spare link activation and logical indices transfer is iterated until at last the spare node S_1 is inserted into operation performing the logical functions previously performed by $N_{3,5}$.

The same procedure is applied for reconfiguration of all the other faults in the example. In the particular case of node $N_{5,5}$, for which no spare could be found, substitution is effected by involving a *cousin* at the same level 5, $N_{6,5}$, and displacements along the level are effected until spare node S_2 is reached.

Fatal failure can occur, with this approach, whenever particular fault distributions are present (independent of the availability of fault-free spares). Examples of such fault distributions are given in figure 16.12. While the fault model relates to faulty *nodes*, the technique allows also to tolerate isolated link failures and isolated switch failures.

VLSI-orientation of the SOFT approach is examined by referring, here also, to

APPENDIX: RELIABILITY PREDICTION

This appendix briefly treats the problem of predicting arrays reliability (i.e., robustness against run-time faults), adopting the well established approach based upon Markov models. Three standard figures of merit will be considered:

- computational availability,
- reliability,
- performability.

The usual procedures of reliability theory will be followed, as described in [MEY80] and [MEY82], and these techniques will be adapted to arrays by following methods common to [GEI83, TRI83], and [COS87].

First, the properties of an isolated PE will be modeled, thus finding the difficulties and conventionalities of modeling a complex digital circuit, and predicting failure rates. Whole arrays will then be modeled, together with the reconfiguration algorithms implemented. This will be easy only for simple arrays and, above all, for simple reconfiguration algorithms.

Only the case of *fault-degradation* for *rectangular meshes* will be dealt with here. Simple degradation of the array following run-time faults (contrary to reconfiguration based upon spares) will allow for reasonably simple models. Every new fault simply determines a reduction of the array size, without forcing new spare PEs to work.

Fault-degradation algorithms and methods treated in [FOR85] and in [GEI83] are particularly suited to simple computations, and will be followed. After a brief survey of Markov models, the single PE will be modeled, introducing the so-called *Fault/Error Handling Model (FEHM)*. From this it will be possible to derive cumulative parameters of the PE and to adopt them in a *Fault Occurrence and Repair Model (FORM)* that will account for both array structure and fault-degradation algorithm.

This analysis will lead to a complex, closed form solution for availability, reliability and performability figures: complexity of procedures, and intrinsic conventionality of some input parameters, will explain why simpler approaches based upon simulations aimed at determining harvesting and survival probability are usually followed.

A.1. Markov models: a brief introduction

The subset of the general Markov models to be adopted is characterized by constant transition rates, and requires two assumptions:

(1) PEs, that are assumed as the elementary modules of the model, fail independently in different moments (i.e., the probability that two or more PEs change

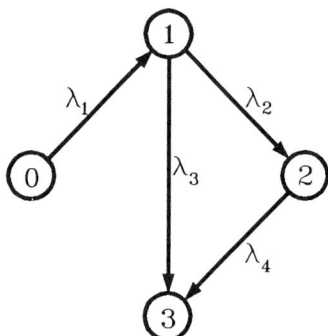

Figure A.1
An example of Markov Model.

their error state in a small finite time interval can be neglected);

(2) the probability of transition between two different error states of a PE depends only upon the two states in question, and its rate is constant (i.e., if the transition rate from state k to state r is λ_{kr} , then the probability of transition from k to r during the interval Δt is simply $\lambda_{kr} \cdot \Delta t$).

Consequently, we can define the *error state* of each PE, and of the array. Error states represent all the past error history of PEs and array, and allow prediction of the future evolution, if all the probability rates of transitions between error states are known.

It also follows that:

- if at time t the probability of being in error state k is $P_k(t)$, then the probability of transition from this state k to state r during the interval Δt is

$$P_k(t) \cdot \lambda_{kr} \cdot \Delta t \qquad [A.1]$$

- from equation [A.1] we can derive a probability balance related to all the transitions involving state k during Δt, and by dividing it by Δt we have the rate of change of $P_k(t)$:

$$\frac{\Delta P_k(t)}{\Delta t} = -\sum_r P_k(t) \cdot \lambda_{kr} + \sum_j P_j(t) \cdot \lambda_{jk} \qquad [A.2]$$

where the first sum expresses probability of leaving state k following all transitions from k to any other error state, and the second gives probability of entering k.

Let us see, by means of an example, how these equations allow computation of all the probabilities $P_k(t)$ for every error state in a simple and known system.

Without any practical loss of generality, assume that the number of error states is finite. A graph like that of figure A.1 can then be built, where each state corresponds to a node, and transitions between states correspond to arcs labeled by the corresponding probability rates.

Let $P_0(t), \cdots, P_3(t)$ be the probabilities of being in state $0, \cdots, 3$ at time t. Then, obviously:

$$P_0(t) + P_1(t) + P_2(t) + P_3(t) = 1 \qquad [A.3]$$

The $P_k(t)$ are the unknown of the problem, to be computed on the basis of equation [A.3] and equation [A.2] and of an initial condition. If we suppose that for $t = 0$ the system is certainly in state 0, then the initial condition is:

$$P_0(0) = 1, \quad P_1(0) = P_2(0) = P_3(0) = 0 \qquad [A.4]$$

For the graph of figure A.1, equation [A.2] becomes ($\Delta t \to dt$):

$$
\begin{aligned}
\frac{dP_0(t)}{dt} &= -\lambda_1 P_0(t) \\
\frac{dP_1(t)}{dt} &= \lambda_1 P_0(t) - \lambda_2 P_1(t) - \lambda_3 P_1(t) \\
\frac{dP_2(t)}{dt} &= \lambda_2 P_1(t) - \lambda_4 P_2(t) \\
\frac{dP_3(t)}{dt} &= \lambda_3 P_1(t) - \lambda_4 P_2(t)
\end{aligned}
\qquad [A.5]
$$

or, using matrix algebra:

$$
\frac{d}{dt}
\begin{bmatrix} P_0(t) \\ P_1(t) \\ P_2(t) \\ P_3(t) \end{bmatrix}
=
\begin{bmatrix}
-\lambda_1 & 0 & 0 & 0 \\
\lambda_1 & -\lambda_2 - \lambda_3 & 0 & 0 \\
0 & \lambda_2 & -\lambda_4 & 0 \\
0 & \lambda_3 & \lambda_4 & 0
\end{bmatrix}
\cdot
\begin{bmatrix} P_0(t) \\ P_1(t) \\ P_2(t) \\ P_3(t) \end{bmatrix}
\qquad [A.6]
$$

Note that the last column of the matrix with all zeroes indicates that the equations are not linearly independent: obviously one equation can be eliminated and one unknown substituted by means of equation [A.3].

A customary way of solving equation [A.6] is based upon Laplace transforms: it is very easy to derive from equation [A.6] the corresponding system linking the Laplace transforms

$$\mathcal{L}(P(t)) = P(s)$$

of the unknown.

Remembering that the transform of a derivative (for $t \in \overline{0 \ \infty}$) is given by

$$\mathcal{L}\left(\frac{dP_k(t)}{dt}\right) = sP_k(s) - P_k(0^+) \qquad [A.7]$$

and that the Laplace transform is a linear operation, we can apply it to equation [A.6] obtaining:

$$\begin{bmatrix} s+\lambda_1 & 0 & 0 & 0 \\ -\lambda_1 & s+\lambda_2+\lambda_3 & 0 & 0 \\ 0 & -\lambda_2 & s+\lambda_4 & 0 \\ 0 & -\lambda_3 & -\lambda_4 & s \end{bmatrix} \cdot \begin{bmatrix} P_0(s) \\ P_1(s) \\ P_2(s) \\ P_3(s) \end{bmatrix} = \begin{bmatrix} P_0(0^+) \\ P_1(0^+) \\ P_2(0^+) \\ P_3(0^+) \end{bmatrix} \qquad [A.8]$$

where the unknown $P_k(t)$ are substituted by their Laplace transforms.

The solution of equation [A.8] is a standard problem of linear algebra. Note that the vector of known terms

$$[P_0(0^+), P_1(0^+), P_2(0^+), P_3(0^+), P_4(0^+)]^T$$

coincides with the initial state, e.g., $[1,0,0,0]^T$ in our example.

By solving equation [A.8] we can compute $P_0(s)$, $P_1(s)$, $P_2(s)$, $P_3(s)$ and then, by inverse transformations, the corresponding time-domain functions $P_k(t)$ for t from 0 to ∞.

If, as it often happens, only the steady state values are required, they can be computed directly from the $P_k(s)$, by means of the Final Value Theorem, that states:

$$P_k(\infty) = \lim_{t \to \infty} P_k(t) = \lim_{s \to 0} sP_k(s) \qquad [A.9]$$

If the application allows us to safely assume fault probabilities so small that following each new fault the system reaches the new steady state, before appearance of a new fault, then the problem can be treated in this way:

(1) a new fault arises in a PE. By means of the PE model (FEHM), starting from the initial state, the steady state probabilities are computed with the Final Value Theorem;

(2) the steady state probabilities previously computed are inserted in the model of array reconfiguration (FORM). The FORM model can be computed giving the $P_k(t)$ for the complete array.

A.2. Fault/Error Handling Model (FEHM)

Adopting the methods of [TRI83], and following [COS87], this section is devoted to the study of a sufficiently simple but significant model for a PE, so as to be able to describe the PE's behavior after a run-time fault arises.

The FEHM of the PE shown in figure A.2 (introduced in [COS87] and adopted here) considers six states for the PE. This model shows how faults become errors, if active, and describes permanent, transient, and intermittent faults/errors.

Input state is state 0, which corresponds to *a fault is active*; the output states are state 4 (a fault/error has been detected, and reconfiguration will follow) and

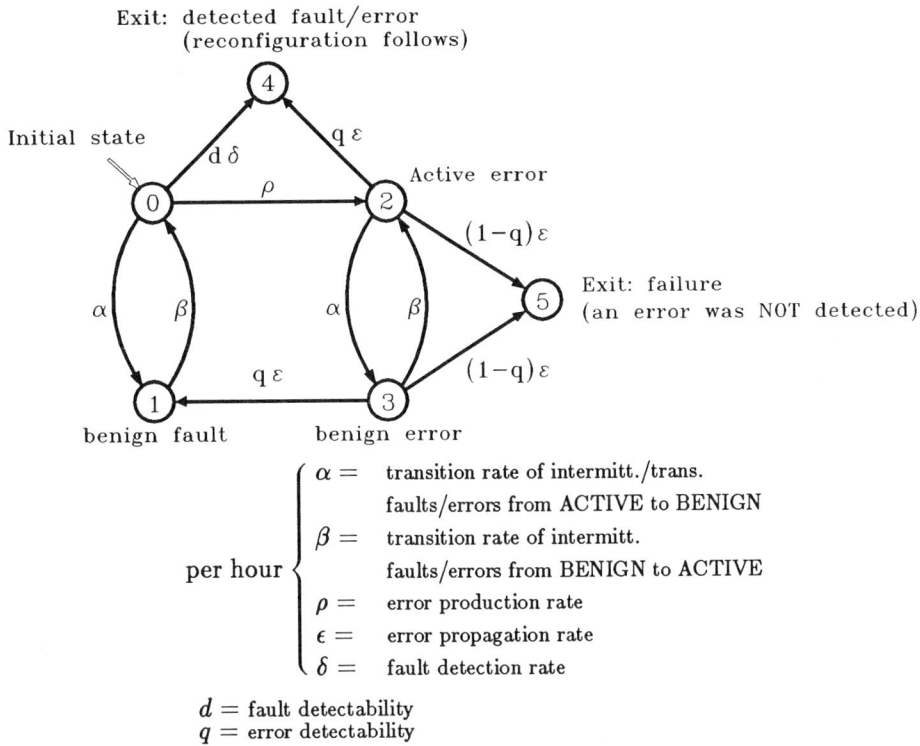

Exit: detected fault/error
(reconfiguration follows)

Initial state

Active error

Exit: failure
(an error was NOT detected)

benign fault   benign error

$\alpha =$ transition rate of intermitt./trans.
faults/errors from ACTIVE to BENIGN

$\beta =$ transition rate of intermitt.
faults/errors from BENIGN to ACTIVE

per hour

$\rho =$ error production rate

$\epsilon =$ error propagation rate

$\delta =$ fault detection rate

$d =$ fault detectability
$q =$ error detectability

Figure A.2

A Fault/Error Handling Model (FEHM) from [COS87].

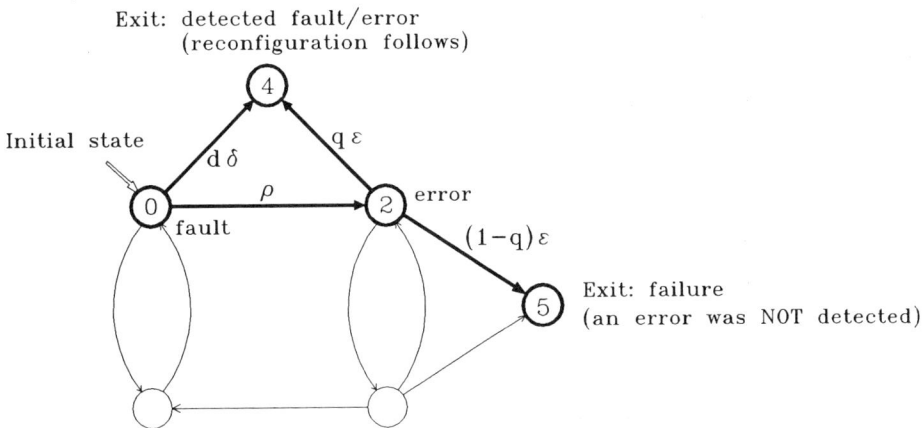

Exit: detected fault/error
(reconfiguration follows)

Initial state

error

fault

Exit: failure
(an error was NOT detected)

Figure A.3

The case of Permanent Faults/Errors.

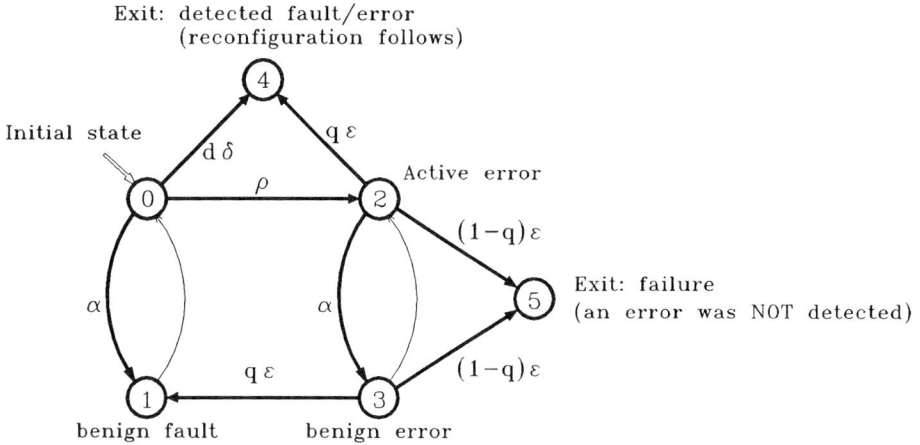

Figure A.4

FEHM allowing also for transient Faults/Errors.

state 5 (failure: an error was not detected; the necessary reconfiguration action is not activated).

If permanent faults/errors alone are of interest, the subgraph shown in figure A.3 can be considered; the subgraph of figure A.4 allows modeling of transient faults/errors (i.e., faults/errors that are active once). The complete graph of figure A.2 shows behavior of intermittent faults/errors.

By means of the approach described in the previous section, adopting Laplace transforms and Final Value Theorem, from this FEHM the following steady state probabilities can be easily derived:

- from the graph of figure A.2:

  coverage probability for intermittent faults (followed by reconfiguration):

  $$C_i = P_4(\infty) = \lim_{s \to 0} sP_4(s) = \frac{d\delta(\alpha + \beta + \epsilon) + \rho q(\beta + \epsilon)}{(\alpha + \beta + \epsilon)(d\delta + \rho) - \alpha\rho q} \qquad [A.10]$$

- from the subgraph of figure A.3:

  coverage probability for permanent faults (followed by reconfiguration):

  $$C_p = P_4(\infty) = \lim_{s \to 0} sP_4(s) = \frac{d\delta + \rho q}{d\delta + \rho} \qquad [A.11]$$

- from the subgraph of figure A.4:

  recovery from transient faults (the PE returns to fault-free state, no reconfiguration needed):

  $$C_t = P_1(\infty) = \lim_{s \to 0} sP_1(s) = \frac{\alpha(\alpha + \epsilon + \rho q)}{(\alpha + \epsilon)(\alpha + d\delta + \rho)} \qquad [A.12]$$

k reconfigured faults

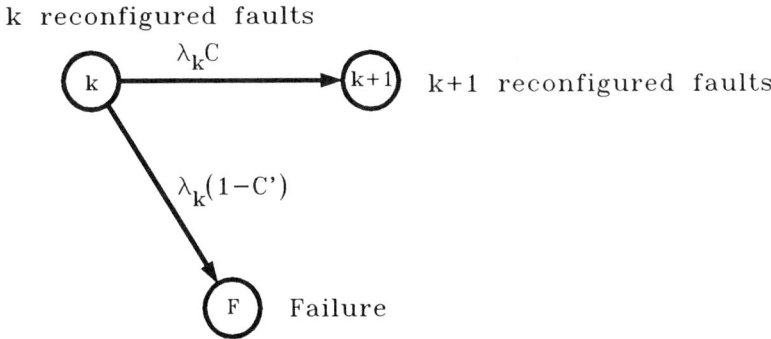

Figure A.5
Array transitions after a new fault.

coverage of transient faults (the fault/error is detected, reconfiguration follows, i.e., the transient fault is treated as permanent, e.g., staticized):

$$C_{tp} = P_4(\infty) = \lim_{s \to 0} sP_4(s) = \frac{d\delta(\alpha + \epsilon) + \epsilon \rho q}{(\alpha + \epsilon)(\alpha + d\delta + \rho)} \qquad [A.13]$$

Note that C_{tp} cannot be derived from C_i by simply substituting $\beta = 0$ in equation [A.10], because [A.10] does not hold for the subgraph in figure A.4.

It is now necessary to mix these probabilities, by adding weights corresponding to the probability that a fault is permanent, transient, or intermittent.

If a new fault has constant probabilities:

h_p of being permanent,

h_t of being transient,

h_i of being intermittent,

then, of course, $h_p + h_t + h_i = 1$, and the following coverage probabilities can be computed:

$$C = h_p C_p + h_i C_i + h_t C_{tp} \qquad [A.14]$$

where C is the steady state probability that, when a fault arises, its effects are identified and the PE is declared faulty (reconfiguration follows); and:

$$C' = C + h_t C_t \qquad [A.15]$$

which is the steady state probability that, when a new fault arises, either the PE is correctly declared faulty or else it becomes again fault-free without propagating errors (i.e., this is the probability of not failing to properly manage the fault and its consequences).

Then, figure A.5 can be drawn, showing how from an error state of the array that survives to *k* past faults, a new fault puts the array in the state corresponding to "surviving to $k+1$ past faults" if the fault is detected and the array is reconfig-ured, or in the state corresponding to failure if produced errors neither disappear nor are detected. It portrays the following conditions.

- Assume that the array is in a state that has been reached after appearence of (and survival to) k faults. When the $(k+1)$-th fault appears:
  - either the fault is detected and reconfiguration is correctly performed, so that a state representing survival to $k+1$ faults is reached;
  - or the failure state is reached, if the new fault is not detected and it does not desappear (i.e., it is not a transient fault).

In figure A.5, the parameter λ_k is the fault rate corresponding to the array con-figuration after k faults: for example, if every PE has the same constant fault rate λ, and the array in state k comprises M_k PEs, then the array fault rate is $\lambda_k = M_k \cdot \lambda$.

A.3. Fault Occurrence and Repair Model (FORM)

It is now possible to study the fault model of the whole array, considering the repair actions as well. Following [COS87], we will treat here only the case of fault-degradation, where each fault at most induces a reduction of the size of the array, and the reduction of the number of used PEs can be easily computed, in order to determine the λ_k previously seen, for each k. By iterating the graph of figure A.5, that of figure A.6 can be obtained. In this graph, each node is labeled with the number of past faults that have been correctly treated. Thus, node 0 represents the original array with no faults; every undetected error puts the system in the failure state $L + 1$; L corresponds to the maximum number of faults that the system can tolerate.

If simple fault-degradation algorithms are considered, as in [FOR85], it is quite simple to determine λ_k as a function of M_k and then of k, by means of the previously seen

$$\lambda_k = M_k \cdot \lambda \qquad [A.16]$$

because M_k can be readily determined as a function of k (as will be seen later, using an example from [FOR85]).

Otherwise, in the case of fault-tolerance algorithms, some hypotheses must be adopted to simplify the computation of M_k and λ_k. Some of the following hypotheses could be useful:

(a) all the spare and not used PEs are either permanently fault-free or faulty;

(b) every reconfiguration action following the k-th fault always configures an array with the same M_k ;

(c) the number L of faults that can be tolerated is independent from the past reconfiguration actions.

In particular, if all these hypotheses hold, the same graph of figure A.5 holds, along with equation [A.16].

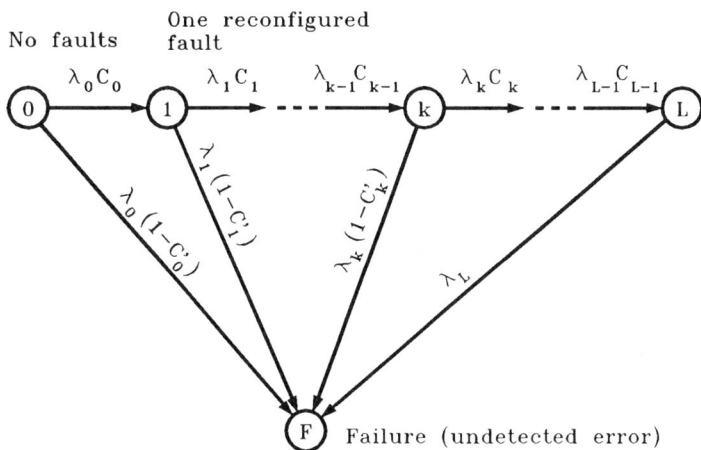

Figure A.6

A Fault Occurrence and Repair Model (FORM).

The FORM of figure A.5 can now be used; in this case the equations for the state probabilities are:

$$\frac{dP_0(t)}{dt} = -(C - C_0' + 1)\lambda_0 P_0(t)$$

$$\vdots$$

$$\frac{dP_k(t)}{dt} = -(C_k - C_k' + 1)\lambda_k P_k(t) + C_{k-1}\lambda_{k-1} P_{k-1}(t) \qquad [A.17]$$

$$\vdots$$

(when $k = L$ it is enough to assume $C_L = C_L'$)

Equations [A.17] can be solved by using Laplace transforms; then, returning to the time domain, the following solution is obtained:

$$P_0(t) = e^{-(C_0 - C_0' + 1)\lambda_0 t}$$

$$P_k(t) = \sum_{j=0}^{k} \frac{\prod_{i=0}^{k-1} C_i \lambda_i}{\prod_{i=0}^{k} \left[(C_i - C_i' + 1)\lambda_i - (C_j - C_j' + 1)\lambda_j \right]} e^{-(C_j - C_j' + 1)\lambda_j t} \qquad [A.18]$$

This closed form solution is suited for computerized analysis.

As a brief and simple example of application, we can consider the array and a fault-degradation algorithm presented in [FOR85], reconsidered also in [COS87]. This array is a 2-dimensions mesh, and the fault-degradation is quite simple: when a PE fails, its entire row is excluded from the array.

Starting from a $N \times N$ square array, after k faults hitting the degrading array, only $M_k = (N-k) \cdot N$ PEs remain connected, because k rows have been eliminated.

If we consider that the array is working when at least one row survives, then the maximum number of tolerated faults is

$$L = N - 1$$

and, after L faults, the array is composed just by a row (i.e., $1 \times N$ PEs). These values, along with equations [A.16] and [A.18], give the state probabilities.

Again, considering the general case, equations [A.18] are then computed, and their values are substituted in the following definitions, giving the usual reliability parameters.

(1) *Reliability*, which gives the probability that, at time t, the array is working, in one of the L possible configured states:

$$R(t) = \sum_{k=0}^{L} P_k(t) \tag{A.19}$$

(2) *Performability*, which gives the probability that the array is working at least at a given level B of performance (e.g., B is a fraction of the number of the initial PEs. In our example, we might add here only the probabilities corresponding to the states with at least a given number of rows, i.e., up to only B faults):

$$Perf(B, t) = \sum_{k} P_k(t) \tag{A.20}$$

(3) *Computational availability*, which gives a probabilistic measure of the number of working PEs at time t, i.e., of the computing power:

$$Ac(t) = \sum_{k=0}^{L} M_k P_k(t) \tag{A.21}$$

Computational availability can also be evaluated in a normalized way, by relating it to the total number of PEs in the architecture.

This model can also be adopted to determine the effect of classes of faults: for example, by first setting

$$h_t = h_i = 0 \quad \text{and} \quad h_p = 1,$$

and then assuming values corresponding to the presence of all kinds of faults, by analyzing the changes in results, the influence of intermittent and transient faults can be determined.

rates: $\alpha, \beta, \rho, \varepsilon, \delta, d, q$

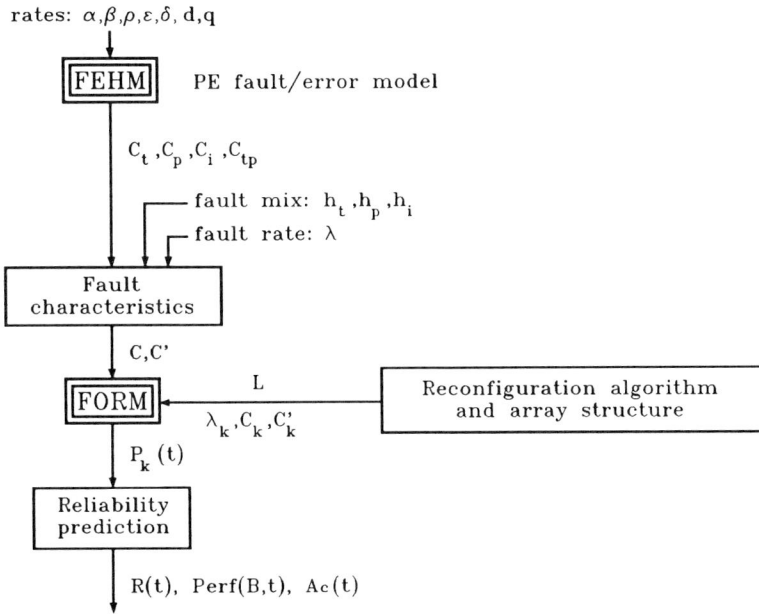

Figure A.7
Procedure for reliability prediction.

In conclusion, the procedure for computing the closed form results is given in figure A.7.

This approach for determining the reliability parameters of arrays shows the usual drawbacks of reliability modeling: in particular, the FEHM model poses, as always, the problem of the choice of realistic figures for rates and weights. These figures often have some intrinsic conventionality, and precise values are difficult to determine from experience. Difficulties in the FORM model also relate to the graph itself, which can be very difficult to determine from the reconfiguration algorithm and array structure. Position of faults inside arrays can often condition the number of PEs that remain connected, or the following capacity of tolerating further faults.

This explains why, in the literature, more pragmatic approaches are usually preferred; typically, simulation of very large numbers of reconfiguration problems corresponding to randomly generated faults allows for significant computation of harvesting and survival probability, instead of reliability parameters.

A.4. References

[BEI87]   V.Bein, C.Costantinescu: *Fault-Tolerant Systolic Arrays for Antialiasing*, Proc. COMPEURO-87, Hamburg, May 1987, 720-723, IEEE.

[COS87]   C.Costantinescu: *Effect of Transient Faults on Gracefully Degrading Processor Arrays*, submitted to IEEE-TC, 1987.

[FOR85]   J.A.B.Fortes, C.S.Raghevendra: *Gracefully Degradable Processor Arrays*, IEEE-TC, Vol. C-34, N. 11, Nov. 1985, 1033-1044.

[GEI83]   R.M.Geist, K.S.Trivedi: *Ultrahigh Reliability Prediction for Fault Tolerant Computer Systems*, IEEE-TC, Vol. C-32, N. 12, Dec. 1983, 1118-1127.

[MEY80]   J.F.Meyer:*On Evaluating the Performability of Degradable Computer Systems*, IEEE-TC, Vol. C-29, N. 8, Aug. 1980, 720-731.

[MEY82]   J.F.Meyer: *Closed-Form Solutions of Performability*, IEEE-TC Vol. C-31, N. 7, July 1982, 648-657.

[TRI83]   K.S.Trivedi, R.M.Geist: *Decomposition in Reliability Analysis of Fault Tolerant Systems*, IEEE-TR, Vol. R-32, N. 5, Dec. 1983, 463-468.

INDEX

AUTHOR INDEX

The MIT Press, with Peter Denning, general consulting editor, and Brian Randell, European consulting editor, publishes computer science books in the following series:

ACM Doctoral Dissertation Award and Distinguished Dissertation Series

Artificial Intelligence, Patrick Winston, Michael Brady, and Daniel Bobrow, editors

Charles Babbage Institute Reprint Series for the History of Computing, Martin Campbell-Kelly, editor

Computer Systems, Herb Schwetman, editor

Exploring with Logo, E. Paul Goldenberg, editor

Foundations of Computing, Michael Garey and Albert Meyer, editors

History of Computing, I. Bernard Cohen and William Aspray, editors

Information Systems, Michael Lesk, editor

Logic Programming, Ehud Shapiro, editor; Fernando Pereira, Koichi Furukawa, and D. H. D. Warren, associate editors

The MIT Electrical Engineering and Computer Science Series

Research Monographs in Parallel and Distributed Processing, Christopher Jesshope and David Klappholz, editors

Scientific Computation, Dennis Gannon, editor